Die Grundlehren der mathematischen Wissenschaften

in Einzeldarstellungen
mit besonderer Berücksichtigung
der Anwendungsgebiete

Band 196

Herausgegeben von
J. L. Doob · A. Grothendieck · E. Heinz · F. Hirzebruch
E. Hopf · W. Maak · S. Mac Lane · W. Magnus · J. K. Moser
M. M. Postnikov · F. K. Schmidt · D. S. Scott · K. Stein

Geschäftsführende Herausgeber
B. Eckmann und B. L. van der Waerden

K. B. Athreya · P. E. Ney

Branching Processes

Springer-Verlag New York Heidelberg Berlin 1972

Krishna B. Athreya

Associate Professor of Mathematics
University of Wisconsin, Madison, Wisconsin

Peter E. Ney

Professor of Mathematics
University of Wisconsin, Madison, Wisconsin

AMS Subject Classifications (1970)
Primary 60J80, 60K99, 60F99
Secondary 60J85, 39A20, 45M05, 60J45, 92A15

ISBN 0-387-05790-0 Springer-Verlag New York Heidelberg Berlin
ISBN 3-540-05790-0 Springer-Verlag Berlin Heidelberg New York

 Library of Congress Catalog Card Number 72-75819. Printed in Germany. Monophoto typesetting and offset printing: Zechnersche Buchdruckerei, Speyer. Bookbinding: Konrad Triltsch, Würzburg.

Dedicated to our parents

Preface

The purpose of this book is to give a unified treatment of the limit theory of branching processes. Since the publication of the important book of T. E. Harris (*Theory of Branching Processes*, Springer, 1963) the subject has developed and matured significantly. Many of the classical limit laws are now known in their sharpest form, and there are new proofs that give insight into the results. Our work deals primarily with this decade, and thus has very little overlap with that of Harris. Only enough material is repeated to make the treatment essentially self-contained. For example, certain foundational questions on the construction of processes, to which we have nothing new to add, are not developed.

There is a natural classification of branching processes according to their criticality condition, their time parameter, the single or multi-type particle cases, the Markovian or non-Markovian character of the process, etc. We have tried to avoid the rather uneconomical and unenlightening approach of treating these categories independently, and by a series of similar but increasingly complicated techniques. The basic Galton-Watson process is developed in great detail in Chapters I and II. In the subsequent treatment of the continuous time (Markov and age-dependent) cases in Chapters III and IV, we try wherever possible to reduce analogous questions to their Galton-Watson counterparts; and then concentrate on the genuinely new or different aspects of these processes. We hope that this gives the subject a more unified aspect. In our development we give a number of new proofs of known results; and also some new results which appear here for the first time.

Although we have included a chapter on applications and special processes (Chapter VI), these are a reflection of our own interests, and there is no attempt to catalogue or even dent the great variety of special models in physics and biology that have been investigated in recent years. Rather, our emphasis is on the basic techniques, and it is our hope that this volume will bring the reader to the point where he can pursue the literature and his own research in the subject. In this connection we have included sections on complements and problems, which may suggest new work.

With the exception of Chapter V which deals with a finite number of distinct particle types, we have concentrated on single type processes. There is an extensive literature on branching processes on general state spaces which is outside the spirit and scope of this book. Readers interested in this subject are referred to the papers of Ikeda, Nagasawa, Watanabe, Jirina, Moyal, Mullikin, and Skorohod listed in the bibliography.

The prerequisites for reading this book are analysis and probability at about the first year graduate level. The reader should, for example, have some familiarity with Markov chains at the level of K. L. Chung's book (Springer, 1967); the martingale theorem; renewal theory and analytical probability at the level of W. Feller, Vol. II (Wiley, 1966).

Some of the more technical sections can be skipped over without impairing continuity. A possible sequence to follow at a first reading is

Chapter I
Chapter III, sections 1—8,
Chapter IV
Chapter V, omitting section 8,
Chapter VI, selections according to taste.

Acknowledgements

We want to express our appreciation to our collegues and teachers J. Chover, T. Harris, A. Joffe, S. Karlin, H. Kesten, and F. Spitzer for their help and influence on our work. Their contributions to the theory of branching processes permeate these pages. We also wish to extend our sincere thanks to J. L. Doob, as editor of this series, for his counsel and valuable suggestions.

Many others have given us assistance. J. Lamperti and J. Williamson read parts of the manuscript and made helpful comments. Among our former and present students, M. Goldstein prepared the original seminar notes from which the book started; J. Foster, K. B. Erickson, W. Esty, and R. Alexander corrected vast numbers of errors; W. Esty prepared the bibliography and index and helped extensively in the preparation of the final manuscript. Judy Brickner did a superb job of typing the manuscript, and then cheerfully saw it through revisions and re-revisions.

Finally, we would like to thank our wives for their moral support throughout this long project.

Madison, Wisconsin *
September, 1971 Krishna B. Athreya · Peter E. Ney

* K. Athreya is now at the Indian Institute of Science in Bangalore, India.

Table of Contents

Chapter I. The Galton-Watson Process

Chapter II. Potential Theory

Chapter III. One Dimensional Continuous Time Markov Branching Processes

Chapter IV. Age-Dependent Processes

Chapter V. Multi-Type Branching Processes

Chapter VI. Special Processes

Chapter I

The Galton-Watson Process

Part A. Preliminaries

1. The Basic Setting

A Galton-Watson process is a Markov chain $\{Z_n; n=0,1,2,\ldots\}$ *on the nonnegative integers. Its transition function is defined in terms of a given probability function* $\{p_k; k=0,1,2,\ldots\}$, $p_k \geqslant 0$, $\sum p_k = 1$, *by*

$$P(i,j) = P\{Z_{n+1}=j \mid Z_n=i\} = \begin{cases} p_j^{*i} & \text{if } i \geqslant 1, \quad j \geqslant 0, \\ \delta_{0j} & \text{if } i=0, \quad j \geqslant 0, \end{cases} \tag{1}$$

δ_{ij} *being the Kronecker delta and* $\{p_k^{*i}; k=0,1,2,\ldots\}$ *being the i-fold convolution of* $\{p_k; k=0,1,2,\ldots\}$.

The probability function $\{p_k\}$ is thus the total datum of the problem.

The process can be thought of as representing an evolving population of particles. It starts at time 0 with Z_0 particles, each of which (after one unit of time) splits independently of the others into a random number of offspring according to the probability law $\{p_k\}$. The total number Z_1 of particles thus produced is the sum of Z_0 random variables, each with probability function $\{p_k\}$. It constitutes the first generation. These go on to produce a second generation of Z_2 particles, and so on. The number of "offspring" produced by a single "parent" particle at any given time is independent of the history of the process, and of other particles existing at the present. The number of particles in the n'th generation is a random variable Z_n. Clearly (1) tells us that if $Z_n=0$, then $Z_{n+k}=0$ for all $k \geqslant 0$. Thus 0 is an absorbing state, and reaching 0 is the same as the process becoming extinct.

The study of branching processes has a long history, which, as might be expected, is closely interwoven with a number of applications in the physical and biological sciences. The original problem, which was introduced by Francis Galton in 1873 [see also Galton (1889, 1891)] and first successfully attacked by the Reverend Henry Watson in that year [Watson and Galton (1874)], was in fact concerned with the extinction of family names in the British peerage. For a most enjoyable historical introduction we refer the reader to D. Kendall (1966); and for a complete early bibliography to T. E. Harris (1963).

Notation. When we want to draw particular attention to the initial number of particles, we will let

$$\{Z_n^{(i)}; n=0, 1, 2, \dots\}$$

denote the branching process with i intitial particles. Since most of the time we will be assuming $Z_0=1$, it will be convenient, unless specified to the contrary, to write $Z_n^{(1)}=Z_n$.

Generating Functions. An important tool in the analysis of the process is the generating function

$$f(s)=\sum_{k=0}^{\infty} p_k s^k, \qquad |s|\leqslant 1, \tag{2}$$

and its iterates

$$f_0(s)=s, \qquad f_1(s)=f(s), \qquad f_{n+1}(s)=f[f_n(s)],$$

where s is complex in general, but will be assumed real throughout most of chapter I.

Observe that,

$$\sum_j P(1,j)s^j=f(s); \qquad \sum_j P(i,j)s^j=[f(s)]^i, \qquad i\geqslant 1. \tag{3}$$

Also, letting $P_n(i,j)$ be the n-step transition probabilities, and using the Chapman-Kolmogorov equation, we get

$$\begin{aligned}\sum_j P_{n+1}(1,j)s^j &= \sum_j \sum_k P_n(1,k)P(k,j)s^j = \sum_k P_n(1,k)\sum_j P(k,j)s^j \\ &= \sum_k P_n(1,k)[f(s)]^k.\end{aligned}$$

Thus if we let $\sum P_n(1,j)s^j=f_{(n)}(s)$, then we have shown that

$$f_{(n+1)}(s)=f_{(n)}[f(s)].$$

Hence it follows by induction that

$$f_{(n)}(s)=f_n(s), \tag{4}$$

a crucial formula.

From (3) and (4) we also have

$$\sum_{j=0}^{\infty} P_n(i,j)\, s^j = [f_n(s)]^i . \tag{5}$$

Probabilistic Setting. Although most of our study of the branching process will be directly in terms of the generating functions $f_n(s)$, it is sometimes important to be aware of the probabilistic structure underlying this analytical setting.

From the definition of $\{Z_n\}$ as a Markov chain with a given transition function, we know from general considerations (the Kolmogorov theorem) that there is a probability space $(\Omega, \mathbb{F}, P)$ on which $\{Z_n(\omega); n \geq 0\}$ are defined, and have the distributions determined by (1). This construction does not, however, tell us enough about Ω to assure us that we can define certain other random variables, such as

$$\begin{aligned} Z_{n,k}^{(j)}(\omega) = {} & \text{the number of } k\text{'th generation offspring of the } j\text{'th} \\ & \text{of the } Z_n(\omega) \text{ particles of the } n\text{'th generation.} \end{aligned} \tag{6}$$

To this end we need a space whose points represent entire "family trees", i.e. specify the generation number, ancestors, and offspring of each particle. The construction of such a space has been carried out by T. E. Harris in chapter VI of his book (1963) (actually he treats the more general age-dependent process, which we study in chapter IV). He gives a labeling scheme in which each sample tree is completely specified in terms of a sequence of finite sequences. The σ-field $\mathbb{F}$ is constructed from the cylinder sets of Ω. A probability measure is defined on this space, from which $\{Z_n(\omega); n \geq 0\}$ can be shown to be a Markov chain with the transition mechanism (1), and random variables like (6) can now be seen to be well defined.

By repeated application of (1), we also obtain the following fundamental feature of the Galton-Watson process:

Additive Property. The process $\{Z_n^{(i)}; n=0,1,2,\ldots\}$ is the sum of i independent copies of the branching process $\{Z_n; n=0,1,2,\ldots\}$. In other words, if P_i denotes the measure on $\mathbb{F}$ corresponding to the initial measure $P\{Z_0=i\}=1$, then P_i is the i-fold convolution of P_1. Thus the joint distribution of $(Z_{n_1}^{(i)}, \ldots, Z_{n_k}^{(i)})$, for integers $1 \leqslant n_1 \leqslant \cdots \leqslant n_k$, is the i-fold convolution of the distribution of $(Z_{n_1}, \ldots, Z_{n_k})$.

An assumption. In order to avoid trivialities, we assume throughout that

$$p_0 + p_1 < 1 ,$$

and

$$p_j \neq 1 \quad \text{for any } j .$$

2. Moments

The moments of the process, when they exist, can be expressed in terms of the derivatives of $f(s)$ at $s=1$. For the mean we have

$$E Z_1 = \sum P(1,j)j = f'(1) \equiv m \qquad \text{(say)},$$

and from the chain rule

$$E Z_n = \sum P_n(1,j)j = f_n'(1) = f_{n-1}'(1) f'(1) = \cdots = [f'(1)]^n = m^n. \tag{1}$$

Similarly, using the fact that

$$f_{n+1}''(1) = f''(1)[f_n'(1)]^2 + f'(1) f_n''(1),$$

one can show that

$$f_n''(1) = f''(1)[m^{2n-2} + m^{2n-3} + \cdots + m^{n-1}],$$

and hence, letting $\sigma^2 = \text{variance } Z_1$, conclude that

$$\operatorname{var} Z_n = \begin{cases} \dfrac{\sigma^2 m^{n-1}(m^n - 1)}{(m-1)} & \text{if } m \neq 1, \\ n\sigma^2 & \text{if } m = 1. \end{cases} \tag{2}$$

Higher moments can be derived similarly.

3. Elementary Properties of Generating Functions

Since all the properties of the transition functions $P_n(i,j)$ are contained in the generating functions $f_n(s)$, and since, in particular, the asymptotic behavior of $\{f_n(s)\}$ can be translated into limit theorems about the $\{Z_n\}$ process, we shall eventually want to develop very refined estimates of these functions. We start, however, by showing that even their simple properties lead to interesting results.

Let t be real. From the definition of f as a power series with nonnegative coefficients $\{p_k\}$ adding to 1, and with $p_0 + p_1 < 1$, we see at once that:

(i) *f is strictly convex and increasing in $[0, 1]$;*

(ii) $f(0) = p_0$; $f(1) = 1$;

(iii) *if $m \leq 1$ then $f(t) > t$ for $t \in [0, 1)$;*

(iv) *if $m > 1$ then $f(t) = t$ has a unique root in $[0, 1)$.*

Let q be the smallest root of $f(t) = t$ for $t \in [0, 1]$. Then (i)—(iv) imply that there is such a root, and furthermore:

Lemma 1. *If $m \leq 1$ then $q = 1$; if $m > 1$ then $q < 1$.*

The situation is illustrated in the following diagrams.

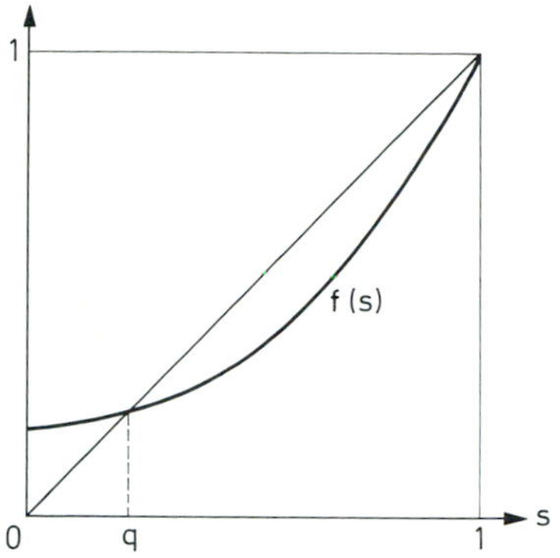

Fig. 1. $m>1$

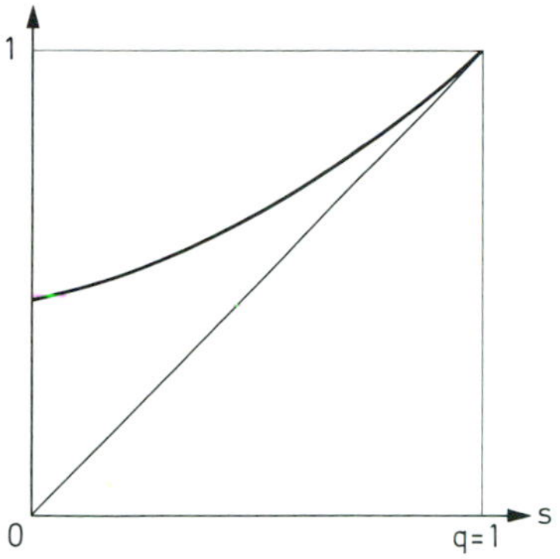

Fig. 2. $m \leqslant 1$

The first order behavior of the iterates $f_n(t)$ is also easy to guess from a picture (see Figs. 3 and 4), and is given in the next lemma.

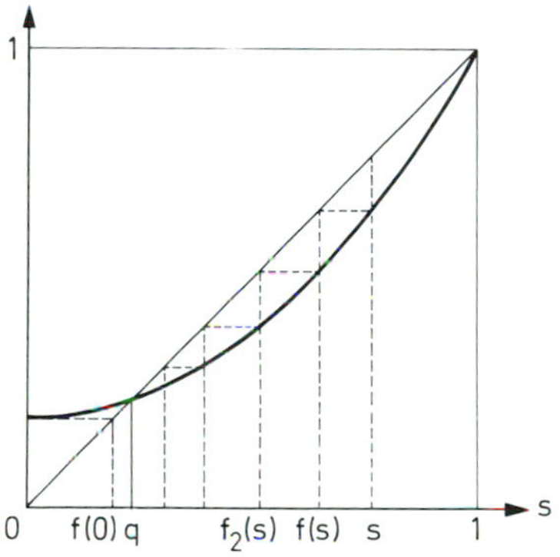

Fig. 3

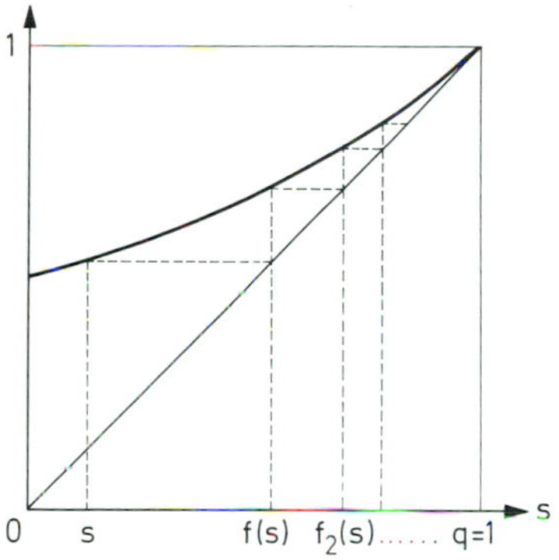

Fig. 4

Lemma 2. *If* $t \in [0, q)$ *then* $f_n(t) \uparrow q$ *as* $n \to \infty$.[1]
If $t \in (q, 1)$ *then* $f_n(t) \downarrow q$ *as* $n \to \infty$.
If $t = q$ *or* 1 *then* $f_n(t) = t$ *for all* n.

Proof. If $0 \leqslant t < q$, then $t < f(t) < f(q)$, and iterating this inequality we have

$$t < f_1(t) < f_2(t) < \cdots < f_n(t) < f_n(q) = q$$

for all $n \geqslant 1$. Thus $f_n(t) \uparrow L \leqslant q$. Since f is continuous we may take limits through the equation $f_{n+1}(t) = f[f_n(t)]$ and conclude that $L = f(L)$. But q is the smallest root in $[0, 1]$, and hence $L = q$.

---

[1] The symbols $\uparrow(\downarrow)$ mean that the convergence is monotone nondecreasing (nonincreasing).

If $q<t<1$, then arguing similarly we show that $1>f_n(t)\downarrow L\geqslant q$, where $L=f(L)$. But by the properties (iii) and (iv) above there are no roots of $t=f(t)$ in $(q,1)$. Hence again $L=q$.

This proves the first two parts of the lemma, and the third is trivial. □

Remark. The convergence $f_n(t)\uparrow q$ for $t\in[0,q)$ is uniform since $f_n(0)\leqslant f_n(t)\leqslant q$.

Arguing as in lemma 2 and using induction we also have

Lemma 3. *The functions $f_n(s)$ are differentiable and converge on $[0,1)$. Moreover for all s in $[q,1)$, $f_n'(s)\leqslant(f'(s))^n$ and for all s in $[0,q)$, $f_n'(s)\geqslant(f'(s))^n$.*

This suggests that $f_n'(s)$ has a geometric rate of decay, a fact which will be established in section 11.

4. An Important Example

There is essentially only one non-trivial example for which the iterates $f_n(s)$ have been explicitly computed. This is called the linear fractional case, because of the form taken by the generating function. (Some other examples can be computed from this one [see e.g. Harris (1963), p. 10].) Suppose

$$p_k=b\,p^{k-1}\qquad k=1,2,\ldots,$$

$$p_0=1-\sum_{i=1}^{\infty}p_i=[1-b-p][1-p]^{-1}.$$

Then one readily computes

$$f(s)=1-\frac{b}{1-p}+\frac{bs}{1-ps},\tag{1}$$

and

$$m=\frac{b}{(1-p)^2}.$$

Now for any two points u,v

$$\frac{f(s)-f(u)}{f(s)-f(v)}=\frac{s-u}{s-v}\cdot\frac{1-pv}{1-pu}.\tag{2}$$

The equation $f(s)=s$ has roots s_0 and 1. If $m>1$ then $s_0<1$; if $m=1$, $s_0=1$; if $m<1$, $s_0>1$. If we take $u=s_0$ and $v=1$ then for $m\neq 1$ the above formula becomes

$$\frac{1-p}{1-ps_0}=\lim_{s\to 1}\left(\frac{f(s)-s_0}{s-s_0}\right)\left(\frac{f(s)-1}{s-1}\right)^{-1}=\frac{1}{m};$$

and hence (2) becomes

$$\frac{f(s)-s_0}{f(s)-1}=\frac{1}{m}\,\frac{s-s_0}{s-1}. \tag{3}$$

Iterating this we get

$$\frac{f_n(s)-s_0}{f_n(s)-1}=\frac{1}{m^n}\,\frac{s-s_0}{s-1} \tag{4}$$

which can be solved explicitly for $f_n(s)$. The answer is

$$f_n(s)=1-m^n\left(\frac{1-s_0}{m^n-s_0}\right)+\frac{m^n\left(\frac{1-s_0}{m^n-s_0}\right)^2 s}{1-\left(\frac{m^n-1}{m^n-s_0}\right)s} \quad \text{if } m\neq 1\,. \tag{5}$$

If $m=1$, then $b=(1-p)^2$ and $s_0=1$. Then

$$f(s)=\frac{p-(2p-1)s}{1-ps},$$

which can be iterated to yield

$$f_n(s)=\frac{np-(np+p-1)s}{1-p+np-nps}. \tag{6}$$

5. Extinction Probability

As a special case of lemma 3.2 we note that $f_n(0)\uparrow q$. But $\lim\limits_{n\to\infty} f_n(0) = \lim\limits_{n} P\{Z_n=0\}=\lim\limits_{n} P\{Z_i=0 \text{ for some } 1\leqslant i\leqslant n\}=P\{Z_i=0 \text{ for some } i\geqslant 1\}$, which is by definition the probability that the process ever becomes extinct. Applying lemma 3.1 we get the classical extinction probability theorem.

Theorem 1. *The extinction probability of the* $\{Z_n\}$ *process is the smallest nonnegative root* (q) *of the equation* $t=f(t)$. *It is* 1 *if* $m\leqslant 1$ *and* <1 *if* $m>1$.

But we can say more [Harris (1963) pp. 8—9]. Note that for $k\geqslant 1$

$$P\{Z_{n+i}\neq k,\ i\geqslant 1\,|\,Z_n=k\}\geqslant\begin{Bmatrix} P(k,0) & \text{if } p_0>0\\ 1-P(k,k) & \text{if } p_0=0\end{Bmatrix}>0\,.$$

Thus all states $k\geqslant 1$ are transient, *and with probability* 1, $Z_n\to 0$ *or* ∞. Together with theorem 1 this implies

Theorem 2. $\lim_{n\to\infty} P\{Z_n=k\}=0$ *for* $k\geq 1$. *Furthermore,*

$$P\left\{\lim_n Z_n=0\right\}=1-P\left\{\lim_n Z_n=\infty\right\}=q\,.$$

Harris (1963), Ch. I, remarks that this instability of the population is contrary to the behavior of biological populations, which tend to reach a state of balance with their environment (though, of course, they may oscillate). The situation is in fact even worse than this theorem suggests. Even in models where particle production depends on present population size, positive states will be transient provided particles reproduce independently of others existing at the present. An unmodified branching process is thus not a satisfactory model for most biological situations. We shall have occasion to return to this point in chapter VI.

When $m<1$, $=1$, or >1, we shall refer to the Galton-Watson process as *subcritical, critical* or *supercritical* respectively.

Remark. We showed in section 3 that $f_n(s)\to q$ for s real and $0\leq s<1$. If s is complex with $|s|<1$ then

$$|f_n(s)-f_n(0)|\leq E\{|s|^{Z_n};\,Z_n\neq 0\}=f_n(|s|)-f_n(0)\to q-q=0\,,$$

and hence we still have

$$f_n(s)\to q \quad \text{for } |s|<1\,.$$

This kind of extension of results about generating functions from real to complex arguments is typical of many results in this chapter. In those cases where the extensions are trivially achieved, we will occasionally (and without further proof) take the liberty of using the complex version of a result which has been proved in only the real case.

Part B. A First Look at Limit Theorems

Our primary objective in this chapter is to prove limit theorems about Z_n. We have already seen that, with probability 1, Z_n does not fluctuate, i.e., Z_n either becomes 0 for large n, or blows up to ∞ as $n\to\infty$.

It is natural to inquire about the nature of this divergence. Part B of this chapter is devoted to answering this question, but imposes some strong hypotheses in order to facilitate the exposition. In Part C we will refine these results and give the best known to date.

6. Motivating Remarks

We start by motivating and outlining the types of limit theorems to be studied. Recall the additive property mentioned in section 1, namely given $Z_n=i$, the stochastic process $\{Z_{n+k}; k=0,1,2,\ldots\}$ is the sum of i independent copies of the process $\{Z_0\equiv 1, Z_1, Z_2, \ldots\}$. Using the Markov property this immediately yields

$$E(Z_{n+k}|Z_n=i_n, Z_{n-1}=i_{n-1}, \ldots Z_1=i_1, Z_0=i_0)$$
$$=E(Z_{n+k}|Z_n=i_n)=i_n E(Z_k|Z_0=1)=i_n m^k.$$

Hence if we set

$$W_n\equiv Z_n m^{-n}, \tag{1}$$

then

$$E(W_{n+k}|W_0, W_1, \ldots, W_n)=W_n \quad \text{a.s.} \tag{2}$$

Thus we get the following result [first observed by J. Doob; see page 13 of Harris (1963)].

Theorem 1. *If* $0<m\equiv f'(1-)<\infty$, $W_n=Z_n m^{-n}$, *and* $\boldsymbol{F}_n$ *is the* σ-*algebra generated by* $Z_0, Z_1, \ldots, Z_n$, *then the sequence* $\{W_n, \boldsymbol{F}_n; n=0,1,2,\ldots\}$ *is a martingale. Furthermore, since* $W_n\geqslant 0$, *there exists a random variable* W *such that*

$$\lim_{n\to\infty} W_n=W \quad \text{a.s.} \tag{3}$$

Proof. Clear from (2) and the fact that any nonnegative martingale converges with probability 1.

From (3) we see that $Z_n(\omega)$ grows like $m^n W(\omega)$. This is the stochastic analogue of the so-called Malthusian law of geometric population growth. Although theorem 1 gives us a neat result with only a very weak hypothesis, it is not very satisfactory in the sense that it does not tell us anything about W. Of course we can conclude by using Fatou's lemma that

$$EW\leqslant \liminf EW_n=EZ_0.$$

This does not, however, rule out the possibility that $W\equiv 0$, in which case theorem 1 says nothing except that m^n is too big a normalizing factor. In fact, if $m\leqslant 1$ we know that with probability 1 Z_n is 0 for sufficiently large n, and hence W is indeed degenerate at 0. Thus the theorem could be meaningful, if at all, only when $m>1$ (the supercritical case). Even in this case it could happen that $P(W=0)=1$. However, if $\sigma^2<\infty$ then we can assert that W is nondegenerate. In fact, we have the following

Theorem 2. *If* $m>1$, $\sigma^2<\infty$, *and* $Z_0\equiv 1$, *then*
(i) $\lim_{n\to\infty} E(W_n-W)^2=0$;

(ii) $EW=1, \quad \operatorname{var} W=\sigma^2/(m^2-m)$;

(iii) $P(W=0)=q\equiv P(Z_n=0 \text{ for some } n)$.

Proof. From (2.2)

$$E\,W_n^2=\frac{E\,Z_n^2}{m^{2n}}=\left\{\frac{\sigma^2(1-m^{-n})}{m^2-m}\right\}+1$$

and hence $\sup_n E\,W_n^2=\lim_n E\,W_n^2=\{\sigma^2/(m^2-m)\}+1<\infty$. Now by standard martingale theory [Doob (1953), p. 319], (i) and (ii) follow.

If $r=P(W=0)$ then $E\,W=1$ implies $r<1$. Furthermore, by conditioning on Z_1, we see that r must satisfy

$$r=\sum_k P(W=0\,|\,Z_1=k)\,P(Z_1=k)=\sum_k p_k[P(W=0)]^k=f(r),$$

and hence must coincide with q. □

Quite a bit more can be said about W and under much weaker hypotheses than in theorem 2 (e.g. $\sigma^2<\infty$ is not necessary). From (1.4)

$$E\,e^{-uW_n}=f_n(e^{-u/m^n})=f\left(f_{n-1}\left(\exp\left\{-\frac{u}{m}\,\frac{1}{m^{n-1}}\right\}\right)\right). \tag{4}$$

Letting $n\to\infty$ we see that $\varphi(u)=E\,e^{-uW}$ satisfies Abel's equation, namely,

$$\varphi(u)=f\left(\varphi\left(\frac{u}{m}\right)\right). \tag{5}$$

This equation is useful only when $P(W=0)<1$, in which case (5) can be used to show that the distribution of W is absolutely continuous on the open interval $(0,\infty)$. We shall prove this fact, and the nondegeneracy of W in part C.

In the subcritical case there are two devices available for a finer study of Z_n for large n. The most obvious is to "assume" ourselves into a nondegenerate situation by conditioning the process on nonextinction. The conditional generating function is given by

$$\sum_{k=0}^{\infty} P(Z_n=k\,|\,Z_n\neq 0)\,s^k=\sum_{k=1}^{\infty}\frac{P(Z_n=k)}{P(Z_n\neq 0)}\,s^k=\frac{f_n(s)-f_n(0)}{1-f_n(0)}. \tag{6}$$

In section 8 we will show that this function converges to a generating function, and will study properties of the limit variable, and some related limit phenomena.

Another way preventing the Z_n process from dying out (instead of conditioning) is to introduce an "immigration" process in some systematic manner. This device has a greater physical appeal than conditioning, and leads to limit variables which are related to the conditioned ones. Immigration processes will be studied in chapter VI.

The critical case plays a distinct role in the theory and leads to some of the nicest results. In this case we will see (in section 9) that the conditional generating function in (6) goes to 0 as $n\to\infty$; i.e. the conditional sequence $\{Z_n|Z_n>0\}$ goes to ∞ in distribution. An idea as to the rate of divergence is given by a simple moment calculation:

$$1=EZ_n=E(Z_n|Z_n>0)P(Z_n>0)+0\cdot P(Z_n=0),$$

implying

$$E(Z_n|Z_n>0) = \frac{1}{P(Z_n>0)}.$$

We will show that $P(Z_n>0)=1-P(Z_n=0)$ is $\sim$ constant/n, and hence the conditional expectation goes up linearly. This suggests studying the conditional process

$$\left\{\frac{Z_n}{n}\,\middle|\, Z_n>0\right\}, \qquad n>0,$$

and indeed we shall see that this process converges to a negative exponential distribution. Again immigration provides an appealing alternative to conditioning.

The above results form the core of the classical limit laws for the Galton-Watson process. In addition to these we will prove a number of variants involving different types of conditioning, and different ways of taking limits (part D). We shall also see that there is a class of conditional limit laws in the supercritical case which are closely related to the subcritical situation.

Before turning to these basic results we will develop (in section 7) some useful properties of the iterates of the Markov transition function $P(i,j)$.

7. Ratio Theorems

In the present section we shall prove theorems about ratios of transition functions. Although the results are of interest in their own right, they are slightly technical in nature, and the main reason for giving them at this time is that they will be useful tools in the proofs of the conditional limit laws in the next section.

The present formulation and proofs are due to Papangelou (1968). For related work see also the survey paper of Seneta (1969).

They key idea is the simple but very useful *monotone ratio lemma.*

Lemma 1. *The j'th derivatives* $(d^j/ds^j)f_n(s)\equiv f_n^{(j)}(s)$ *satisfy*

$$f_n^{(j)}(s)=a_{n,j}(s)+f'[f_{n-1}(s)]\,f_{n-1}^{(j)}(s), \qquad n,j\geqslant 1, \tag{1}$$

where $a_{n,j}(s)$ *is a power series with nonnegative coefficients.*

Proof. If $j=1$ then (1) holds with $a_{n,j}(s)=0$. Suppose (1) is true for some fixed j. Then differentiation shows it to be true for $j+1$ with

$$a_{n,j+1}(s)=a'_{n,j}(s)+f''[f_{n-1}(s)]f'_{n-1}(s)f^{(j)}_{n-1}(s),$$

and thus the lemma is true by induction. □

Lemma 2. *(Monotone Ratio). Assume that* $p_1 \neq 0$. *Then*

$$\frac{P_n(1,j)}{P_n(1,1)} \uparrow \pi_j \leqslant \infty, \qquad j \geqslant 1. \tag{2}$$

Proof. The hypothesis implies that $P_n(1,1)>0$ for all n. Thus

$$\frac{P_{n+1}(1,j)}{P_{n+1}(1,1)}=\frac{1}{j!}\frac{f^{(j)}_{n+1}(0)}{f'_{n+1}(0)}=\frac{1}{j!}\frac{a_{n+1,j}(0)+f'[f_n(0)]f^{(j)}_n(0)}{f'[f_n(0)]f'_n(0)}$$
$$\geqslant\frac{1}{j!}\frac{f^{(j)}_n(0)}{f'_n(0)}=\frac{P_n(1,j)}{P_n(1,1)},$$

where the inequality follows from lemma 1. Hence the ratios converge to a limit (to be shown $<\infty$), which we denote by π_j. □

The π_j's defined above have interesting probabilistic interpretations. We proceed to develop some of their properties. From now on let

$$\gamma=f'(q), \tag{3}$$

and

$$\mathscr{P}(s)=\sum_{n=1}^{\infty}\pi_n s^n \tag{4}$$

whenever the series converges.

Assumption. Throughout the rest of this section we assume that $p_1>0$.

Theorem 1. $\{\pi_j; j=1,2,\ldots\}$ *always satisfy the equations*

$$\gamma\pi_j=\sum_{k=1}^{\infty}\pi_k P(k,j), \qquad j\geqslant 1; \tag{5}$$

and $\mathscr{P}(\cdot)$ *satisfies*

$$\mathscr{P}[f(s)]=\gamma\mathscr{P}(s)+\mathscr{P}(p_0), \tag{6}$$

whenever s *and* p_0 *are in the region of convergence of* $\mathscr{P}$.

Remark. We will see in the next theorem that the radius of convergence of $\mathscr{P}$ is $\geqslant 1$. The role of the π's as stationary measures will be discussed in chapter II.

Proof. Note that

$$P_{n+1}(1,1)=f'_{n+1}(0)=f'[f_n(0)]f'_n(0)=f'[f_n(0)]P_n(1,1),$$

and hence

$$\frac{P_{n+1}(1,1)}{P_n(1,1)}\uparrow\gamma. \tag{7}$$

From

$$P_{n+1}(1,j)=\sum_{k=0}^{\infty}P_n(1,k)P(k,j)$$

we get

$$\frac{P_{n+1}(1,j)}{P_{n+1}(1,1)}\cdot\frac{P_{n+1}(1,1)}{P_n(1,1)}=\sum_{k=1}^{\infty}\frac{P_n(1,k)}{P_n(1,1)}P(k,j).$$

Taking limits of both sides of this expression, (5) follows from the monotone ratio lemma and the monotone convergence theorem; and taking generating functions of both sides then yields (6). □

Theorem 2. *$\mathscr{P}(s)$ converges for $0\leqslant s<1$. Furthermore, $\sum\pi_i$ converges if $m<1$ and diverges if $m\geqslant 1$.*

Proof. Let $\mathsf{S}=\{j\geqslant 1: P_n(1,j)>0$ for some $n\geqslant 1\}$. Due to (5) and the hypothesis $p_1>0$, we can infer that (see complements)

$$\pi_k<\infty,\qquad k\geqslant 1, \tag{8}$$

and

$$\pi_k>0\quad\text{for } k\in\mathsf{S}.$$

Applying (5) again we see that

$$\gamma=\gamma\pi_1=\sum_{k=1}^{\infty}\pi_k P(k,1)=\sum_{k=1}^{\infty}\pi_k k p_0^{k-1}p_1\geqslant\frac{p_1}{p_0}\sum_{k=1}^{\infty}\pi_k p_0^k=\frac{p_1}{p_0}\mathscr{P}(p_0).$$

Thus $\mathscr{P}(p_0)<\infty$, and (6) then implies that $\mathscr{P}[f_n(p_0)]<\infty$ for all $n>0$, which in turn implies

$$\mathscr{P}(s)<\infty\quad\text{for } 0\leqslant s<q. \tag{9}$$

Suppose $m>1$ and $q\leqslant s<1$. Then

$$\mathscr{P}(s)=\sum_{j=1}^{\infty}\pi_j s^j=\lim_{n\to\infty}\sum_{j=1}^{\infty}\frac{P_n(1,j)}{P_n(1,1)}s^j=\lim_{n\to\infty}\frac{f_n(s)-f_n(0)}{f'_n(0)}\leqslant\lim_{n\to\infty}\frac{f'_n(s)}{f'_n(0)}. \tag{10}$$

We will see in theorem 11.1 that $\gamma^{-n}f'_n(s)$ converges to a finite positive function. Modulo this fact, the proof of the first part is complete.

Now returning to (6) iteration yields

$$\mathscr{P}[f_n(s)]=\gamma^n\mathscr{P}(s)+(\gamma^{n-1}+\cdots+\gamma+1)\mathscr{P}(p_0). \tag{11}$$

If $m=1$ then $\gamma=1$ and the right side $\uparrow\infty$ as $n\uparrow\infty$. Since $f_n(s)\uparrow 1$ in this case we have $\mathscr{P}(1)=\infty$. If $m<1$ the right side converges and $\mathscr{P}(1)<\infty$.

If $m>1$ then $\gamma<1$, and we can write

$$\mathscr{P}(s)=\frac{1}{\gamma^n}\left(\mathscr{P}[f_n(s)]-\frac{1-\gamma^n}{1-\gamma}\mathscr{P}(p_0)\right).$$

Substituting q in (6) we easily check that

$$\frac{\mathscr{P}(p_0)}{1-\gamma}=\mathscr{P}(q)$$

and hence

$$\mathscr{P}(s)=\frac{1}{\gamma^n}(\mathscr{P}[f_n(s)]-(1-\gamma^n)\mathscr{P}(q))\geqslant\frac{\mathscr{P}[f_n(s)]-\mathscr{P}(q)}{\gamma^n}.$$

But clearly we can find a sequence $s_n\uparrow 1$ as $n\uparrow\infty$ such that $f_n(s_n)\uparrow 1>q$. Since $\gamma^n\downarrow 0$ we will have $\mathscr{P}(s_n)\uparrow\infty$. Thus $\mathscr{P}(1)=\infty$. This proves the theorem. □

Theorem 3 *(Uniqueness). The solution of* (5) *is unique up to a multiplicative constant. That of* (6) *on* $[0,q)$ *is unique (up to a multiplicative constant) among power series vanishing at zero and having non-negative coefficients.*

Proof. We prove the uniqueness assertion for (6). This will then imply that for (5). Suppose $\mathscr{R}$ is another solution in the stated class. Then $\mathscr{R}$ will also satisfy (11), and differentiating the latter we see that both $\mathscr{P}$ and $\mathscr{R}$ will satisfy

$$\mathscr{P}'[f_n(s)]f_n'(s)=\gamma^n\mathscr{P}'(s). \tag{12}$$

Now for any $s\in[0,q)$ there will be a k such that

$$f_k(0)\leqslant s\leqslant f_{k+1}(0),$$

and hence by (12)

$$\frac{\mathscr{R}'(s)}{\mathscr{P}'(s)}=\frac{\mathscr{R}'[f_n(s)]}{\mathscr{P}'[f_n(s)]}\leqslant\frac{\mathscr{R}'[f_{n+k+1}(0)]}{\mathscr{P}'[f_{n+k}(0)]}=\frac{\mathscr{R}'[f_{n+k+1}(0)]}{\mathscr{P}'[f_{n+k+1}(0)]}\cdot\frac{\mathscr{P}'[f_{n+k+1}(0)]}{\mathscr{P}'[f_{n+k}(0)]}.$$

But by the first equality above

$$\frac{\mathscr{R}'[f_n(0)]}{\mathscr{P}'[f_n(0)]}=\frac{\mathscr{R}'(0)}{\mathscr{P}'(0)}=\text{constant},\qquad n\geqslant 1,$$

and hence

$$\begin{aligned}\frac{\mathscr{R}'(s)}{\mathscr{P}'(s)}&\leqslant\frac{\mathscr{R}'(0)}{\mathscr{P}'(0)}\cdot\frac{\mathscr{P}'[f_{n+k+1}(0)]}{\mathscr{P}'[f_{n+k}(0)]}\\&=\frac{\mathscr{R}'(0)}{\mathscr{P}'(0)}\cdot\frac{f_{n+k}'(0)}{f_{n+k+1}'(0)}\gamma\qquad\text{(applying (12) again)}\\&=\frac{\mathscr{R}'(0)}{\mathscr{P}'(0)}\cdot\frac{\gamma}{f'(f_{n+k}(0))}.\end{aligned}$$

Letting $n \to \infty$ we get $\mathscr{R}'(s)/\mathscr{P}'(s) \leqslant \mathscr{R}'(0)/\mathscr{P}'(0)$. The converse inequality is established similarly. Since $\mathscr{R}(0) = \mathscr{P}(0)$, this implies the theorem. □

We conclude with the main limit theorem. Let $\tau_i = i q^{i-1}$. Note that the τ's satisfy $\gamma \tau_i = \sum_{j=1}^{\infty} P(i,j)\tau_j$, and thus play a dual role to the π's.

Theorem 4. *Ratio Theorem.*

$$\lim_{n\to\infty} \frac{P_{n+m}(i,j)}{P_n(k,l)} = \gamma^m \frac{\tau_i \pi_j}{\tau_k \pi_l}.$$

Proof. For $0 \leqslant s < 1$ we have

$$\begin{aligned}
\sum_{j=1}^{\infty} \left\{\frac{P_n(i,j)}{P_n(1,1)}\right\} s^j &= \frac{f_n^i(s) - f_n^i(0)}{P_n(1,1)} \\
&= \frac{f_n(s) - f_n(0)}{P_n(1,1)} \cdot [f_n^{i-1}(s) + f_n^{i-2}(s) f_n(0) + \cdots + f_n^{i-1}(0)] \\
&= [f_n^{i-1}(s) + f_n^{i-2}(s) f_n(0) + \cdots + f_n^{i-1}(0)] \sum_{j=1}^{\infty} \left\{\frac{P_n(1,j)}{P_n(1,1)}\right\} s^j \\
&\to i q^{i-1} \sum_{j=1}^{\infty} \pi_j s^j < \infty
\end{aligned}$$

due to lemma 2 and theorem 2. Thus by the continuity theorem for generating functions, $\dfrac{P_n(i,j)}{P_n(1,1)} \to \tau_i \pi_j$. But by (7) $\dfrac{P_{n+m}(1,1)}{P_n(1,1)} \uparrow \gamma^m$. Combining these two facts proves the theorem. □

8. Conditioned Limit Laws

When $m < 1$ the process dies out with probability one (section 5). To describe its asymptotic behavior in these cases Kolmogorov (1938) and Yaglom (1947) introduced the device of conditioning Z_n on the event $\{Z_n > 0\}$. The main result of this section is the corollary of theorem 1, which states that when $m < 1$ the distribution of $\{Z_n | Z_n > 0\}$ converges to a proper distribution.

This result was first proved by Yaglom (1947) under moment restrictions. The proof was simplified and the moment assumptions removed by Joffe (1967) and Seneta, Vere-Jones (1967).

Theorem 1 is not restricted to $m < 1$, and brings out the symmetry between the sub and super-critical processes. It was observed by Papangelou (1968), Seneta (1969) (and probably others).

Let $\{\pi_j\}$ be the sequence whose existence was proved in lemma 7.2, and recall that $\mathsf{S}=\{j\geq 1: P_n(1,j)>0 \text{ for some } n\geq 1\}$. When $p_1>0$ and $q>0$ we can define

$$\tilde{b}_j=\frac{\pi_j q^j}{\sum_{j=1}^{\infty}\pi_j q^j}\geq 0\,. \tag{1}$$

Then by theorem 7.2 if $m\neq 1$

$$\tilde{b}_j>0 \quad \text{for } j\in\mathsf{S} \quad \text{and } \sum\tilde{b}_j=1\,; \tag{2}$$

while if $m=1$

$$\tilde{b}_j=0 \quad \text{for } j\geq 1\,.$$

Let T denote the extinction time of the Galton-Watson process, i.e.

$$T=k \Leftrightarrow Z_{k-1}>0, \qquad Z_k=0\,. \tag{3}$$

Theorem 1. *Assume that* $q>0$. *Then*

(i) $$\lim_{n\to\infty} P\{Z_n=j|n<T<\infty\}=b_j \quad \textit{exists}. \tag{4}$$

(ii) *If* $m\neq 1$ *then* b_j *is a probability function and its generating function* $\mathscr{B}(s)=\sum b_j s^j$ *is the unique solution of the equation*

$$\mathscr{B}\left(\frac{f(sq)}{q}\right)=\gamma\,\mathscr{B}(s)+(1-\gamma) \tag{5}$$

among generating functions vanishing at 0.

(iii) *If* $p_1>0$, *then* $b_j=\tilde{b}_j$ *and* $\mathscr{B}(s)=\mathscr{P}(qs)/\mathscr{P}(q)$.

Proof. Suppose first that $m\leq 1$. Then $P\{T<\infty\}=1$, and hence

$$\mathscr{B}_n(s)\equiv E(s^{Z_n}|n<T<\infty)=E(s^{Z_n}|n<T)=\frac{f_n(s)-f_n(0)}{1-f_n(0)}=1-\frac{1-f_n(s)}{1-f_n(0)}\,.$$

Let $G_n(s)=(1-f_n(s))/(1-f_n(0))$, and $\Gamma(s)=(1-f(s))/(1-s)$. Then

$$G_n(s)=\frac{\Gamma[f_{n-1}(s)]}{\Gamma[f_{n-1}(0)]}\,G_{n-1}(s)\,. \tag{6}$$

Now $\Gamma(s)$ and $f_n(s)$ are increasing in s and hence $G_n(s)$ is increasing in n. Thus $\lim_{n\to\infty} G_n(s)\equiv G(s)$ and $\lim_{n\to\infty}\mathscr{B}_n(s)=\mathscr{B}(s)$ exist, and $\mathscr{B}(s)=1-G(s)$. This proves (i) when $m\leq 1$.

Now from the definitions of b_n and Γ

$$G_n(f(s))=G_{n+1}(s)\,\Gamma(f_n(0))\,. \tag{7}$$

Since $f_n(0)\uparrow 1$ as $n\uparrow\infty$ and $\Gamma(x)\uparrow m$ as $x\uparrow 1$, the above relation implies that

$$G(f(s))=mG(s) \tag{8}$$

or

$$\mathscr{B}(f(s))=m\mathscr{B}(s)+(1-m). \tag{9}$$

If $m<1$ then $\gamma=m$ and $q=1$, and thus (5) is established in this case. Letting $s\uparrow 1$ in (8) yields $G(1-)=mG(1-)$, and when $m<1$ this implies that $G(1-)=0$ or $\mathscr{B}(1-)=1$. Thus $\{b_j\}$ is a probability function. The proof of uniqueness in (5) is exactly the same as in theorem 7.3.

If $m>1$, then

$$\sum_{j=1}^{\infty} s^j P\{Z_n=j|n<T<\infty\}=\frac{\sum_{j=1}^{\infty} s^j P\{Z_n=j, n<T<\infty\}}{P\{n<T<\infty\}}$$

$$=\frac{\sum P\{Z_n=j\}q^j s^j}{\sum P\{Z_n=j\}q^j}=\frac{f_n(sq)-f_n(0)}{q-f_n(0)}$$

$$=\frac{f_n^*(s)-f_n^*(0)}{1-f_n^*(0)}$$

where $f_n^*(s)$ is the n-fold iterate of $f^*(s)=q^{-1}f(qs)$. Note that $f^*(s)$ is a generating function with mean

$$m^*=f^{*\prime}(1)=f'(q)=\gamma<1. \tag{10}$$

Hence the above argument for the subcritical case applies identically with $f(s)$ replaced by $f^*(s)$. (For a probabilistic interpretation of f^*, see theorems 12.3 and 12.4.) Thus $\{b_j\}$ is again a probability function, and substituting f^* for f in (9) yields (5). This completes the proof of (i) and (ii).

When $p_1>0$, then by lemma 7.2

$$\begin{aligned}P\{Z_n=j|n<T<\infty\}&=\frac{P_n(1,j)q^j}{\sum_{k=1}^{\infty} P_n(1,k)q^k}\\&=\frac{\left[\dfrac{P_n(1,j)}{P_n(1,1)}\right]q^j}{\sum_k\left[\dfrac{P_n(1,k)}{P_n(1,1)}\right]q^k}\to\frac{\pi_j q^j}{\sum_k \pi_k q^k}=\tilde{b}_j.\end{aligned} \tag{11}$$

(Note that the denominator converges by theorem 7.2). The generating functions $\mathscr{B}(s)$ and $\mathscr{P}(s)$ are related by

$$\mathscr{B}(s) = \frac{\mathscr{P}(qs)}{\mathscr{P}(q)} \tag{12}$$

and this with (7.6) gives another proof of (5). □

Corollary 1 *(Yaglom's Theorem). If* $m<1$ *then* $P\{Z_n=j|Z_n>0\}$ *converges (as* $n\to\infty$*) to a probability function whose generating function* $\mathscr{B}$ *satisfies* (9).

Remark. We will show in section 11 that if $m<1$ then the mean of the limit distribution $\{b_j\}$ is finite if and only if $EZ_1 \log Z_1 < \infty$; and in section 12 that if $m>1$ then $\sum j^r b_j < \infty$ for all $r>0$.

Although (4) is valid even when $m=1$, it is not very interesting in this case since all the limits are 0. ($\mathscr{B}[f(s)]=\mathscr{B}(s)$ and $f(s)>s$ imply $\mathscr{B}(s)=\text{constant}=\mathscr{B}(0)=0$). However if instead of conditioning on eventual extinction we condition on extinction *at the next stage*, then we get a nondegenerate result. The following result was first noted by Seneta (1967a), but it is Papangelou's lemma that again leads to the trivial proof.

Theorem 2. *If* $m=1$, *and* $p_1>0$ *then*

$$\lim_{n\to\infty} P\{Z_n=j|Z_n>0, Z_{n+1}=0\} = \theta_j = \frac{\pi_j p_0^j}{\sum \pi_i p_0^i} \tag{13}$$

where $\theta_j \geqslant 0$, $\sum \theta_j = 1$. *The generating function* $\Theta(s)=\sum \theta_j s^j$ *is related to* $\mathscr{P}(s)$ *by*

$$\Theta(s) = \frac{\mathscr{P}(p_0 s)}{\mathscr{P}(p_0)}. \tag{14}$$

Proof.

$$P\{Z_n=j|Z_n>0, Z_{n+1}=0\} = \frac{P_n(1,j)\,p_0^j}{\sum_{i=1}^{\infty} P_n(1,i)\,p_0^i}.$$

Dividing numerator and denominator by $P_n(1,1)$ and letting $n\to\infty$, the right side becomes (by the monotone ratio lemma)

$$\frac{\pi_j p_0^j}{\sum \pi_i p_0^i} = \theta_j.$$

The denominator converges by theorem 7.2, proving (13). Formula (14) is immediate. □

Remark. In the critical case we thus have a probabilistic interpretation of π_j as

$$\pi_j = p_0^{-j} \lim_{n\to\infty} P\{Z_n=j|Z_n>0, Z_{n+1}=0\} \cdot \mathscr{P}(p_0).$$

9. The Exponential Limit Law for the Critical Process

The case $m=1$ is in a sense the most interesting. Theorem 5.2 tells us that $Z_n \to 0$ with probability 1. On the other hand the limit probabilities b_j of the sequence of conditional distributions of $\{Z_n|Z_n>0\}$ are 0 (see section 8), and hence this process is diverging to ∞.

We shall see that a further normalization is needed to make the conditioned process converge to a nondegenerate limit. The suitably modified process will then converge to a universal limit law (which will have the same form for all f's). This is in sharp contrast to the noncritical cases studied till now. (Also compare with section 10).

Our approach will depend on a more careful analysis of the asymptotic behavior of $f_n(t)$ than we have required till now. We have seen in section 3 that $f_n(t)\uparrow 1$, and now need to know the rate of convergence. To that end the following lemma will be important.

Basic Lemma. *If* $m=EZ_1=1$ *and* $\sigma^2=\operatorname{var} Z_1<\infty$, *then*

$$\lim_{n\to\infty} \frac{1}{n}\left[\frac{1}{1-f_n(t)}-\frac{1}{1-t}\right]=\frac{\sigma^2}{2} \tag{1}$$

uniformly for $0\leqslant t<1$.

We will prove this lemma later; in fact, two proofs will be given because of the importance of their methods.

First, however, we examine some consequences of the lemma.

We already know that $P\{Z_n>0\}\to 0$ as $n\to\infty$ in the present case. Setting $t=0$ in the lemma, and noting that $P\{Z_n>0\}=1-f_n(0)$, we obtain the following estimate of the rate of convergence to 0 (first proved by Kolmogorov (1938) under a third moment assumption).

Theorem 1. *If* $m=1$ *and* $\sigma^2<\infty$ *then as* $n\to\infty$

$$[1-f_n(0)]=P\{Z_n>0\}\sim\frac{2}{n\sigma^2}. \tag{2}$$

Remark. If $\sigma^2=\infty$ the result remains correct with the interpretation that $\lim nP\{Z_n>0\}=0$. The rate of convergence to 0 has also been studied when $\{p_k;k=0,1,\dots\}$ is in the domain of a stable law. (See Slack (1968).)

We have remarked that the conditioned process $\{Z_n|Z_n>0\}$ diverges, and theorem 1 can be used to give a first idea of its growth rate. As was already observed (see section 6)

$$E(Z_n|Z_n>0)=\frac{1}{P(Z_n>0)},$$

and thus

$$E(Z_n|Z_n>0) \sim \frac{n\sigma^2}{2}. \tag{3}$$

Since the mean of the conditioned process is growing at the rate n, it is reasonable to cut the process down by a factor of n. This leads to the following result (originally proved by Yaglom (1947)—again under a third moment assumption).

Theorem 2. *If* $m=1$ *and* $\sigma^2<\infty$ *then*

$$\lim_{n\to\infty} P\left\{\frac{Z_n}{n} > z|Z_n>0\right\} = \exp\left\{-\frac{2z}{\sigma^2}\right\}, \qquad z\geqslant 0. \tag{4}$$

Remark. Again if $\sigma^2=\infty$ then this theorem is still correct with the interpretation that the limit in (4) equals 1, and there is an appropriate refinement for $\{p_k\}$ in the domain of a stable law [Slack (1968)].

Proof of theorem 2. We shall show that

$$\lim_{n\to\infty} E\left[e^{-\alpha\left(\frac{Z_n}{n}\right)}|Z_n>0\right] = \frac{1}{1+\frac{\alpha\sigma^2}{2}}. \tag{5}$$

This is sufficient since the right side of (5) is the Laplace transform of the right side of (4). But the conditional expectation in (5) equals

$$\frac{f_n\left(e^{-\frac{\alpha}{n}}\right)-f_n(0)}{1-f_n(0)} = 1 - \frac{\{n[1-f_n(0)]\}^{-1}}{\left\{n\left[1-f_n\left(e^{-\frac{\alpha}{n}}\right)\right]\right\}^{-1}}. \tag{6}$$

By the basic lemma

$$\lim_{n\to\infty} \frac{1}{n[1-f_n(0)]} = \frac{\sigma^2}{2},$$

and (using the uniform convergence)

$$\lim_{n\to\infty} \frac{1}{n\left[1-f_n\left(e^{-\frac{\alpha}{n}}\right)\right]} = \frac{\sigma^2}{2} + \lim_{n\to\infty} \frac{1}{n\left(1-e^{-\frac{\alpha}{n}}\right)} = \frac{\sigma^2}{2}+\frac{1}{\alpha};$$

which with (6) implies (5) and the theorem. □

Two Proofs of the Basic Lemma.

Proof 1. (Kesten, Ney, Spitzer (1966)). This proof involves a direct estimation procedure. We note that by L'Hospital's rule

$$\lim_{t\uparrow 1} \frac{f(t)-t}{(1-t)^2} = \lim_{t\uparrow 1} \frac{f'(t)-1}{2(t-1)} = \frac{f''(1)}{2} \equiv a \qquad \text{(say)}.$$

Let

$$\varepsilon(t)=a-\frac{f(t)-t}{(1-t)^2}.$$

Then it can be checked (see complements) that

$$\varepsilon(t)\geqslant 0 \quad \text{and} \quad \varepsilon(t)\downarrow 0 \quad \text{as} \quad t\uparrow 1. \tag{7}$$

Thus

$$\frac{f(t)-t}{(1-t)^2}\leqslant a.$$

Now let

$$\delta(t)=a-\left[\frac{1}{1-f(t)}-\frac{1}{1-t}\right]. \tag{8}$$

But $t\leqslant f(t)$ implies that

$$\frac{1}{1-f(t)}-\frac{1}{1-t}\geqslant\frac{f(t)-t}{(1-t)^2},$$

and hence

$$\delta(t)\leqslant\varepsilon(t). \tag{9}$$

Substituting $f_i(t)$ for t in (9), summing over i, and using (7), we see that

$$an-\left[\frac{1}{1-f_n(t)}-\frac{1}{1-t}\right]=\sum_{i=0}^{n-1}\delta(f_i(t))\leqslant\sum_{i=0}^{n-1}\varepsilon[f_i(t)]\leqslant\sum_{i=0}^{n-1}\varepsilon[f_i(0)]=o(n), \tag{10}$$

where the last equality holds since $\varepsilon[f_i(0)]\to 0$ as $i\to\infty$. To get a bound in the other direction we observe that

$$\delta(s)=\left(a\cdot\frac{1-f(s)}{1-s}-\frac{f(s)-s}{(1-s)^2}\right)\left(\frac{1-f(s)}{1-s}\right)^{-1}\geqslant a\,\frac{s-f(s)}{1-f(s)}$$

due to (7). But by definition of $\varepsilon(s)$, $\dfrac{f(s)-s}{1-s}=(1-s)(a-\varepsilon(s))$, and hence the above inequality becomes

$$\begin{aligned}\delta(s)&\geqslant -a\,\frac{f(s)-s}{1-s}\,\frac{1-s}{1-f(s)}=-a(1-s)(a-\varepsilon(s))\frac{1-s}{1-f(s)}\\&\geqslant -a^2(1-s)\left[\frac{1-s}{1-f(s)}\right]\geqslant -a^2(1-s)\cdot\frac{1}{1-f(0)},\end{aligned}$$

since $\dfrac{1-s}{1-f(s)}$ is decreasing and positive on $[0,1]$. Hence

$$\sum_{k=0}^{n-1}\delta[f_k(t)]\geqslant-\frac{a^2}{1-f(0)}\sum_{k=0}^{n-1}[1-f_k(t)]\geqslant\frac{-a^2}{1-f(0)}\sum_{k=0}^{n-1}[1-f_k(0)]=o(n). \tag{11}$$

Dividing (10) and (11) through by n and letting $n \to \infty$ then implies the lemma. □

Proof 2. The following proof is due to F. Spitzer (unpublished). Although less direct than the proof we have just given, it is based on a very ingenious idea which will be suggestive in another setting later on.

We have seen one (and only one) example for which $f_n(s)$ could be explicitly evaluated, namely the linear fractional. The present proof starts by observing that (1) is trivially true when $f(s)$ is linear fractional. This follows from (4.6) since the latter implies

$$\frac{1}{1-f_n(t)} - \frac{1}{1-t} = n\frac{p}{1-p} = n\frac{\sigma^2}{2}, \qquad 0 \leqslant t < 1. \tag{12}$$

The idea then is to bound an arbitrary generating function f, $f'(1)=1$, $f''(1)<\infty$, between two linear fractional generating functions. Knowledge of the asymptotic behavior of the bounding functions then yields the result for the original function. The basis of this argument is the following

Comparison Lemma (Spitzer). *If $f^{(1)}(t)$ and $f^{(2)}(t)$ are the generating functions of critical branching processes with variances $\sigma_1^2 < \sigma_2^2 \leqslant \infty$, then there exist integers n_1, n_2 such that*

$$f_{n+n_1}^{(1)}(t) \leqslant f_{n+n_2}^{(2)}(t) \tag{13}$$

for $0 \leqslant t \leqslant 1$, $n = 0, 1, 2, \ldots$.

Proof. From (7) and the fact that $\sigma_1^2 < \sigma_2^2$ it follows that there is a $0 < t_0 < 1$ such that

$$f^{(1)}(t) \leqslant f^{(2)}(t) \quad \text{for } t_0 \leqslant t \leqslant 1, \tag{14}$$

and hence that

$$f_n^{(1)}(t) \leqslant f_n^{(2)}(t), \qquad t_0 \leqslant t \leqslant q, \qquad 1 \leqslant n. \tag{15}$$

Now choose n_1 and then $n_2 > n_1$ so that

$$t_0 \leqslant f_{n_1}^{(1)}(0) \quad \text{and} \quad f_{n_1}^{(1)}(t_0) \leqslant f_{n_2}^{(2)}(0). \tag{16}$$

(This choice is possible since the $f_n^{(i)}(t) \uparrow 1$ as $n \to \infty$, $i = 1, 2$).

Now if $t_0 \leqslant t \leqslant 1$ then (13) clearly holds. If $0 \leqslant t \leqslant t_0$, then

$$t_0 \leqslant f_{n_1}^{(1)}(0) \leqslant f_{n_1}^{(1)}(t) \leqslant f_{n_1}^{(1)}(t_0) \leqslant f_{n_2}^{(2)}(0) \leqslant f_{n_2}^{(2)}(t), \tag{17}$$

and since all the above terms are $\geqslant t_0$, (13) again follows. □

Now return to the proof of the basic lemma. Recall that $f''(1) = \sigma^2$ = the variance of the given distribution. Choose an arbitrary $\varepsilon > 0$.

Take $f^{(1)}$ in the comparison lemma to be the given generating function f, and $f^{(2)}$ to be the linear fractional generating function with variance $(1+\varepsilon)\sigma^2$. Then the lemma tells us that there are integers n_1, n_2 such that

$$f_{n_1+n}(t) \leqslant f^{(2)}_{n_2+n}(t), \qquad n \geqslant 0, \qquad 0 \leqslant t \leqslant 1,$$

or

$$\begin{aligned}\frac{1}{1-f_{n_1+n}(t)} - \frac{1}{1-t} &\leqslant \frac{1}{1-f^{(2)}_{n+n_2}(t)} - \frac{1}{1-t} \\ &= (n+n_2)(1+\varepsilon)\frac{\sigma^2}{2}, \qquad 0 \leqslant t < 1,\end{aligned} \tag{18}$$

where the equality on the right is due to (12). Similarly the comparison lemma can be used to get a lower bound since it assures us of the existence of integers m_1, m_2 such that

$$(n+m_1)(1-\varepsilon)\frac{\sigma^2}{2} \leqslant \frac{1}{1-f_{m_2+n}(t)} - \frac{1}{1-t}, \qquad 0 \leqslant t < 1. \tag{19}$$

Finally (18) and (19) imply (1). □

We conclude this section by stating two useful corollaries of the above work. From the basic lemma it follows that (see complements)

Corollary 1. *If* $m=1$ *and* $\sigma^2 < \infty$ *then*

$$\lim_{n\to\infty} n^2 [f_{n+1}(s) - f_n(s)] = \frac{2}{\sigma^2}, \qquad s < 1. \tag{20}$$

Note that

$$P\{Z_n = j | Z_0 = 1, Z_n > 0, Z_{n+1} = 0\} = \frac{P_n(1,j)\, p_0^j}{f_{n+1}(0) - f_n(0)}.$$

If $p_1 > 0$, then theorem 8.2 and the above corollary imply that

$$\lim_{n\to\infty} n^2 P_n(1,j) = \frac{2\theta_j}{\sigma^2 p_0^j} = \frac{2\pi_j}{\sigma^2 \mathscr{P}(p_0)}. \tag{21}$$

Applying theorem 7.4 we thus get

Corollary 2. *If* $p_1 > 0$, $m = 1$ *and* $\sigma^2 < \infty$ *then for* $j \geqslant 1$

$$\lim_{n\to\infty} n^2 P_n(i,j) = c\, i\, \pi_j,$$

where $c = 2/\sigma^2 \mathscr{P}(p_0)$.

Part C. Finer Limit Theorems

10. Strong Convergence in the Supercritical Case

We will now go deeper into the behavior of Z_n, and develop more refined limit properties; usually under weaker assumptions than in part B.

We start with the random variable W, to which the sequence $W_n = Z_n/m^n$ converges strongly (i.e. a.s.). We know that $P\{W=0\}=1$ when $m \leqslant 1$. When $m>1$ we have already seen (theorem 6.2) that $\sigma^2<\infty$ is sufficient to make $P\{W=0\}<1$. Levinson (1959) was the first to observe that if $\sigma^2=\infty$ then $P\{W=0\}$ could be 1, but nothing as strong as $\sigma^2<\infty$ is needed for the nondegeneracy of W. The necessary and sufficient condition given below (Theorem 1) is only slightly stronger than the existence of the mean. As for the distribution of W, it turns out to be absolutely continuous on the positive reals and has, in fact, a continuous, strictly positive density function (corollary 1 and theorem II.5.2).

The present sharp result was first given for a more general situation (multidimensional Galton-Watson process) by Kesten and Stigum (1966a, b). The proof given here is different from theirs and is a sharpening of Levinson's argument.

Throughout this section we shall *assume that* $Z_0 \equiv 1$, $m>1$ *and* $p_j \neq 1$ *for any* j.

Theorem 1. *If*

$$E Z_1 \log Z_1 < \infty \tag{1}$$

then

$$E W = 1 . \tag{2}$$

If $E Z_1 \log Z_1 = \infty$, *then*

$$E W = 0, \quad \textit{or equivalently} \quad P\{W=0\}=1 . \tag{3}$$

Remark. This theorem tells us that if $E Z_1 \log Z_1 = \infty$ then

$$P\left\{\lim \frac{Z_n}{m^n} = 0\right\} = 1 .$$

It leaves open the possibility that there exists another normalizing sequence $\{C_n\}$, such that $\{Z_n/C_n\}$ converges to a non-degenerate limit. This question was settled by Seneta (1969), who showed that such a normalization always exists.

Theorem 1 can thus be obtained as a special case of Seneta's result. However there is still a fair amount of work needed to show that $C_n \sim m^n$ when $E Z_1 \log Z_1 < \infty$. Also the method developed in the direct proof of theorem 1 will be applied to more general processes in chapters 4 and 5. For this reason we treat theorem 1 independently, and then generalize it in theorem 3.

Note that since W_n converges to W almost everywhere

$$\varphi_n(u) \equiv E e^{-uW_n} \to E e^{-uW} \equiv \varphi(u), \qquad u \geqslant 0, \tag{4}$$

where $\varphi(u)$ has been shown to satisfy the fundamental functional equation

$$\varphi(u) = f\left(\varphi\left(\frac{u}{m}\right)\right). \tag{5}$$

The key ingredients for the proof of theorem 1 are (5) and the following lemma, which, with its two corollaries, will also be useful in later sections.

Lemma 1. *Let X be a nonnegative random variable with $0 < m = EX < \infty$. Then for any $a > 0$*

$$\int_0^a \frac{1}{u^2}\left[E\left(e^{-\frac{uX}{m}}\right) - e^{-u}\right] du < \infty \tag{6}$$

if and only if

$$E X |\log X| < \infty. \tag{7}$$

Proof.

$$E\left(e^{-\frac{uX}{m}}\right) - e^{-u} = E\left(e^{-\frac{uX}{m}} - 1 + \frac{uX}{m}\right) + 1 - u - e^{-u}.$$

Since for $u \geqslant 0$ one has $0 \leqslant e^{-u} - 1 + u \leqslant u^2/2$, it suffices to show that

$$\int_0^a \frac{1}{u^2} E\left(e^{-\frac{uX}{m}} - 1 + \frac{uX}{m}\right) du < \infty \tag{8}$$

if and only if (7) holds. Writing $F(x) = P\{X < mx\}$ the left side of (8) equals

$$\int_0^a \left(\int_0^\infty u^{-2}[e^{-ux} - 1 + ux]\, dF(x)\right) du.$$

On interchanging orders of integration (the integrand is nonnegative), the above expression becomes

$$\int_0^\infty \left(\int_0^a u^{-2}[e^{-ux} - 1 + ux]\, du\right) dF(x). \tag{9}$$

But

$$\int_0^a u^{-2}[e^{-ux} - 1 + ux]\, du = x \int_0^{ax} u^{-2}[e^{-u} - 1 + u]\, du$$

and hence

$$\lim_{x\to\infty}\left(\int_0^a u^{-2}[e^{-ux}-1+ux]\,du\right)(x\log x)^{-1}=1\,.$$

The lemma is now immediate from (9). □

Corollary 1. *Let* $f(s)=\sum_{j=0}^{\infty} p_j s^j$ *be a probability generating function. Let* $0<m\equiv f'(1-)<\infty$. *Then*

$$\int_0^1 u^{-2}\left[f\left(e^{-\frac{u}{m}}\right)-e^{-u}\right]du<\infty$$

if and only if

$$\sum_{j=0}^{\infty} p_j(j\log j)<\infty\,.$$

Proof. Take X to be a nonnegative integer valued random variable with $f(s)$ as its probability generating function. Apply lemma 1. □

Let

$$A(u)=\begin{cases} m-u^{-1}[1-f(1-u)] & \text{for } 0<u\leqslant 1,\\ 0 & \text{for } u=0\,.\end{cases} \tag{10}$$

Corollary 2. *Under the same hypothesis as in Corollary* 1 $A(u)$ *is nonnegative and nondecreasing. Furthermore for any* r *and* c *in* (0, 1),

$$\sum_{n=0}^{\infty} A(cr^n)<\infty \quad \text{iff} \quad \sum_{j=0}^{\infty} p_j(j\log j)<\infty\,.$$

Proof. Since $\lim_{u\downarrow 0} A(u)=0$ and f is convex, $A(u)\geqslant 0$ and $A'(u)\geqslant 0$ for all $u\geqslant 0$. Thus the first part is obvious. An alternative argument would be to note that

$$A(u)=\sum_{j=1}^{\infty} p_j\left[\sum_{r=0}^{j-1}(1-(1-u)^r)\right].$$

As for the last part, by monotonicity of A we see that

$$\sum_{n=0}^{\infty} A(cr^n)<\infty \quad \text{if and only if} \int_0^{\infty} A(cr^t)\,dt<\infty\,,$$

which (by setting $cr^t=v$) is true if and only if

$$\int_0^1 v^{-1}A(v)\,dv<\infty\,.$$

But

$$\begin{aligned} v^{-1}A(v)&=[f(1-v)-1+mv]v^{-2}\\ &=\left[f\left(e^{-\frac{u}{m}}\right)-e^{-u}+m\left(1-e^{-\frac{u}{m}}-\frac{u}{m}\right)+(e^{-u}-1+u)\right]u^{-2}(uv^{-1})^2, \end{aligned}$$

where $1-v=e^{-u/m}$. Now $x(1-e^{-x})^{-1}$ is bounded and $\geqslant 1$ in any finite interval and $|1-x-e^{-x}|x^{-2}\leqslant\frac{1}{2}$ for $x\geqslant 0$. Thus

$$\int_0^1 v^{-1}A(v)\,dv<\infty \quad \text{iff} \int_0^{c'}\left(f\left(e^{-\frac{u}{m}}\right)-e^{-u}\right)u^{-2}\,du<\infty,$$

where c' is some positive constant. Now use Corollary 1. □

We now proceed with the proof of theorem 1.

Step 1. (1)$\Rightarrow$(2). First we note that $\varphi_n(u)$ is nondecreasing in n for each fixed u. This follows from Jensen's inequality and the fact that W_n is a martingale; namely

$$E(e^{-uW_{n+1}}|Z_n)\geqslant e^{-uE(W_{n+1}|Z_n)}=e^{-uW_n} \qquad \text{a.s.}$$

Taking expectations of both sides

$$E(e^{-uW_{n+1}})\geqslant E(e^{-uW_n}).$$

Now set

$$\psi_{n+1}(u)\equiv\frac{\varphi_{n+1}(u)-\varphi_n(u)}{u}, \qquad u>0.$$

Noting that $\varphi_{n+1}(u)=f(\varphi_n(u/m))$ and using the mean value theorem on $f(s)$ for $0\leqslant s\leqslant 1$ we get the recurrence relation

$$\psi_{n+1}(u)\leqslant\psi_n\left(\frac{u}{m}\right), \tag{11}$$

and iterating,

$$\psi_{n+1}(u)\leqslant\psi_1\left(\frac{u}{m^n}\right). \tag{12}$$

Since $\varphi_0(u)$ equals e^{-u} by definition,

$$0\leqslant\frac{\varphi(u)-e^{-u}}{u}=\lim_{n\to\infty}\frac{\varphi_{n+1}(u)-\varphi_0(u)}{u}=\sum_{n=1}^{\infty}\psi_n(u). \tag{13}$$

We shall show that under (1)

$$g(u)\equiv\sum_{n=1}^{\infty}\psi_n(u)<\infty \quad \text{for each } u>0 \tag{14}$$

and

$$\lim_{u\downarrow 0} g(u)=0. \tag{15}$$

It follows from (13) and (15) that

$$\lim_{u\downarrow 0}\frac{[1-\varphi(u)]}{u}=\lim_{u\downarrow 0}\left[\frac{1}{u}(1-e^{-u})+\frac{1}{u}(e^{-u}-\varphi(u))\right]=1.$$

Thus step 1 will be complete if we establish (15).

Using (12) and the monotonicity of ψ_1 for small u (say $u \leqslant u_0$) we get

$$0 \leqslant g(u) \leqslant \sum_{n=0}^{\infty} \psi_1\left(\frac{u}{m^{n-1}}\right) \leqslant \int_0^{\infty} \psi_1\left(\frac{u}{m^t}\right) dt. \tag{16}$$

Setting $u/m^t = v$ we get for $u \leqslant u_0$

$$0 \leqslant g(u) \leqslant \frac{1}{\log m} \int_0^u \left(\frac{\psi_1(v)}{v}\right) dv. \tag{17}$$

But $\psi_1(v) = v^{-1}[f(e^{-v/m}) - e^{-v}]$. and thus by Corollary 1 we have (under (1))

$$\int_0^u \left(\frac{\psi_1(v)}{v}\right) dv < \infty \quad \text{for } u \leqslant u_0 .$$

Hence (15) holds. This completes step 1.

Step 2. If (1) does not hold we shall show that

$$E(W) = \lim_{u \downarrow 0} \frac{1 - \varphi(u)}{u} = 0. \tag{18}$$

Since W is a nonnegative random variable it will follow that $P\{W=0\}=1$. Suppose $EW > 0$. Then there exist α, β in $(0,1)$ such that

$$\frac{1-\varphi(u)}{u} \geqslant 2\beta \quad \text{for } 0 \leqslant u \leqslant \alpha. \tag{19}$$

Let

$$\lambda_n(u) \equiv \frac{1-\varphi_n(u)}{u}. \tag{20}$$

Then since $\varphi_n(u) \to \varphi(u)$ for all $n > 0$, we see that there exists an n_0 such that

$$\lambda_n(\alpha) > \beta \quad \text{for } n \geqslant n_0 . \tag{21}$$

But for any $n, \lambda_n(u)$ is a decreasing function of $u > 0$ (just note that $(1-e^{-x})/x$ is a decreasing function of x for $x > 0$). Thus we get

$$\lambda_n(u) > \beta \quad \text{for } 0 \leqslant u \leqslant \alpha, \qquad n \geqslant n_0 . \tag{22}$$

From the recursive relation $\varphi_{n+1}(u) = f(\varphi_n(u/m))$ it follows that

$$\lambda_{n+1}(u) = \frac{\left[1 - f\left(\varphi_n\left(\frac{u}{m}\right)\right)\right]}{u} = \lambda_n\left(\frac{u}{m}\right)\left\{1 - \left(\frac{1}{m}\right) A\left[\left(\frac{u}{m}\right)\lambda_n\left(\frac{u}{m}\right)\right]\right\}, \tag{23}$$

where $A(\cdot)$ is as defined in (10).

Since $A(u)$ is a nonnegative, nondecreasing function of u in $[0,1]$, (21), (23) and the fact that for $x \geqslant 0$, $1-x \leqslant e^{-x}$, imply that for $n \geqslant n_0$

$$\begin{aligned}\lambda_{n+1}(u) &\leqslant \lambda_n\left(\frac{u}{m}\right)\left[1-\left(\frac{1}{m}\right)A\left(\frac{\beta u}{m}\right)\right] \\ &\leqslant \lambda_n\left(\frac{u}{m}\right)\exp\left[-\left(\frac{1}{m}\right)A\left(\frac{\beta u}{m}\right)\right] \quad \text{for } 0 \leqslant u \leqslant \alpha.\end{aligned} \tag{24}$$

Iterating (24) yields

$$\lambda_{n_0+k}(u) \leqslant \exp\left[-\frac{1}{m}\sum_{r=1}^{k} A\left(\frac{\beta u}{m^r}\right)\right] \tag{25}$$

since

$$\lambda_{n_0}\left(\frac{u}{m^k}\right) \leqslant \lim_{x\downarrow 0}\lambda_{n_0}(x) = E\,W_{n_0} = 1\,.$$

By Corollary 2, if (3) does not hold then

$$\lim_{k\to\infty}\sum_{r=1}^{k} A\left(\frac{\beta u}{m^r}\right) = \infty\,,$$

and thus letting $k \to \infty$ in (25) this would imply that $\lim_{n\to\infty}\lambda_n(u)=0$ for $0 \leqslant u \leqslant \alpha$. This contradicts (22) and hence implies (18), thus completing the proof of theorem 1. □

The trick used above can be imitated and exploited further to prove an interesting result about the functional equation (5). Let C denote the class of completely monotone functions on $[0,\infty]$, namely,

$$\mathsf{C} = \left\{\varphi(\cdot):\varphi(u)=\int_0^\infty e^{-ux}\,dF(x),\right.$$

$$\left. F \text{ a distribution function, } F(0-)=0,\ F(0+)<1\right\}.$$

Theorem 2. *The functional equation* (5)

$$\varphi(u) = f\left[\varphi\left(\frac{u}{m}\right)\right], \qquad u \geqslant 0\,,$$

always has a unique solution (up to a scale change) in C. *Further, for any* $p \geqslant 0$

$$\int_0^\infty t|\log t|^{\mathrm{p}}\,dF(t) < \infty$$

if and only if

$$\sum_1^\infty p_j j(\log j)^{\mathrm{p}+1} < \infty\,.$$

For the proof see Athreya (1971 a). Letting $u \to \infty$ in the above functional equation and noting that $\varphi(\infty) = F(0+) < 1$, we conclude that

Corollary 3. $F(0+) = q$.

We now turn to the case when $\sum_j p_j j \log j = \infty$. This condition, we know from theorem 1, implies that m^n is not the right normalization for Z_n to yield convergence to a limit law which is not degenerate at 0. Seneta (1968 b) was the first to show that there always exists a sequence of constants $\{C_n\}$ such that $C_n^{-1} Z_n$ *does* converge in distribution to a nondegenerate limit. Heyde (1970) strengthened this result using a martingale argument and established the almost sure convergence. We give a synthesis of their arguments.

Theorem 3. *Let $\{Z_n: n = 0, 1, 2, \ldots\}$ be a Galton-Watson process with $1 < m < \infty$. Then, there always exists a sequence of constants $\{C_n\}$ with $C_n \to \infty$ and $C_n^{-1} C_{n+1} \to m$ as $n \to \infty$, such that the random variables $W_n = C_n^{-1} Z_n$ converge almost surely to a random variable W with $P(W > 0) = 1 - q$. Furthermore if $\varphi(z) = E(e^{-zW})$ for* $\operatorname{Re} z \geqslant 0$ *then $\varphi(z)$ satisfies* (5),

$$\varphi(z) = f\left(\varphi\left(\frac{z}{m}\right)\right).$$

We shall break up the proof of this theorem into a number of lemmas. First some preliminaries.

Let $k(s) = -\log f(e^{-s})$ be the so-called cumulant generating function of $\{p_j\}$. It is immediate that $k(s)$ is well-defined for s in $[0, \infty)$ and is strictly increasing, continuous and concave. The range of $k(s)$ is the interval $[0, -\log p_0)$. We can therefore define a unique inverse function $h(s)$ of $k(s)$ for s in $[0, -\log q)$, where q is the extinction probability of the process. One readily verifies the following facts about k, h, and their iterates:

$$k_n(s) = -\log f_n(e^{-s}); \qquad h_n(s) \text{ is the inverse of } k_n(s).$$

Let s_0 be a fixed number in the open interval $(0, -\log q)$. For $n \geqslant 1$ define

$$\begin{aligned}
C_n &= (h_n(s_0))^{-1}, \\
W_n &= C_n^{-1} Z_n, \\
Y_n &= \exp(-W_n), \\
\boldsymbol{F}_n &= \sigma\text{-algebra generated by } Z_0, Z_1, \ldots, Z_n.
\end{aligned}$$

We start with some estimates on $\{C_n\}$.

Lemma 2. $\lim_{n\to\infty} C_n^{-1}=0$ *and* $\lim_{n\to\infty} C_n C_{n+1}^{-1}=m^{-1}$.

Proof. Suppose there exists a subsequence $\{n_j\}$ such that $\{h_{n_j}(s_0)\}$ is bounded away from 0. Then the sequence $\{\exp(-h_{n_j}(s_0))$ will be bounded away from 1. This, by our results on extinction probability in section 5, implies that

$$f_{n_j}(\exp(-h_{n_j}(s_0)))\to q,$$

and hence that

$$k_{n_j}(h_{n_j}(s_0))\to -\log q.$$

But by definition of k and h the left side above equals s_0 for all j, and by our choice $s_0<-\log q$. This contradiction proves the first part. The second part is obvious since

$$\frac{C_n}{C_{n+1}}=\frac{h(h_n(s_0))}{h_n(s_0)}$$

and

$$\lim_{x\to 0}\frac{h(x)}{x}=\lim_{x\to 0}\left[\frac{k(x)}{x}\right]^{-1}=m^{-1}. \qquad \square$$

Our next step is the key to the almost sure convergence.

Lemma 3. *The family* $\{Y_n, \boldsymbol{F}_n: n=0,1,2,\ldots\}$ *is a martingale. Hence*

$$\lim_{n\to\infty} Y_n=Y$$

exists almost surely, and

$$\lim_{n\to\infty} E\,Y_n^u=E\,Y^u \quad \textit{for all } u>0.$$

Proof.

$$\begin{aligned} E\{Y_{n+1}|\mathbf{F}_n\}&=E\{e^{-C_{n+1}^{-1}Z_{n+1}}|\boldsymbol{F}_n\}\\ &=(f(e^{-C_{n+1}^{-1}}))^{Z_n} \qquad \text{a.s.}\\ &=e^{-Z_n k(C_{n+1}^{-1})}=Y_n \end{aligned}$$

since

$$k(C_{n+1}^{-1})=k(h_{n+1}(s_0))=k(h(h_n(s_0)))=h_n(s_0)=C_n^{-1}.$$

This establishes the martingale property. The almost sure convergence is now automatic since Y_n is nonnegative. The convergence of the moments follows by the bounded convergence theorem since $0\leqslant Y_n\leqslant 1$ a.s. for all n. $\square$

The next lemma enables us to conclude that $W\equiv -\log Y$ is finite valued and not degenerate at 0.

Lemma 4. *Let* Y *be as in lemma* 3. *Then*

$$P(Y>0)=1 \quad \textit{and} \quad P(Y=1)=q.$$

Proof. From lemma 3 and the martingale property of Y_n we note that

$$E\,Y = \lim_{n\to\infty} E\,Y_n = E\,Y_1 = e^{-s_0}.$$

Now if we set $G_n(u, s_0) = E(Y_n^u)$ (for $u>0$) and $G(u, s_0) = E(Y^u)$ then Lemma 3 says that

$$G_n(u, s_0) \to G(u, s_0) \quad \text{for } u>0.$$

But using the definition of G_n and C_n, one easily verifies that

$$G_n(u, s_0) = f\left[G_{n-1}(u, h(s_0))\right],$$

and letting $n\to\infty$ we arrive at the functional equation

$$G(u, s_0) = f\left[G(u, h(s_0))\right]. \tag{26}$$

From the definition of $G(u, s_0)$ as $E(Y^u)$ it follows that

$$P(Y>0) = \lim_{u\downarrow 0} G(u, s_0)$$

and

$$P(Y=1) = \lim_{u\uparrow\infty} G(u, s_0).$$

However,

$$P(Y>0) = P\left(\lim_n W_n < \infty\right) \quad \text{and} \quad P(Y=1) = P\left(\lim_n W_n = 0\right).$$

Thus,

$$\lim_{u\downarrow 0} G(u, s_0) = P\left(\lim_n (h_n(s_0))\, Z_n < \infty\right)$$

and

$$\lim_{u\uparrow\infty} G(u, s_0) = P\left(\lim_n (h_n(s_0))\, Z_n = 0\right).$$

Similarly,

$$\lim_{u\downarrow 0} G(u, h(s_0)) = P\left(\lim_n (h_{n+1}(s_0))\, Z_n < \infty\right),$$

and

$$\lim_{u\uparrow\infty} G(u, h(s_0)) = P\left(\lim_n (h_{n+1}(s_0))\, Z_n = 0\right).$$

From lemma 2 we know that the following set identities hold.

$$\left\{\lim_n h_n(s_0)\, Z_n < \infty\right\} = \left\{\lim_n h_{n+1}(s_0)\, Z_n < \infty\right\}$$

and

$$\left\{\lim_n h_n(s_0)\, Z_n = 0\right\} = \left\{\lim_n h_{n+1}(s_0)\, Z_n = 0\right\}.$$

Hence, $\lim_{u\downarrow 0} G(u, s_0)$ and $\lim_{u\uparrow\infty} G(u, s_0)$ retain their values when s_0 is replaced by $h(s_0)$. Thus by (26) we see that $\alpha \equiv \lim_{u\downarrow 0} G(u, s_0)$ and $\beta \equiv \lim_{u\uparrow\infty} G(u, s_0)$ satisfy the equation $x = f(x)$. Since $s_0>0$, $\beta = P\{Y=1\} \leqslant E\,Y = e^{-s_0} < 1$, and hence $\beta = q$. Furthermore $s_0 < -\log q$, and thus

$\alpha = P\{Y>0\} \geqslant EY = e^{-s_0} > \beta$. Since $\alpha = f(\alpha)$ we can conclude that $\alpha = 1$. □

From lemmas 3 and 4 it is clear that $W_n \to W$ a.s., and that $P(W>0) = 1-q$. Finally use lemma 2 to conclude that for $\operatorname{Re} z \geqslant 0$

$$\varphi(z) = \lim_{n\to\infty} E(e^{-zW_n}) = \lim_{n\to\infty} f_{n+1}(e^{zC_{n+1}^{-1}})$$

$$= \lim_{n\to\infty} f(f_n(e^{zC_n^{-1}C_{n+1}^{-1}C_n})) = f\left(\varphi\left(\frac{z}{m}\right)\right). \quad \square$$

Remarks. 1. We see from the above that for any s in $(0, -\log q)$ the sequence $(h_n(s))Z_n$ converges in law to a nondegenerate limit law. By Khinchine's law [Gnedenko and Kolmogorov (1949)] on the uniqueness of normalizing constants it follows that

$$\lim_{n\to\infty} \frac{h_n(s)}{h_n(s_0)} = H(s)$$

exists. This was proved in a purely analytic way by Seneta (1968b) who used it to prove the convergence of $Z_n(h_n(s_0))$.

2. If $\sum_j p_j j \log j < \infty$ then we know from theorem 1 that $Z_n m^{-n}$ converges in law to a nondegenerate limit. Again by Khinchine's law it follows that $C_n \sim \text{const} \cdot m^n$. Conversely, if $C_n \sim \text{const} \cdot m^n$ then $Z_n m^{-n}$ converges in law to a nondegenerate limit (by theorem 3) which implies (by theorem 1) that $\sum p_j j \log j < \infty$. We further know from theorem 2 that $EW < \infty$ if and only if $\sum p_j j \log j < \infty$. Thus we arrive at the implications

$$EW < \infty \quad \text{iff } \sum p_j j \log j < \infty \quad \text{iff } C_n \sim \text{const} \cdot m^n .$$

This was proved first by Seneta (1969).

3. The functional equation (5) is often impossible to solve explicitly for $\varphi(z)$ and even then it is rarely possible to invert $\varphi(z)$ to obtain the distribution function of W. However, we can use (5) to determine all the integral moments of W of order $\geqslant 2$ (when they exist).

4. Suppose $\{Z_n^{(m)}; n=0,1,2,\ldots\}$, $m=2,3,\ldots$, is a sequence of branching processes indexed by the mean m and defined on the same probability space. Assume all the processes have second moments. Suppose $Z_1^{(m)}/\sigma_m$ converges as $m\to\infty$ to a random variable Z in mean square, where $\sigma_m = \operatorname{var}(Z_1^{(m)}|Z_0 \equiv 1)$. Let $W^{(m)}$ denote the a.s. limit of $Z_n^{(m)}/m^n$ as $n\to\infty$. Then

$$E\left(\frac{mW^{(m)}}{\sigma_m} - Z\right)^2 \leqslant 2E\left(\frac{mW^{(m)}}{\sigma_m} - \frac{Z_1^{(m)}}{\sigma_m}\right)^2 + 2E\left(\frac{Z_1^{(m)}}{\sigma_m} - Z\right)^2 \to 0 \quad \text{as } m\to\infty$$

since

$$E\left(\frac{m W^{(m)}}{\sigma_m}-\frac{Z_1^{(m)}}{\sigma_m}\right)^2=\frac{m}{m^2-m}.$$

Thus if $Z_1^{(m)}/\sigma_m \to Z$ in mean square, $m W^{(m)}/\sigma_m \to Z$ in mean square also. Thus for large m the limit random variable $W^{(m)}$ behaves like $Z_1^{(m)}/m$.

5. We have discussed so far only the case $1<m<\infty$. It has been shown by Seneta (1969) that when $m=\infty$ there exists no normalizing sequence $\{C_n\}$ such that $Z_n C_n^{-1}$ converges in law to a proper limit distribution. However, Darling (1970) has shown that if $g(s)$ is the inverse function of $1-f(1-s)$ and

$$g'(s)=a s^{b-1}(1+0(s^\delta)) \quad \text{as } s\to 0$$

for some $a>0$, $b>1$, $\delta>0$, then

$$b^{-n}\log(Z_n+1)$$

converges in law to a proper distribution.

We conclude this section with

Theorem 4. *If $q=0$ then the random variable W in Theorem 3 has an absolutely continuous distribution* $(0, \infty)$.

This result was first proved by Harris (1963) with second moment assumption, by Stigum (1966) with $\sum p_j j \log j<\infty$ and by Athreya (1971 a) without any hypothesis.

The hypothesis $q=0$ will be removed in corollary 12.1.

The proof is somewhat longer than those encountered so far, and the reader may wish to skip over it on a first reading of the book.

Lemma 5. *The distribution of W is not concentrated at one point.*

Proof. If W is degenerate at c then $\varphi(u)$ must coincide with e^{-uc}. Now (5) implies

$$e^{-uc}=f\left(e^{-\frac{uc}{m}}\right) \quad \text{for all } u\geq 0.$$

That is, if $s=e^{-uc/m}$ then $s^m=f(s)$ for $0\leq s\leq 1$, implying that m is an integer and $p_m=1$. This contradicts our initial assumption. □

We use lemma 5 to establish

Lemma 6. *If $\varphi(-it)=E(e^{itW})$ is the characteristic function of W, then*

$$|\varphi(-it)|<1 \quad \text{for all } t\neq 0. \tag{27}$$

Proof. Since the distribution of W is not degenerate, there exists a $\delta>0$ such that

$$|\varphi(-it)|<1 \quad \text{for all } |t|<\delta, \quad t\neq 0. \tag{28}$$

This is an elementary result from the theory of characteristic functions and is proved, for example, in Feller (Vol. II. p. 475).

We know from Theorem 3 that

$$\varphi(-it)=f\left(\varphi\left(-\frac{it}{m}\right)\right) \quad \text{for } -\infty<t<\infty\,. \tag{29}$$

Let $|t|<m\delta$ and $t\neq 0$. Then from (28)

$$|\varphi(-it)|\leqslant f\left(\left|\varphi\left(-\frac{it}{m}\right)\right|\right)<1 \tag{30}$$

since $|t/m|<\delta$, $t\neq 0$. Thus $|\varphi(-it)|<1$ for $|t|<m\delta$, $t\neq 0$, and hence for $|t|<m^n\delta$, $t\neq 0$ for any n. Since $m>1$ this implies (27). □

The following lemma is a key step.

Lemma 7. *Let* $\delta=(-\log f'(q))(\log m)^{-1}$. *Then,*

(i) *for any* $0<\delta_0<\delta$, $\sup_t |\varphi(it)-q|\cdot|t|^{\delta_0}<\infty$;

(ii) *the additional hypothesis* $\sum p_j j\log j<\infty$ *implies, for any* $0<\delta_0<\delta$, $\sup_t |\varphi'(it)|\,|t|^{1+\delta_0}<\infty$.

Proof. For any t, (29) yields on iteration

$$\varphi(im^n t)=f_n(\varphi(it))\,. \tag{31}$$

We know from section 5 that $f_n(\varphi(it))\to q$ uniformly for t such that $|\varphi(it)|\leqslant\beta'<1$. By lemma 6 and the continuity of $\varphi(it)$ we know that

$$\beta\equiv\sup_{1\leqslant|t|\leqslant m}|\varphi(it)|<1\,. \tag{32}$$

Let $\gamma_0=m^{-\delta_0}$ and let $\varepsilon>0$ be such that $f'(q+\varepsilon)<\gamma_0$. Such an ε exists since $0<\delta_0<\delta$ implies $f'(q)<\gamma_0<1$. Now choose k such that

$$\sup_{1\leqslant|t|\leqslant m}|f_n(\varphi(it))-q|\leqslant\varepsilon \quad \text{for all } n\geqslant k\,. \tag{33}$$

Next, an application of the mean value theorem yields for $|s-q|\leqslant\varepsilon$

$$|f(s)-f(q)|\leqslant\gamma_0|s-q|\,,$$

and by iteration we get (using (31) and (33))

$$\sup_{1\leqslant|t|\leqslant m}|\varphi(im^{n+k}t)-q|\leqslant\varepsilon\gamma_0^n\,. \tag{34}$$

This yields

$$\sup_{1\leqslant|t|\leqslant m}|m^{n+k}t|^{\delta_0}|\varphi(im^{n+k}t)-q|\leqslant\varepsilon\gamma_0^{-(k+1)}, \qquad n\geqslant 0\,;$$

and hence

$$\sup_{m^{k+n}\leqslant|t|\leqslant m^{k+n+1}}|\varphi(it)-q|\,|t|^{\delta_0}<\infty, \qquad n\geqslant 0\,.$$

This establishes (i).

Turning to (ii) we first note that the additional hypothesis $\sum p_j j \log j < \infty$ is needed to ensure that $EW < \infty$ which in turn makes φ differentiable.

Differentiating (31) yields

$$m^n \varphi'(i m^n t) = f_n'(\varphi(it))\varphi'(it). \tag{35}$$

But for $0 \leqslant x < 1$,

$$f_n'(x) = \prod_{j=1}^{n} f'(f_{j-1}(x))$$

and hence

$$f_{n+k}'(\beta) \leqslant f_k'(\beta)\gamma_0^n, \tag{36}$$

where β and k are as in (32) and (33).

It is clear from (35) and (36) that for all $n \geqslant 0$

$$\sup_{1 \leqslant |t| \leqslant m} |m^{n+k} t|^{1+\delta_0} |\varphi'(i m^{n+k} t)| \leqslant m f_k'(\beta) \gamma_0^{-k-1} |\varphi'(0)| < \infty.$$

Then, as before,

$$\sup_{m^{k+n} \leqslant |t| \leqslant m^{k+n+1}} |t|^{1+\delta_0} |\varphi'(it)| < \infty \quad \text{for } n \geqslant 0,$$

thus finishing the proof of (ii). □

We shall determine the precise rate of convergence of $f_n(s)$ to q in the next section and use it in Chapter 2 to improve the assertion of the above lemma to a precise rate of decay.

An easy consequence of lemma 7 (ii) is the following.

Corollary 4. *If* $\sum_j p_j j \log j < \infty$, *then* W *is absolutely continuous on* $(0, \infty)$ *with a continuous density function.*

Proof. Define for x in $(0, \infty)$

$$F(x) = (1-q)^{-1} P\{0 < W \leqslant x\}.$$

Then, $F(x)$ is a distribution function and

$$\int_0^\infty e^{-itx} dF(x) = (1-q)^{-1}(\varphi(it) - q).$$

Also, since $\sum p_j j \log j < \infty$ we note by remark 2 following Theorem 3 that $EW = \int_0^\infty x\, dF(x) < \infty$. Let $d\tilde{F}(x) = x\, dF(x)$. Then, $\int_0^\infty d\tilde{F}(x) < \infty$ and $\int_0^\infty e^{-itx} d\tilde{F}(x) = (1-q)^{-1} \varphi'(it)$. By lemma 7 (ii), $\varphi'(it)$ is integrable. Thus, by the standard inversion theorem there exists a continuous function $w(x)$ on $(0, \infty)$ such that

$$x\, dF(x) = d\tilde{F}(x) = w(x)\, dx. \quad \square$$

To prove the absolute continuity without the hypothesis $\sum p_j \log j < \infty$ we use lemma 7 (i) and lemmas 8 and 9 below. It is here that the assumption $q=0$ is used.

If $\delta > 1$ then $\varphi(it)$ will be integrable and W will have a uniformly continuous density $w(x)$.

When $\delta < 1$ note that since $\delta > 0$, $\varphi^k(it)$ will be integrable if $k\delta > 1$. This means that without any moment conditions there always exists a positive integer k such that the k-fold convolution of W is absolutely continuous and has a uniformly continuous density. It takes a little argument to show that this implies the absolute continuity of W. We do this in two more lemmas.

Lemma 8. *Let* $\{W_j : j=0,1,2,\ldots\}$ *be a sequence of independent random variables with the same distribution as* W. *Let the sequence also be independent of the branching process* $\{Z_n\}$. *Then for each n the two random variables* W_0 *and* $m^{-n}\sum_{j=1}^{Z_n} W_j$ *have the same distribution.*

Proof. The characteristic function of the two random variables in question are respectively $\varphi(it)$ and $f_n(\varphi(itm^{-n}))$, and these are clearly identical as can be seen by iterating the basic functional equation $\varphi(it) = f(\varphi(it/m))$ n times. □

Lemma 9. *Let* E *be a set of Lebesgue measure zero in* $(0, \infty)$. *Then* $P(W \in \mathsf{E}) = 0$.

Proof. By lemma 4

$$P(W_0 \in \mathsf{E}) = P\left(\sum_{j=1}^{Z_n} W_j \in m^n \mathsf{E}\right) = \sum_{r=1}^{\infty} P\left(Z_n = r, \sum_{j=1}^{Z_n} W_j \in m^n \mathsf{E}\right)$$

$$= \sum_{r=1}^{\infty} P(Z_n = r) P\left(\sum_{j=1}^{r} W_j \in m^n \mathsf{E}\right).$$

But for $r \geqslant k, P\left(\sum_{j=1}^{r} W_j \in m^n \mathsf{E}\right) = 0$ where k is an integer such that $k\delta > 1$. Thus,

$$P(W_0 \in \mathsf{E}) \leqslant P(Z_n < k).$$

Now for any fixed k, $P(Z_n < k) \to 0$ as $n \to \infty$. □

Thus we conclude that W is absolutely continuous on $(0, \infty)$. □

S. Dubuc (1971) has independently established these results and, by a direct estimation of the inversion formula, has in fact shown that the density function of W is continuous.

11. Geometric Convergence of $f_n(s)$ in the Noncritical Cases[2]

We know that for any $s<1, f_n(s)\to q$ as $n\to\infty$, and we shall now study the rate of convergence. In the critical case we already saw that $1-f_n(s)$ was of order $1/n$, and we will show that when $m\neq 1$ the rate is geometric; namely $f_n(s)-q\sim\gamma^n$, where $\gamma=f'(q)<1$. An easy consequence of this result will be a similar estimate on the decay rate of $P_n(i,j)$, which will be useful later. Further symmetries between the super and subcritical cases will also become apparent in this section.

We start by studying

$$Q_n(s)=\gamma^{-n}[f_n(s)-q]. \tag{1}$$

A symmetry argument will then also yield the asymptotic behavior of $m^n[g_n(s)-1]$, where $g_n(s)$ is the n-fold iterate of $g(s)=$ the inverse function of $f(s)$; and this leads to another proof of theorem 10.1. We conclude this section with some applications to conditioning, and to an approximation for large populations.

Let $Q_n'(s)=$ the derivative of $Q_n(s)$. Clearly

$$Q_n'(s)=\frac{f_n'(s)}{\gamma^n}=\prod_{j=0}^{n-1}\frac{f'(f_j(s))}{\gamma}=\prod_{j=0}^{n-1}\left\{1-\left(1-\frac{f'(f_j(s))}{\gamma}\right)\right\}. \tag{2}$$

We shall show first that

$$\lim_{n\to\infty}Q_n'(s)=Q'(s) \tag{3}$$

exists for all $0\leqslant s<1$. Then, integration with respect to s will yield the convergence of $Q_n(s)$. When $m<1$ we will need an additional hypothesis to ensure that $Q'(s)$ is not identically zero, namely $\sum p_j j\log j<\infty$.

If $m<1$ then $q=1$ and so $\gamma=f'(q)=m$. For $0\leqslant s<1$, $Q_n'(s)$ is monotone in n and $0\leqslant Q_n'(s)\leqslant 1$. Thus (3) is immediate in this case.

When $m>1, q<1$ and for any $0\leqslant s<1$, $f_j(s)\to q$. Fix s in $[0,1)$ and set $a_j=|f'(q)-f'(f_j(s))|$. We shall show that

$$\sum_{j=0}^{\infty}a_j<\infty. \tag{4}$$

This clearly implies (3) and also guarantees that $Q'(s)\neq 0$. Let $\varepsilon>0$ be such that $0\leqslant q<q+\varepsilon<1$, and $\gamma'=f'(q+\varepsilon)<1$. Choose j_0 such that $f_j(s)<q+\varepsilon$ for all $j\geqslant j_0$. By the mean value theorem

$$a_{j+j_0}\leqslant f''(q+\varepsilon)|q-f_{j+j_0}(s)|.$$

---

[2] Results in this section have been independently obtained by a number of authors; including B. I. Selivanov (1969), E. Seneta (1969), S. Karlin and J. McGregor (1968a), and A. Joffe and F. Spitzer (1967).

Another application of the mean value theorem yields the estimate

$$|q-f_{j+j_0}(s)|=|f_{j+j_0}(q)-f_{j+j_0}(s)|\leq\gamma'|f_{j+j_0-1}(q)-f_{j+j_0-1}(s)|,$$

and on iteration

$$a_{j+j_0}\leq \text{const}\cdot(\gamma')^j.$$

But $0<\gamma'<1$ and this implies (4).

Now return to the case $m<1$, and observe that a necessary and sufficient for $Q'(s)$ to be nonvanishing is

$$\sum_{j=1}^{\infty}(m-f'(f_j(s)))<\infty. \tag{5}$$

But

$$m-f'(f_j(s))=\sum_{k=1}^{\infty}kp_k(1-f_j^{k-1}(s)),$$

and hence

$$\sum_{j=1}^{\infty}(m-f'(f_j(s)))=\sum_{k=1}^{\infty}kp_k\sum_{j=1}^{\infty}(1-f_j^{k-1}(s)) \tag{6}$$

(all the terms involved are nonnegative).

By the convexity of f

$$(1-p_0)(1-s)\leq 1-f(s)\leq m(1-s).$$

Iterating this we get

$$(1-s)(1-p_0)^j\leq 1-f_j(s)\leq m^j. \tag{7}$$

Now write

$$\sum_{j=1}^{\infty}(1-f_j^{k-1}(s))=\sum_{j=1}^{\infty}[1-\{1-(1-f_j(s))\}^{k-1}].$$

Observe that for $0<r<1$, $\sum_{j=1}^{\infty}[1-(1-r^j)^k]<\infty$ if and only if $\int_0^{\infty}[1-(1-r^t)^k]dt=(\log r)^{-1}\int_0^1[1-(1-v)^k]v^{-1}dv<\infty$, and that the last integral is $\sim\log k$. Hence there exist constants k_1, k_2 such that

$$0<k_1<(\log k)^{-1}(-\log r)\sum_{j=1}^{\infty}[1-(1-r^j)^{k-1}]<k_2<\infty. \tag{8}$$

From (6), (7), and (8) we see that (5) holds if and only if

$$\sum p_k k\log k<\infty.$$

We thus conclude that either $Q'(s)\equiv 0$; or that $Q'(s)>0$ for all $s\in[0,1)$, and $\lim_{s\to q}Q'(s)=1$. To see the last point use the bounded convergence theorem and the monotonicity of $f(s)$. Namely note from (3) that

$$Q'(s) = \prod_{j=0}^{\infty} \frac{f'(f_j(s))}{\gamma}.$$

Then,

$$-\log Q'(s) = \sum_{j=0}^{\infty} \left(-\log \frac{f'(f_j(s))}{\gamma}\right) < \sum_{j=0}^{\infty} \left(-\log \frac{f'(f_j(0))}{\gamma}\right) < \infty,$$

and by the dominated convergence theorem

$$\lim_{s\to q} \sum_{j=0}^{\infty} \left(-\log \frac{f'(f_j(s))}{\gamma}\right) = 0.$$

Summarizing this discussion we have

Theorem 1. *Without any assumptions*

$$\lim_{n\to\infty} Q_n'(s) = Q'(s) \tag{9}$$

exists for $0 \leqslant s < 1$. *The function* $Q'(s)$ *has the following properties:*

(i) $Q'(s) \equiv 0$ *in* $[0, 1)$ *if and only if* $m < 1$ *and* $\sum p_j j \log j = \infty$;

(ii) *in all other cases* $Q'(s) > 0$ *in* $[0, 1)$ *and* $\lim_{s\to q} Q'(s) = 1$.

Now define

$$Q(s) = \int_q^s Q'(x)\,dx, \qquad 0 \leqslant s < 1.$$

Then we have

Corollary 1. *If* $m \neq 1$ *then*

$$\lim_{n\to\infty} Q_n(s) = Q(s) \quad \text{for } 0 \leqslant s < 1.$$

Proof. By the bounded convergence theorem

$$Q_n(s) = Q_n(s) - Q_n(q) = \int_q^s Q_n'(t)\,dt \to Q(s). \qquad \square$$

Throughout the rest of this section we shall assume that $\sum p_j j \log j < \infty$ whenever $m < 1$.

Theorem 2. *The function* $Q(s)$ *defined in theorem 1 is the unique solution of the functional equation*

$$Q(f(s)) = \gamma Q(s) \quad \textit{for } 0 \leqslant s < 1 \tag{10}$$

satisfying

$$Q(q) = 0 \quad \textit{and } \lim_{s\to q} Q'(s) = 1. \tag{11}$$

Proof. That Q satisfies (10) can be seen by substituting $f(s)$ for s in the definition of $Q_n(s)$, and taking limits. As for uniqueness, note that if $Q^{(1)}(s)$ and $Q^{(2)}(s)$ are two solutions of (10) satisfying (11) then

$$\begin{aligned} |Q^{(1)}(s)-Q^{(2)}(s)| &= \gamma^{-1}|Q^{(1)}(f(s))-Q^{(2)}(f(s))| \\ &= \gamma^{-n}|Q^{(1)}(f_n(s))-Q^{(2)}(f_n(s))| \\ &\leq |Q_n(s)|\left\{\left|1-\frac{Q^{(1)}(f_n(s))}{f_n(s)-q}\right|+\left|1-\frac{Q^{(2)}(f_n(s))}{f_n(s)-q}\right|\right\}. \end{aligned} \tag{12}$$

Now for any $0 \leq s < 1$, $f_n(s) \to q$ and

$$\lim_{n\to\infty} \frac{Q^{(i)}(f_n(s))}{f_n(s)-q} = \lim_{s\to q} Q^{(i)\prime}(s) = 1, \qquad i=1,2.$$

Thus from (12) $|Q^{(1)}(s)-Q^{(2)}(s)|=0$ for any s. □

Since $Q(s)$ is a limit of power series, we may write

$$Q(s) = \sum_0^\infty v_j s^j, \qquad 0 \leq s < 1. \tag{13}$$

Clearly $v_0 < 0$ and $v_j > 0$ for $j \geq 1$. The following result asserting the geometric decay of $P_n(i,j)$ is now immediate. For some related matter see theorem 2.2 of Seneta (1969b), and the spectral representation of Karlin and McGregor (1966a, b).

Theorem 3. *For fixed i and j (both ≥ 1)*

$$\lim_{n\to\infty} \left(\frac{P_n(i,j)}{\gamma^n}\right) = i q^{i-1} v_j. \tag{14}$$

Proof. First observe that

$$\sum_{j=0}^\infty s^j P_n(i,j) = f_n^i(s).$$

Furthermore

$$\frac{f_n^i(s)-q^i}{\gamma^n} = \frac{f_n(s)-q}{\gamma^n}\left(f_n^{i-1}(s)+f_n^{i-2}(s)q+\cdots+q^{i-1}\right),$$

and hence for $0 \leq s < 1$

$$\lim_{n\to\infty} \frac{f_n^i(s)-q^i}{\gamma^n} = Q(s)\, i q^{i-1}.$$

Comparing coefficients on both sides implies (14). □

The function $Q(s)$ is very useful in the study of $f_n(s)$ and the spectral representation of $P_n(i,j)$. This has been developed by Karlin and

McGregor who exploited the symmetry between the super and sub-critical cases. Since $Q'(s)>0$ for s in $[0, 1)$, the function $Q(s)$ possesses an inverse which we shall denote by $P(s)$. Iterating (10) one gets

$$Q(f_n(s)) = \gamma^n Q(s),$$

or

$$f_n(s) = P(\gamma^n Q(s)). \tag{15}$$

Although it is difficult to find P and Q explicitly, (15) is useful in studying the iterates $f_n(s)$. Karlin and McGregor (1966a) have shown that $P(s)$ has a power series expansion in a neighborhood of 0. Writing

$$P^i(w) = \sum_{r=0}^{\infty} b_{ir} w^r \quad \text{for } |w| < \rho,$$

and

$$Q^r(s) = \sum_{j=0}^{\infty} Q_{rj} s^j \quad \text{for } |s| < 1,$$

they get from (15) that for n sufficiently large

$$f_n^i(s) = \sum_{r=0}^{\infty} b_{ir} \left(\sum_{j=0}^{\infty} a_{rj} s^j \right) \gamma^{nr} = \sum_{j=0}^{\infty} s^j \left(\sum_{r=0}^{\infty} a_{rj} b_{ir} \gamma^{nr} \right).$$

Hence one obtains the expansion

$$P_n(i,j) = \sum_{r=0}^{\infty} \gamma^{nr} a_{rj} b_{ir}.$$

The reader is also referred to Karlin, McGregor (1968a, b) for further applications of the functions P and Q.

Techniques similar to those above can be used to give another derivation of the necessary and sufficient conditions for nondegeneracy of W, the limit random variable studied in the last section. The idea is to use an analogue of theorems 1 and 2 for the inverse function g of f and its iterates g_n. This is given in theorem 4, which further clarifies the symmetry between the super and sub-critical cases.

Theorem 4. *If* $p_0 = 0$, *then*

(i) $\lim_{n\to\infty} (g_n(s) - 1) m^n = Q^*(s)$ *exists for* $0 < s \leq 1$.

(ii) *If* $\sum_j (j \log j) p_j = \infty$, *then* $Q^*(s) \equiv -\infty$ *for* $s < 1$.

(iii) *If* $\sum_j (j \log j) p_j < \infty$ *then* $-\infty < Q^*(s) \leq 0$ *for* $0 < s < 1$, *and*

$$\lim_{s \uparrow 1} Q^*(s) = Q^*(1) = 0,$$

$$Q^{*\prime}(s) > 0 \quad \textit{for } s > 0.$$

Furthermore, $Q^*(s)$ *satisfies*

$$Q^*(f(s)) = m\,Q^*(s), \qquad 0 < s \leqslant 1$$

and is the unique solution of this functional equation among those satisfying

$$Q^*(1) = 0, \qquad Q^{*\prime}(1) \equiv \lim_{x \uparrow 1} \frac{Q^*(x)}{x-1} = 1 .$$

Remark. We will see in section 12 that when $m > 1$, the case $p_0 \neq 0$ can be reduced to $p_0 = 0$.

The proof is along the same lines as in theorems 1 and 2 and so we omit it. Using this theorem we shall show that $\sum_j p_j(j \log j) < \infty$ iff W is not degenerate at zero.

Alternate Proof of Theorem 10.1: From theorem 4 we get that if $\sum_j p_j(j \log j) < \infty$, then $Q^{*\prime}(s) > 0$, and hence there exists a unique inverse function $P^*(s)$ of $Q^*(s)$. As in (15) we get

$$f_n(s) = P^*(m^n Q^*(s)) . \tag{16}$$

Let $s = e^{-(\lambda/m^n)}$. Then

$$m^n Q^*\left(e^{-\frac{\lambda}{m^n}}\right) = (-\lambda) \frac{Q^*\left(e^{-\frac{\lambda}{m^n}}\right)}{\left(e^{-\frac{\lambda}{m^n}} - 1\right)} \cdot \frac{\left(e^{-\frac{\lambda}{m^n}} - 1\right)}{\left(-\frac{\lambda}{m^n}\right)}$$

$$\to -\lambda \quad \text{as } n \to \infty \quad \text{by theorem 4} .$$

Since $P^*(s)$ is continuous (because $Q^*(s)$ is)

$$\lim_{n \to \infty} f_n\left(e^{-\frac{\lambda}{m^n}}\right) = P^*(-\lambda) .$$

Thus $\varphi(\lambda) \equiv E(e^{-\lambda W})$ coincides with $P^*(-\lambda)$. Under the condition $\sum_j p_j(j \log j) < \infty$, $Q^*(s)$ and hence $P^*(s)$ are strictly monotone. But $\varphi(\lambda)$ is strictly monotone if and only if W has positive mass on $(0, \infty)$.

To see the converse let W be nontrivial. Define $B^*(-\lambda) = \varphi(\lambda) \equiv E(e^{-\lambda W})$ for λ in $[0, \infty)$. Since W is nontrivial $B^*(x)$ is strictly monotone increasing in $(-\infty, 0]$. Let $A^*(x)$ be the inverse function of $B^*(x)$. Since φ is continuous so is B^* and hence A^*. We know φ satisfies the functional equation

$$\varphi(m\lambda) = f(\varphi(\lambda))$$

which implies

$$f(B^*(-\lambda)) = B^*(-\lambda m) . \tag{17}$$

Operating with A^* on both sides we get

$$A^*(f(x)) = m A^*(x) \quad \text{for } 0 \leqslant x \leqslant 1,$$

which can be rewritten

$$A^*(x) = m A^*(g(x)). \tag{18}$$

Iterating (18)

$$A^*(x) = m^n A^*(g_n(x)),$$

so that

$$\frac{g_n(x)-1}{m^{-n}} = \frac{[B^*(m^{-n} A^*(x)) - 1]}{m^{-n} A^*(x)} A^*(x).$$

For any $x > 0$, $A^*(x) > -\infty$ and hence $m^{-n} A^*(x) \to 0$. Since $\varphi'(0) = -m$ we conclude that

$$\lim_{n\to\infty} \frac{B^*(m^{-n} A^*(x)) - 1}{m^{-n} A^*(x)} = m,$$

yielding the conclusion that, for $0 < x \leqslant 1$,

$$\lim_{n\to\infty} \frac{g_n(x)-1}{m^{-n}} \quad \text{exists and } = m A^*(x) > -\infty.$$

By theorem 4 this implies that $\sum_j p_j(j \log j) < \infty$. □

We close this section with some more applications of theorem 1. We first prove the result announced in remark 8.1 about the finiteness of the mean of the limit distribution in the Yaglom theorem. This will follow at once from the lemma below.

Lemma 1. *Let* $\{b_j; j = 1, 2, \ldots\}$ *be the limit function in theorem 8.1. Then*

$$\sum_j j b_j = -q[Q(0)]^{-1}.$$

Proof. Recall that $\mathscr{B}(s) = \sum_j b_j s^j$, and that

$$\mathscr{B}'(1) = \lim_{s \uparrow 1} \frac{1 - \mathscr{B}(s)}{1-s} \quad (\text{note that } \mathscr{B}(1) = 1)$$

$$= \lim_{k\to\infty} \frac{1 - \mathscr{B}\left[\frac{f_k(0)}{q}\right]}{1 - \frac{f_k(0)}{q}}. \tag{19}$$

But by (8.6)

$$\mathscr{B}(s) = \lim_{n\to\infty} \frac{f_n(qs) - f_n(0)}{q - f_n(0)}, \tag{20}$$

and hence

$$1-\mathscr{B}\left[\frac{f_k(0)}{q}\right]=\lim_{n\to\infty}\frac{q-f_{k+n}(0)}{q-f_n(0)}=\lim_{s\uparrow q}\frac{q-f_k(s)}{q-s}=\gamma^k. \tag{21}$$

Substituting (21) in (19) and applying theorem 1 yields

$$\mathscr{B}'(1)=\lim_{k\to\infty}\frac{\gamma^k}{1-\dfrac{f_k(0)}{q}}=-\frac{q}{Q(0)}, \tag{22}$$

proving the lemma. □

Corollary 2. *When $m<1$, then $\sum_j j b_j<\infty$ if and only if $\sum_j p_j(j\log j)<\infty$. When $m>1$ then $\sum_j j b_j<\infty$ (unless $q=0$, in which case b_j is not defined).*

Proof. The first assertion follows from lemma 1 [and theorem 1]; the second from the lemma and the fact that $Q(0)<0$ when $m>1$ and $q>0$.

A second application occurs in the treatment of "large populations". Consider a subcritical process. An alternative to starting with $Z_0=1$ and conditioning on nonextinction is to start with $Z_0=\eta$ large, and not conditioning at all. Suppose that $N_n(\eta)$ of the families descending from the original η parents still survive at time n; and let $X_j, j=1,2,\ldots,N_n(\eta)$, denote the number of particles in the jth of these. Then the total number of particles at time n is

$$Z_n(\eta)=X_1+\cdots+X_{N_n(\eta)}.$$

Since the X_j's are independent of each other and for large n and η "approximately" independent of N_n, the generating function of Z_n will be roughly the composition of the p.g.f. of N_n with that of X_1. Now N_n will have a binomial distribution with parameters (η, p_n) where p_n is the survival probability of any one of the original η particles. If we choose $\eta=\eta(n)$ in such a way that $\eta(n)p_n\to\lambda=$ constant as $n\to\infty$, then Z_n should converge in distribution to a compound Poisson variable. This is the idea behind the following theorem.

Theorem 5 (Joffe and Spitzer (1967)). *If $m<1$ and $\sum(k\log k)p_k<\infty$, then*

$$\lim_{n\to\infty}P\{Z_n=k|Z_0=cm^{-n}\textstyle\sum jb_j+o(m^{-n})\}=d_j(c)$$

where $d_j(c)\geqslant 0$ and $\sum_j d_j(c)=1$, and

$$\sum_j d_j(c)s^j=e^{-c(1-\mathscr{B}(s))}. \tag{23}$$

Proof.

$$f_n(s) = 1 - (1 - f_n(0))\left\{1 - \frac{f_n(s) - f_n(0)}{1 - f_n(0)}\right\},$$

which by theorem 1 and lemma 1

$$= 1 - \left(\frac{m^n + o(m^n)}{\mathscr{B}'(1)}\right)[1 - \mathscr{B}(s) + o(1)]$$

$$= 1 - \frac{m^n}{\mathscr{B}'(1)}(1 - \mathscr{B}(s)) + o(m^n).$$

Hence

$$\lim_{n \to \infty} [f_n(s)]^{\{cm^{-n}\mathscr{B}'(1) + o(m^{-n})\}} = e^{-c(1 - \mathscr{B}(s))}. \quad \square$$

We remark that a result symmetric to the above can of course be proved for the supercritical case by conditioning on $T < \infty$.

Summary of Relations among $\mathscr{P}, Q, \mathscr{B}$. We have seen a close relation between the sequences $\{\pi_j\}$, $\{v_j\}$ and $\{b_j\}$, and their generating functions $\mathscr{P}(s)$, $Q(s)$ and $\mathscr{B}(s)$. In fact they are really only different versions of the same limit law. For easy reference, we summarize the relevant relations *for the case* $m \neq 1$ *and* $p_1 > 0$.

$$\mathscr{P}[f(s)] = \gamma \mathscr{P}(s) + \mathscr{P}(p_0), \tag{24}$$

$$Q[f(s)] = \gamma Q(s), \tag{25}$$

$$\mathscr{B}\left[\frac{f(qs)}{q}\right] = \gamma \mathscr{B}(s) + (1 - \gamma), \tag{26}$$

$$\mathscr{B}(s) = \frac{\mathscr{P}(qs)}{\mathscr{P}(q)}, \tag{27}$$

$$Q(s) = Q(0) + \frac{\mathscr{P}(s)}{\mathscr{P}'(q)}. \tag{28}$$

Proof of (28). From (7.11)

$$\mathscr{P}(s) = \frac{\mathscr{P}[f_n(s)] - \mathscr{P}(q)}{\gamma^n} + \mathscr{P}(q)$$

$$= \frac{\mathscr{P}[q + \{f_n(s) - q\}] - \mathscr{P}(q)}{f_n(s) - q} \cdot \frac{f_n(s) - q}{\gamma^n} + \mathscr{P}(q)$$

$$\to \mathscr{P}'(q) Q(s) + \mathscr{P}(q)$$

$$\Rightarrow Q(s) = -\frac{\mathscr{P}(q)}{\mathscr{P}'(q)} + \frac{\mathscr{P}(s)}{\mathscr{P}'(q)}.$$

Since $\mathscr{P}(0)=0$

$$Q(0)=-\frac{\mathscr{P}(q)}{\mathscr{P}'(q)}.$$

There is also a sort of dual to (8.12) which will have a probabilistic interpretation later. In the subcritical case

$$1-\frac{Q_n(s)}{Q_n(0)}\to\mathscr{B}(s), \tag{29}$$

and hence

$$\mathscr{B}(s)=1-\frac{Q(s)}{Q(0)}. \tag{30}$$

But the limit in (29) exists for all f (provided $Q(0)\neq 0$). Call it

$$\hat{\mathscr{B}}(s)=1-\frac{Q(s)}{Q(0)}.$$

Substituting in (25) this yields

$$\hat{\mathscr{B}}[f(s)]=\gamma\hat{\mathscr{B}}(s)+(1-\gamma). \tag{31}$$

Part D. Further Ramifications

12. Decomposition of the Supercritical Branching Process

In the supercritical case the extinction probability q is strictly less than 1, and if $f(0)>0$ then (and only then) q will be strictly positive. Many results in the supercritical case are most easily proved for the case $q=0$, and then reduced from the case $q>0$ to $q=0$ by a transformation of Harris (1948). In this section we give a probabilistic interpretation of this device and exploit it to yield a decomposition of the branching process.

If $f(0)>0$ define

$$\hat{f}(s)\equiv\frac{f((1-q)s+q)-q}{1-q} \quad \text{for } 0\leqslant s\leqslant 1. \tag{1}$$

Clearly

$$\hat{f}(0)=\frac{f(q)-q}{1-q}=0.$$

Since $f(s)$ is a power series with nonnegative coefficients, so are $f((1-q)s+q)$ and $\hat{f}(s)$. Further $\hat{f}(1)=1$, and thus $\hat{f}(s)$ is a probability generating function. Lastly $\hat{f}'(s)=f'((1-q)s+q)$, and letting $s\uparrow 1$ we get $\hat{f}'(1)=f'(1)=m$.

The generating function $\hat{f}(s)$ can be easily identified from a graph of $f(s)$. Suppose the latter is as in the figure below

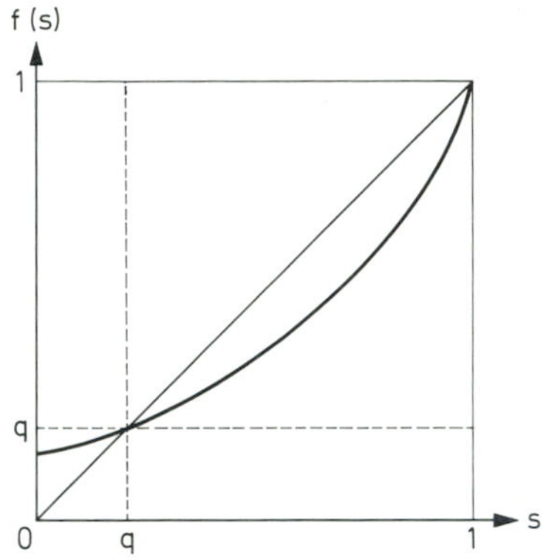

Fig. 5

Now take the square with opposite corners (q,q) and $(1,1)$, and "stretch" it into the unit square, mapping (q,q) into $(0,0)$. The resulting curve will be $\hat{f}(s)$.

Similarly if the square with opposite corners $(0,0)$ and (q,q) has its scale multiplied by q^{-1}, mapping (q,q) into $(1,1)$, then $f(s)$ will go into $f^*(s)=q^{-1}f(qs)$, a generating function we have also encountered (see e.g. the proof of theorem 8.1).

Let $\{Z_n; n=0,1,2,\ldots\}$ and $\{\hat{Z}_n; n=1,2,3,\ldots\}$ be branching processes on the same space $(\Omega, \mathbb{F})$ with generating functions $f(s)$ and $\hat{f}(s)$ respectively. Let P and $\hat{P}$ be the corresponding measures on $(\Omega, \mathbb{F})$. Since $\hat{f}(0)=0$ it follows that $\hat{q}=0$ (where $\hat{q}=$ the extinction probability for the $\{\hat{Z}_n\}$ process).

It can be verified by straightforward computation that the iterates of $\hat{f}(s)$ are

$$\hat{f}_n(s) \equiv \frac{f_n((1-q)s+q)-q}{1-q} \quad \text{for } 0 \leqslant s \leqslant 1,\ n=1,2,\ldots. \tag{2}$$

Many results will be true for the Z_n process if and only if they are also true for the $\hat{Z}_n$ process. But the latter has 0 extinction probability, and hence it will suffice to prove these results for the case $q=0$.

Due to the transience of the non-zero states we know that

$$P(\mathsf{A} \cup \mathsf{B})=1, \tag{3}$$

where

$$\mathsf{A}=\{\omega: Z_n(\omega)\to\infty \text{ as } n\to\infty\},$$

and

$$\mathsf{B}=\{\omega: Z_n(\omega)=0 \text{ for some } n\geqslant 1\}.$$

By definition, $P(\mathsf{B})=q$. From our assumptions we have $0<P(\mathsf{B})<1$ and hence $0<P(\mathsf{A})<1$.

For any ω, $Z_n(\omega)$ is the number of particles in the nth generation of the family tree corresponding to ω. Define

$$Z_n^{(1)}(\omega)=0, \quad \text{if } \omega\in\mathsf{B}, \tag{4}$$
$$= \textit{the number of particles among the } Z_n(\omega) \textit{ of } n\textit{th generation which have infinite line of descent, if } \omega\in\mathsf{A}.$$

The term "infinite line of descent" is self-explanatory.

Let

$$Z_n^{(2)}(\omega)\equiv Z_n(\omega)-Z_n^{(1)}(\omega)\,. \tag{5}$$

It is an easy consequence of these definitions that

$$\begin{gathered} Z_0^{(1)}(\omega)=1 \quad \text{on } \mathsf{A}, \qquad P\{Z_0^{(1)}(\omega)=1\}=1-q\,, \\ \text{and for any} \quad \omega\in\mathsf{A}, \qquad Z_n^{(1)}(\omega)\geqslant Z_{n-1}^{(1)}(\omega)\,. \end{gathered} \tag{6}$$

Let $\mathbb{F}_{\mathsf{A}}$ be the σ-algebra on A obtained by intersecting the sets of $\mathbb{F}$ with A. For any $\mathsf{E}\in\mathbb{F}_{\mathsf{A}}$ we define $P_{\mathsf{A}}\{\mathsf{E}\}\equiv P\{\mathsf{E}\}/P\{\mathsf{A}\}$. Then P_{A} defines a probability measure on $\mathbb{F}_{\mathsf{A}}$ and thus $(\mathsf{A}, \mathbb{F}_{\mathsf{A}}, P_{\mathsf{A}})$ becomes a probability space. With this notation we have the following

Theorem 1. *The stochastic processes* $\{\hat{Z}_n; n=0,1,2,\ldots\}$ *on* $(\Omega, \mathbb{F}, \hat{P})$ *and* $\{Z_n^{(1)}; n=0,1,2,\ldots\}$ *on* $(\mathsf{A}, \mathbb{F}_{\mathsf{A}}, P_{\mathsf{A}})$ *are equivalent in the sense of finite dimensional distributions.*

Proof. We shall prove the theorem by showing that the $Z_n^{(1)}$ process on $(\mathsf{A}, \mathbb{F}_{\mathsf{A}}, P_{\mathsf{A}})$ is Markov with the same transition probabilities as the $\hat{Z}_n$ process. This is clearly enough since $Z_0^{(1)}(\omega)\equiv 1$ on A, and hence $P_{\mathsf{A}}\{Z_0^{(1)}(\omega)=1\}=1=\hat{P}\{\hat{Z}_0=1\}$.

Consider the situation at the nth step. On A, $Z_n^{(1)}(\omega)$ is not 0 for any n. All these particles, and only these out of the $Z_n(\omega)$ particles that constitute the nth generation, are going to have infinite lines of descent. Each one of these $Z_n^{(1)}(\omega)$ particles will create a random number of offspring, at least one of which must have an infinite line of descent. By the basic character of the branching process, the $Z_n^{(1)}(\omega)$ particles create independently of each other and of everything else in the past including $Z_j^{(1)}(\omega)$ for $j=0,1,2,\ldots,n-1$. Further $Z_{n+1}^{(1)}(\omega)$, the number of particles among the $Z_{n+1}(\omega)$ constituting the $(n+1)st$ generation having infinite lines of descent, comes solely from the offspring of $Z_n^{(1)}(\omega)$. Thus the $Z_n^{(1)}$ process is Markovian, and given $Z_n^{(1)}=i$, $Z_{n+1}^{(1)}$ can be written as

$$Z_{n+1}^{(1)}=\sum_{j=1}^{i} N_j\,, \tag{7}$$

where N_j, for $j=1,2,\dots,i$, denotes the number of particles (among the offspring of the jth particle belonging to the $Z_n^{(1)}(\omega)$) having infinite lines of descent. As was said earlier, the random variables N_j are mutually independent, and hence it remains only to show that the N_j's have the same distribution as $\hat{Z}_1$. But since the N_j's have the same distribution as $Z_1^{(1)}$ under P_{A}, it suffices to show that

$$P_{\mathsf{A}}\{Z_1^{(1)}=k\}=\hat{P}\{\hat{Z}_1=k\} \quad \text{for } k=1,2,\dots. \tag{8}$$

By definition

$$P_{\mathsf{A}}\{Z_1^{(1)}=k\}=\frac{P\{(Z_1^{(1)}=k)\cap \mathsf{A}\}}{P\{\mathsf{A}\}},$$

where P refers to our original unconditional probability measure. But for $k\geqslant 1$

$$\begin{aligned} P\{(Z_1^{(1)}=k)\cap \mathsf{A}\}=P\{Z_1^{(1)}=k\}&=\sum_{l=k}^{\infty} P\{Z_1^{(1)}=k\mid Z_1=l\}\,P\{Z_1=l\}\\ &=\sum_{l=k}^{\infty}\binom{l}{k}(1-q)^k q^{l-k}P\{Z_1=l\}. \end{aligned}$$

Therefore, for any s in $[0,1]$

$$\begin{aligned} \sum_{k=1}^{\infty} s^k P_{\mathsf{A}}\{Z_1^{(1)}=k\}&=(1-q)^{-1}\sum_{k=1}^{\infty} s^k\sum_{l=k}^{\infty}\binom{l}{k}(1-q)^k q^{l-k}P\{Z_1=l\}\\ &=(1-q)^{-1}\sum_{l=1}^{\infty}P\{Z_1=l\}\sum_{k=1}^{l}\binom{l}{k}(1-q)^k s^k q^{l-k}\\ &=(1-q)^{-1}\sum_{l=1}^{\infty}P\{Z_1=l\}\,[((1-q)s+q)^l-q^l]\\ &=(1-q)^{-1}\sum_{l=0}^{\infty}P\{Z_1=l\}\,[((1-q)s+q)^l-q^l]\\ &=(1-q)^{-1}[f((1-q)s+q)-q]\\ &\equiv\hat{f}(s)\equiv\sum_{k=1}^{\infty}s^k\hat{P}\{\hat{Z}_1=k\}, \end{aligned}$$

thus establishing (8). □

Remark. The same argument also yields

$$E_{\mathsf{A}}(s^{Z_n^{(1)}})=[f_n((1-q)s+q)-q](1-q)^{-1}. \tag{9}$$

But $\hat{E}(s^{\hat{Z}_n})=\hat{f}_n(s)$, and hence by (2) we get $E_{\mathsf{A}}(s^{Z_n^{(1)}})=\hat{E}(s^{\hat{Z}_n})$, as it should be.

We will now discuss the behavior of $Z_n^{(1)}(\omega)/Z_n(\omega)$ on A.

Lemma 1. *For every* $\varepsilon>0$

$$\lim_{n\to\infty} P\left\{\omega:\omega\in\mathsf{A},\left|\frac{Z_n^{(1)}(\omega)}{Z_n(\omega)}-(1-q)\right|>\varepsilon\right\}=0. \tag{10}$$

Proof. Let $\mathsf{E}_n=\{\omega:|(Z_n^{(1)}(\omega)/Z_n(\omega))-(1-q)|>\varepsilon\}$, and

$$\mathsf{F}_{nk}=\{\omega:Z_n(\omega)=k\}, \quad \text{and} \quad \mathsf{F}_n=\bigcup_{k=1}^{\infty}\mathsf{F}_{nk}=\{\omega:Z_n(\omega)>0\}.$$

Clearly,

$$\begin{aligned} P\{\mathsf{A}\,\mathsf{E}_n\}\leqslant P\{\mathsf{F}_n\,\mathsf{E}_n\}&=\sum_{k=1}^{\infty}P\{\mathsf{E}_n\,\mathsf{F}_{nk}\}\\ &=\sum_{k=1}^{\infty}P\{\mathsf{E}_n|\,\mathsf{F}_{nk}\}P\{\mathsf{F}_{nk}\}. \end{aligned} \tag{11}$$

Using the Markov property we see that, given $Z_n=k$ for $k\geqslant 1$, the random variable $Z_n^{(1)}$ is distributed as the number of "successes" in k independent and identical binomial trials with probability for "success" $(1-q)$.

Thus, using Chebychev's inequality,

$$P\{\mathsf{E}_n|\,\mathsf{F}_{nk}\}\leqslant\varepsilon^{-2}q(1-q)k^{-1},$$

and hence from (11)

$$P\{\mathsf{A}\,\mathsf{E}_n\}\leqslant\varepsilon^{-2}q(1-q)E\left(\frac{1}{Z_n};\,\mathsf{F}_n\right),$$

where for any random variable Y on $(\Omega,\mathbb{F},P)$ we use the notation $E(Y;\,\mathsf{A})$ for $\int_{\mathsf{A}}Y(\omega)\,dP(\omega)$. But

$$E\left(\frac{1}{Z_n};\,\mathsf{F}_n\right)=E\left(\frac{1}{Z_n};\,\mathsf{F}_n\cap\mathsf{A}\right)+E\left(\frac{1}{Z_n};\,\mathsf{F}_n\cap\mathsf{B}\right)=a_n+b_n \qquad \text{(say)}.$$

From the transience of the states $k, k\neq 0$, we know that $Z_n(\omega)\to\infty$ on A, and hence $1/Z_n$, which is bounded by 1 on F_n, goes to 0 on $\mathsf{F}_n\,\mathsf{A}$. Thus $a_n\to 0$. Also $b_n\leqslant P\{\mathsf{F}_n\,\mathsf{B}\}$, and since on B $Z_n(\omega)=0$ for large n, $b_n\to 0$. □

One can use theorem 10.3 to strengthen the convergence in probability in the above lemma to convergence a.s.

Theorem 2. *Let W be as in theorem* 10.3. *Then, on* A,

$$\lim_{n\to\infty}\frac{Z_n^{(1)}(\omega)}{Z_n(\omega)}=(1-q) \qquad \text{a.s.}.$$

Proof. Applying theorem 10.3 to the $\hat{Z}_n$ process we note that there exists a sequence $\hat{C}_n$ such that

$$\hat{P}\left\{\lim_{n\to\infty}\hat{Z}_n\hat{C}_n^{-1}\equiv\hat{W}\ \text{exists}\right\}=1,$$

where $P(\hat{W}>0)=1$, and $\hat{W}$ is nondegenerate.

This, by theorem 1, implies that

$$P_{\mathsf{A}}\left\{\lim_{n\to\infty} Z_n^{(1)}\hat{C}_n^{-1} \equiv W^{(1)} \text{ exists}\right\} = 1\,, \tag{12}$$

where $W^{(1)}$ on $(\mathsf{A}, \mathbb{F}_{\mathsf{A}}, P_{\mathsf{A}})$ has the same distribution as $\hat{W}$ on $(\Omega, \mathbb{F}, \hat{P})$. We also know from lemma 1 that $Z_n^{(1)} C_n^{-1} = (Z_n^{(1)}/Z_n) Z_n C_n^{-1}$ converges in probability to $(1-q)W$ on A. By Khinchine's theorem on the uniqueness of normalising constants for convergence to a nondegenerate limit distribution, we may conclude that

$$\lim_{n\to\infty} \hat{C}_n C_n^{-1} = C \quad \text{exists, and } 0 < C < \infty\,.$$

One can choose $\hat{C}_n$ so that $C=1$.

This and (12) imply that on A

$$\frac{Z_n^{(1)}}{Z_n} = \frac{Z_n^{(1)}\hat{C}_n^{-1}}{Z_n C_n^{-1}} C_n^{-1}\hat{C}_n \to \frac{W^{(1)}}{W} \qquad \text{a.s.}$$

By lemma 1, $W^{(1)}/W = (1-q)$ a.s. □

An obvious consequence is that the conditional distribution of W on $(0, \infty)$ is the same as that of $W^{(1)}$ under P_{A}. Hence, by theorem 1 it is also the same as that of $\hat{W}$ under $\hat{P}$. But, since $\hat{f}(0)=0$, it follows from corollary 10.4 that the distribution of $\hat{W}$ is absolutely continuous on $(0, \infty)$. Thus we obtain

Corollary 1. *Let W be as in theorem* 10.3. *Then W has an absolutely continuous distribution on* $(0, \infty)$.

This removes the hypothesis $q=0$ from theorem 10.4.
Again using the same arguments and theorem 10.2 we obtain

Corollary 2. *For $\alpha \geqslant 1, \sum \hat{p}_j j(\log j)^\alpha < \infty$ if and only if $\sum p_j j(\log j)^\alpha < \infty$.*

Of course, one can verify this fact independently of the relation between W and $\hat{W}$.

In our decomposition (5), $Z_n^{(1)}(\omega)$ never vanished on A and vanishes identically on B, the complement of A. So far we have considered the behavior of $Z_n^{(1)}$ and $Z_n^{(2)}$ only on A. Now we look at B. Here Z_n coincides with $Z_n^{(2)}$, and we have the following result.

Theorem 3. *The stochastic process $Z_n(\omega)$ on $(\mathsf{B}, \mathbb{F}_{\mathsf{B}}, P_{\mathsf{B}})$ is equivalent to a subcritical Galton-Watson branching process $(Z_n^*; n=0, 1, 2, \ldots)$ on a probability space $(\Omega, \mathbb{F}, P^*)$ with*

$$P^*\{Z_0^* = 1\} = 1 \quad \textit{and} \quad E^*(s^{Z_1^*}) = f^*(s) \equiv q^{-1} f(sq)\,.$$

Proof. That the process $Z_n(\omega)$ on $(\mathsf{B}, \mathbb{F}_{\mathsf{B}}, P_{\mathsf{B}})$ is still Markov is obvious. That it has the same transition mechanism as $\{Z_n^*\}$ is verified by showing that

$$\sum_{j=0}^{\infty} P\{Z_{n+1}=j|Z_n=i;\ \mathsf{B}\}\, s^j=[f^*(s)]^i,$$

by arguing as in the proof of the supercritical part of theorem 8.1.

That the process is subcritical follows from (8.10). □

We can now state a strengthening of theorem 8.1 for the supercritical case.

Theorem 4. *Let* $m>1$ *and* $q>0$. *Then* $\mathscr{B}(s)$ *as defined in* (8.4) *is analytic for* $|s|<1/q$ *and hence* $\sum_{j=0}^{\infty} j^r b_j<\infty$ *for all* $r>0$.

Proof. Let

$$\mathscr{B}_n(s)=\frac{f_n^*(s)-f_n^*(0)}{1-f_n^*(0)}.$$

From theorem 3 we have

$$\lim_{n\to\infty} \mathscr{B}_n(s)=\mathscr{B}(s) \quad \text{for } 0\leqslant s\leqslant 1$$

and

$$\mathscr{B}(1-)=1=\mathscr{B}(1).$$

Let $1<s_0<1/q$. Then

$$\mathscr{B}_n(s_0)=\frac{f_n(s_0 q)-f_n(0)}{q-f_n(0)}=1-\frac{f_n(s_0 q)-q}{f_n(0)-q}=1-\frac{(f_n(s_0 q)-q)\gamma^{-n}}{(f_n(0)-q)\gamma^{-n}} \qquad (\gamma=f'(q)).$$

Since $m>1$, we see from section 11 that

$$\lim_{n\to\infty} (f_n(s)-q)\gamma^{-n}=Q(s)\neq 0 \quad \text{for } 0\leqslant s<1.$$

Since $s_0 q<1$, this implies $\sup_n |\mathscr{B}_n(s_0)|<\infty$ and hence in the domain $\mathsf{D}=\{s:|s|\leqslant s_0\}$ the family of analytic functions $\mathscr{B}_n(s)$ is normal. (See e.g. Ahlfors (1953) p. 168.) Further, on the open disk we know $\mathscr{B}_n(s)$ converges to the analytic function $\mathscr{B}(s)$. Thus from a well known theorem on normal families of analytic functions (Ahlfors) we conclude that in every compact subdomain of D, $\mathscr{B}_n(s)$ converges uniformly to a limit function which is analytic in that domain and coincides with $\mathscr{B}(s)$ in D. But s_0 being arbitrary in $(1, 1/q)$, this is equivalent to the theorem. □

13. Second Order Properties of Z_n/m^n

In section 10 we saw that with no restriction at all (except the indispensable assumption of finite mean) the sequence $\{Z_n/m^n; n=0,1,2,\ldots\}$, being a nonnegative martingale, converges almost surely to a random variable W. Thus it seems reasonable to approximate Z_n by $m^n W$ for large enough n. This raises the question of the order of magnitude of the difference $Z_n - m^n W$. In this section we shall develop a simple representation for $Z_n - m^n W$ and use it to answer this question.

The probability space $(\Omega, \mathbb{F}, P)$ still denotes the basic space on which the branching process $\{Z_n; n=0,1,2,\ldots\}$ is defined. Assume for definiteness that $P\{Z_0=1\}=1$.

Theorem 1. *Let* $p_0=0$, *and* $\sum_{j=1}^{\infty} p_j(j\log j)<\infty$. *Then for each n there exists a set* $\mathsf{A}_n \in \mathbb{F}$ *with* $P\{\mathsf{A}_n\}=1$, *and random variables* $W_n^{(j)}(\omega)$ *for* $j=1,2,\ldots,Z_n(\omega)$, *such that for each* ω *in* A_n,

$$Z_n(\omega) - m^n W(\omega) = \sum_{j=1}^{Z_n(\omega)} (1 - W_n^{(j)}(\omega)). \tag{1}$$

The vector $\{W_n^{(j)}(\omega); j=1,2,\ldots,Z_n(\omega)\}$ *(conditioned on* $Z_n(\omega)$*), consists of mutually independent and identically distributed random variables with the same distribution as W.*

Remark. We need the hypothesis $\sum_{j=1}^{\infty} p_j(j\log j)<\infty$ (section 10) to ensure that W is not almost surely 0. Since we are interested only in the set $\{\omega: W(\omega)\neq 0\}$, we see from section 12 that it is no loss of generality to add the condition $p_0=0$. Thus under our hypothesis, $P\{W=0\}=0$.

Proof. We know that there exists a set $\mathsf{A}\in\mathbb{F}$ such that $P\{\mathsf{A}\}=1$ and

$$\lim_{n\to\infty} \frac{Z_n(\omega)}{m^n} \quad \text{exists and} \quad = W(\omega) \quad \text{for } \omega\in\mathsf{A}.$$

Thus

$$W(\omega) = \lim_{k\to\infty} \frac{Z_{n+k}(\omega)}{m^{n+k}}, \qquad \omega\in\mathsf{A}. \tag{2}$$

Let $Z_{n,k}^{(j)}$ = the population size of the kth generation offspring of the jth particle among the Z_n particles existing in the nth generation, where $j=1,2,\ldots,Z_n$. Recall (section 1) that Ω and $\mathbb{F}$ have been chosen so that the $Z_{n,k}^{(j)}$ are well defined random variables. Clearly they satisfy the relation

$$Z_{n+k}(\omega) = \sum_{j=1}^{Z_n(\omega)} Z_{n,k}^{(j)}(\omega). \tag{3}$$

Now observe that

(i) Z_n takes only a countable number of values and is finite with probability 1.

(ii) The conditional probability measure on $(\Omega, \mathbb{F})$ given Z_n is a Z_n-fold convolution of P.

We know from section 10 that the above facts imply the existence of a set A_n such that $P\{\mathsf{A}_n\}=1$, and such that on A_n

$$\lim_{k\to\infty} Z_{n,k}^{(j)}(\omega)\, m^{-k} = W_n^{(j)}(\omega) \quad \text{for } j=1,2,\ldots,Z_n(\omega)$$

exists and has the same distribution as W. Clearly from (2) and (3)

$$\begin{aligned} Z_n(\omega) - m^n W(\omega) &= \lim_{k\to\infty} \sum_{j=1}^{Z_n} (1 - Z_{n,k}^{(j)}(\omega)\, m^{-k}) \\ &= \sum_{j=1}^{Z_n} (1 - W_n{}^{(j)}(\omega)). \quad \square \end{aligned}$$

The following result is a consequence of theorem 1.

Corollary 1. *Let* $p_0=0$ *and* $\sum_{j=1}^{\infty} p_j(j\log j)<\infty$. *Then there exists a set* A *in* $\mathbb{F}$ *such that* $P\{A\}=1$, *and random variables* $W_n{}^{(j)}(\omega)$ *for* $j=1,2,\ldots,$ $Z_n(\omega)$; $n=0,1,2,\ldots$ *(as defined in theorem 1), such that for* ω *in* A

$$Z_n(\omega) - m^n W(\omega) = \sum_{j=1}^{Z_n} (1 - W_n{}^{(j)}(\omega)), \quad \textit{for all } n. \tag{4}$$

Proof. Set $\mathsf{A} = \bigcap_{n=1}^{\infty} \mathsf{A}_n$ where A_n has the same meaning as in theorem 1. $\square$

We now use (4) to partially answer the question raised earlier about the order of magnitude of $Z_n - m^n W$.

Theorem 2. *Let* $p_0=0$, $\sum_{j=1}^{\infty} j^2 p_j < \infty$. *Then*

$$\frac{Z_n - m^n W}{\sqrt{Z_n}} \xrightarrow{d} N\left(0, \frac{\sigma^2}{m^2-m}\right)$$

($\xrightarrow{d}$ stands for convergence in distribution; $N(0,\tau^2)$ stands for the normal distribution with mean 0 and variance τ^2; $\sigma^2 = \operatorname{var}(Z_1)$.) (See Athreya (1968) and Heyde (1971a).)

Proof. From corollary 1 we note that $Z_n - m^n W$ is distributed as $\sum_{j=1}^{Z_n} (1 - W^{(j)})$, where $W^{(j)}, j=1,2,\ldots,Z_n$, are independent copies of W given Z_n. Since $\sum_{j=1}^{\infty} j^2 p_j < \infty$ we have

$$E\, W^{(j)} = 1 \quad \text{and} \quad \operatorname{var}(1 - W^{(j)}) = \frac{\sigma^2}{m^2 - m}.$$

Now appeal to Anscombe's generalization of the classical central limit theorem (see Chung (1966)). □

Remarks. 1. It seems plausible that one could make a stronger use of theorem 1 to assert something stronger than theorem 2. For instance, one would expect a law of the iterated logarithm, namely

$$P\left\{\omega:\limsup\frac{Z_n-m^n W}{\sqrt{Z_n\sigma\log\log Z_n}}=1\right\}=1.$$

This has been proved by C. C. Heyde (1971 b).

2. Our representation in theorem 1 used a method which did not critically depend on the process being in discrete time or being one dimensional. In the case of continuous time one uses in addition the right continuity of sample paths. There will be more on this in chapter III.

3. Note further that in theorem 2 we could assert that $\sum_{j=1}^{\infty} j^2 p_j<\infty$ implied $EW^2<\infty$, and so were able to use the central limit theorem and the representation in corollary 1 to prove the asymptotic normality of $Z_n-m^n W$. Lamperti (1967 b) has shown that W has a stable tail, iff $\{p_j\}$ has a stable tail. Thus $Z_n-m^n W$, when appropriately normalized, should converge to a stable law. This has also been proved by Heyde (1971 a).

4. Although we stated and proved Theorem 1 only with the hypothesis $\sum p_j j\log j<\infty$ we can see from theorem 10.3 that the conclusion of Theorem 1 is valid even when $\sum p_j j\log j=\infty$. This is because the constants C_n are such that $C_{n+k}C_k^{-1}\to m^n$ as $k\to\infty$.

5. An invariance principle for branching processes, extending theorem 2 to a result on convergence of processes, has been established by Brown and Heyde (1971).

14. The Q-Process

We have previously conditioned the Galton-Watson process Z_n on the event $\{n<T<\infty\}$, where T is the extinction time. It is meaningful, more generally, to condition on $\{n+k<T<\infty\}$, $k\geqslant 0$; namely, the event that the process is not extinct at time $n+k$ but does eventually die out. We remark again that when $m\leqslant 1$ this is the same as conditioning on $\{n+k<T\}$.

Proceeding as in the proof of theorem 8.1 we see that when $p_1\neq 0$ and $q\neq 0$

$$P\{Z_n=j|n+k<T<\infty\} = \frac{P_n(1,j)\sum_{l=1}^{\infty} P_k(j,l)\,q^l}{\sum_{l=1}^{\infty} P_{n+k}(1,l)\,q^l}$$
$$= \frac{\left[\frac{P_n(1,j)}{P_n(1,1)}\right]\sum_l P_k(j,l)\,q^l}{\sum_l\left[\frac{P_{n+k}(1,l)}{P_{n+k}(1,1)}\right]q^l}\;\frac{P_n(1,1)}{P_{n+k}(1,1)}, \tag{1}$$

$$\to \frac{\pi_j\sum P_k(j,l)\,q^l}{\gamma^k\mathscr{P}(q)} = \frac{\pi_j[f_k^j(q)-f_k^j(0)]}{\gamma^k\mathscr{P}(q)} = \frac{\pi_j[q^j-f_k^j(0)]}{\gamma^k\mathscr{P}(q)} = b_j(k) \quad \text{(say)}, \tag{2}$$

which by (7.11) is a probability function when $m\neq 1$. As in theorem 8.1, $b_j(k)=0$ if $m=1$. However theorem 8.2 has an analogous generalization. Again using the monotone ratio lemma (with $m=1$)

$$P\{Z_n=j|T=n+k\} = \frac{P_n(1,j)\,[f_k^j(0)-f_{k-1}^j(0)]}{\sum_{i=1}^{\infty} P_n(1,i)\,[f_k^i(0)-f_{k-1}^i(0)]}$$
$$\to \frac{\pi_j[f_k^j(0)-f_{k-1}^j(0)]}{\sum\pi_i[f_k^i(0)-f_{k-1}^i(0)]} \equiv \theta_j(k)\,. \tag{3}$$

By theorem 7.2 the denominator converges. Summarizing we have

Theorem 1. If $p_1>0$ *and* $q>0$, *then*

$$\lim_{n\to\infty} P\{Z_n=j|n+k<T<\infty\}=b_j(k)\geqslant 0\,,$$

where $\sum_j b_j(k)=1$ *if* $m\neq 1$ *and* $b_j(k)=0$ *if* $m=1$. *If* $m=1$, *then*

$$\lim_{n\to\infty} P\{Z_n=j|T=n+k\}=\theta_j(k)\geqslant 0\,,$$

where $\sum_j \theta_j(k)=1$.

An alternative derivation of the above result (not requiring $p_1\neq 0$) from theorem 8.1 goes as follows.

Consider the case $m<1$. Let

$$\mathscr{B}_{n,k}(s)\equiv E(s^{Z_n}|n+k<T)\,.$$

Then

$$\mathscr{B}_{n,k}(s)=[E(s^{Z_n};\,n+k<T)](1-f_{n+k}(0))^{-1}$$
$$=[f_n(s)-f_n(s\,f_k(0))](1-f_{n+k}(0))^{-1}$$
$$=[\{f_n(s)-f_n(0)\}-\{f_n(s\,f_k(0))-f_n(0)\}]$$
$$\times(1-f_n(0))^{-1}\frac{[1-f_n(0)]}{[1-f_n(f_k(0))]}\,.$$

Since $(f_n(s)-f_n(0))(1-f_n(0))^{-1}\to\mathscr{B}(s)$ (theorem 8.1), we get, on letting $n\to\infty$ (with k fixed),

$$\lim_{n\to\infty}\mathscr{B}_{n,k}(s)=\frac{[\mathscr{B}(s)-\mathscr{B}(s\,f_k(0))]}{1-\mathscr{B}(f_k(0))}.$$

Hence

$$b_j(k)=\frac{b_j(1-f_k^j(0))}{1-\mathscr{B}(f_k(0))},$$

where the b_j's are as in theorem 8.1. □

Of course these are only trivial extensions of the results of section 8. A new situation arises, however, if we hold n fixed and let $k\to\infty$. *Assume for the rest of this section that* $q>0$ *and* $p_1>0$. Then the multivariate analogue of (1) is easily computed to be

$$P\{Z_{n_1}=i_1,\ldots,Z_{n_\alpha}=i_\alpha|n_\alpha+k<T<\infty\}$$

$$=P\{Z_{n_1}=i_i,\ldots,Z_{n_\alpha}=i_\alpha\}\frac{\sum_{j=1}^{\infty}P_k(i_\alpha,j)q^j}{\sum_{j=1}^{\infty}P_{n_\alpha+k}(1,j)q^j},$$

(where $n_1\leqslant\cdots\leqslant n_\alpha$, $i_1,\ldots,i_\alpha$ are positive integers)

$$=P\{Z_{n_1}=i_1,\ldots,Z_{n_\alpha}=i_\alpha\}\frac{P_k(1,1)}{P_{n_\alpha+k}(1,1)}\frac{\sum_{j=1}^{\infty}\frac{P_k(i_\alpha,j)}{P_k(1,1)}q^j}{\sum_{j=1}^{\infty}\frac{P_{n_\alpha+k}(1,j)}{P_{n_\alpha+k}(1,1)}q^j}.$$

By theorem 7.4 this

$$\begin{aligned}&\to P\{Z_{n_1}=i_1,\ldots,Z_{n_\alpha}=i_\alpha\}\frac{i_\alpha q^{i_\alpha-1}}{\gamma^{n_\alpha}}\quad\text{as }k\to\infty\\&\equiv\hat{\hat{P}}(i_1,\ldots,i_\alpha)\quad\text{(say)}.\end{aligned}\tag{4}$$

Since

$$\frac{\sum P_n(1,i)\,i\,q^{i-1}}{\gamma^n}=\frac{(f_n(q))'}{(f'(q))^n}=1,$$

the $\hat{\hat{P}}'s$ are probability functions and determine a process which we denote by $\{\hat{\hat{Z}}_n;n=0,1,\ldots\}$. Since $\{Z_n\}$ is a Markov chain it follows from (4) that $\{\hat{\hat{Z}}_n\}$ is also; and its n-fold transition function is computed from (4) to be

$$Q_n(i,j)=P\{\hat{\hat{Z}}_{n+k}=j|\hat{\hat{Z}}_k=i\}=P_n(i,j)\frac{j\,q^{j-i}}{i\,\gamma^n},\qquad i,j\geqslant1.\tag{5}$$

We shall refer to $\{\hat{\hat{Z}}_n\}$ as the Q-process associated with $\{Z_n\}$. (It was introduced by F. Spitzer (unpublished) and in Lamperti-Ney (1968).) It can be roughly thought of as the Z_n process conditioned on *not being* extinct in the distant future *and* on *being* extinct in the even more distant future; Note that when the original process is aperiodic and irreducible the Q-process is aperiodic and irreducible in the same sense.

Theorem 2.
(i) *If* $m>1$ *then the Q-process is positive recurrent.*
(ii) *If* $m=1$ *then the Q-process is transient.*
(iii) *If* $m<1$ *then the Q-process is positive recurrent if and only if* $\sum (k \log k) p_k < \infty$,
(iv) *In the positive recurrent cases the stationary measure for Q is*

$$\mu_j = j q^{j-1} v_j, \quad j \geq 1 \quad (\text{i.e. } \mu_j = \sum \mu_i Q(i,j)). \tag{6}$$

Proof. If $m \neq 1$ then by theorem 11.3 and (11.28)

$$\lim_{n\to\infty} Q_n(i,j) = \lim_{n\to\infty} P_n(i,j) \frac{j q^{j-i}}{i \gamma^n} = j q^{j-1} v_j = \mu_j. \tag{7}$$

But $\sum j q^{j-1} v_j = Q'(q) = 1$.

Applying theorem 11.1 proves parts (i), (iii), and (iv). When $m=1$,

$$\sum_n Q_n(1,j) = j \sum_n P_n(1,j),$$

and the right hand side converges since the original process is transient. This proves (ii). $\square$

As in the Galton-Watson process, the case $m=1$ plays a special role in the Q-process. Due to its transience, $\hat{Z}_n \to \infty$ with probability 1; but $\hat{Z}_n/n$ will converge to a nondegenerate limit law.

Theorem 3[3]**.** *If* $m=1$ *and* $\sigma^2 < \infty$, *then* $(2/\sigma^2)(\hat{Z}_n/n)$ *converges in distribution to a random variable with density*

$$h(x) = \begin{cases} x e^{-x} & \text{if } x \geq 0, \\ 0 & \text{if } x < 0. \end{cases}$$

(Note that h is a convolution of two exponentials.)

Heuristic proof.

$$\begin{aligned} E\exp\left\{-\alpha\left(\frac{\hat{Z}_n}{n}\right)\Big| Z_0 = 1\right\} &= \sum_j Q_n(i,j) e^{\frac{-\alpha j}{n}} \\ &= \sum_j j e^{\frac{-\alpha j}{n}} P_n(1,j) = -\frac{\partial}{\partial \alpha}\left\{n \sum_j e^{\frac{-\alpha j}{n}} P_n(1,j)\right\} \\ &= \frac{\partial}{\partial \alpha}\left\{n\left[1 - f_n\left(e^{\frac{-\alpha}{n}}\right)\right]\right\}. \end{aligned} \tag{8}$$

[3] Observed by Harris (1951).

But we know (see the proof of theorem 9.2) that

$$\lim_{n\to\infty} n\left[1-f_n\left(e^{\frac{-\alpha}{n}}\right)\right]=\left(\frac{\sigma^2}{2}+\frac{1}{\alpha}\right)^{-1}.$$

Thus if we justify taking the limit through the derivative on the right side of (8) we would be finished. Rather than do this, however, we give an independent proof. We shall use the following fact, which is a consequence of Helly's theorem.

Lemma. *If F_n are distributions on $[0,\infty)$ with means μ_n, converging (weakly) to a distribution F with mean $\mu>0$ and $\mu_n\to\mu$ then*

$$G_n(x)=\frac{1}{\mu_n}\int_0^x t\,dF_n(t)$$

converge to the distribution

$$G(x)=\frac{1}{\mu}\int_0^x t\,dF(t).$$

Proof of theorem 3. Let

$$F_n(x)=P\left\{\frac{2}{\sigma^2}\frac{Z_n}{n}\leqslant x\,\middle|\,Z_0=1,\ Z_n>0\right\}.$$

Then (see (9.3))

$$\mu_n=\frac{\frac{2}{\sigma^2}E\left(\frac{Z_n}{n};Z_n>0\right)}{P\{Z_n>0\}}=\frac{2}{\sigma^2}\,\frac{1}{nP\{Z_n>0\}}\to 1,$$

and

$$G_n(x)=\frac{1}{\mu_n}\sum_{j=1}^{\left[\frac{nx\sigma^2}{2}\right]}\left(\frac{2}{\sigma^2}\cdot\frac{j}{n}\right)\frac{P_n(1,j)}{1-P_n(1,0)}=(1+o_n(1))\sum_{j=1}^{\left[\frac{nx\sigma^2}{2}\right]}Q_n(1,j)$$

($[x]=$ the integral part of x)

$$=[1+o(1)]\,P\left\{\frac{2}{\sigma^2}\frac{\hat{Z}_n}{n}\leqslant x\,\middle|\,Z_0=1\right\}.$$

By the lemma

$$G_n(x)\to G(x)=\int_0^x t\,dF(t)=\int_0^x t\,e^{-t}\,dt,$$

proving the theorem. □

15. More on Conditioning; Limiting Diffusions

We have studied the process $\{Z_n|n+k<T<\infty\}$ and its limits, first as $n\to\infty$ (14.2) and then as $k\to\infty$ (the Q-process). As a final result in this direction, we shall let k and $n\to\infty$ simultaneously. The only interesting situation arises in the critical case, and hence we will drop the condition $T<\infty$. Since we shall want n and $n+k$ to get large proportionately, it will be convenient to write our process as $\{Z_{[nt]}|Z_n>0\}$, where $0\leqslant t\leqslant 1$. For simplicity we shall drop the square bracket, but nt will always be understood to mean $[nt]$.

Theorem 1. (Spitzer (unpublished), Lamperti and Ney (1968)). *If $m=1$ and $f''(1)<\infty$, then, for each fixed $t<1$, $\{Z_{nt}/n|Z_n>0\}$ converges in distribution (as $n\to\infty$) to a random variable $U+V$, where U and V are independent random variables having exponential densities with parameters $2/t\sigma^2$ and $2/(t(1-t)\sigma^2)$ respectively.*

Proof. Observe that

$$\begin{aligned}
&P\{Z_{nt}=k|Z_n>0,\, Z_0=1\}\\
&=\frac{P\{Z_{nt}=k,\, Z_n>0|Z_0=1\}}{P\{Z_n>0|Z_0=1\}}\\
&=\frac{P\{Z_{nt}=k|Z_0=1\}\,P\{Z_{n(1-t)}>0|Z_0=k\}}{P\{Z_n>0|Z_0=1\}}\\
&=\frac{P\{Z_{nt}=k|Z_{nt}>0,\, Z_0=1\}\,P\{Z_{nt}>0|Z_0=1\}\,P\{Z_{n(1-t)}>0|Z_0=k\}}{P\{Z_n>0|Z_0=1\}}.
\end{aligned}$$

Hence

$$E\left\{\exp\left\{-\alpha\frac{Z_{nt}}{n}\right\}\middle| Z_n>0,\, Z_0=1\right\}$$

$$\begin{aligned}
&=\frac{P\{Z_{nt}>0|Z_0=1\}}{P\{Z_n>0|Z_0=1\}}\,E\left\{\left[\exp\left\{-\alpha t\frac{Z_{nt}}{nt}\right\}\right][1-f_{n(1-t)}^{Z_{nt}}(0)]\middle|Z_{nt}>0,\, Z_0=1\right\}\\
&=\left[\frac{1}{t}+O_n(1)\right]\;E\left\{\exp\left(-\alpha t\,\frac{Z_{nt}}{nt}\right)\right.\\
&\qquad\left.-\exp\left(-\frac{Z_{nt}}{nt}[\alpha t-nt\log f_{n(1-t)}(0)]\right)\middle|Z_{nt}>0,\, Z_0=1\right\}.
\end{aligned}$$

But

$$\lim_{n\to\infty}[\alpha t-nt\log f_{n(1-t)}(0)]=\alpha t+\left(\frac{t}{1-t}\right)\frac{2}{\sigma^2},$$

and hence by (9.5)

$$\lim_{n\to\infty} E\left[\exp\left\{-\alpha\frac{Z_{nt}}{n}\right\}\middle| Z_n>0, Z_0=1\right]=\frac{1}{t}\left[\frac{1}{1+\alpha t\frac{\sigma^2}{2}}-\frac{1}{1+\alpha t\frac{\sigma^2}{2}+\frac{t}{1-t}}\right]$$
$$=\left(1+\alpha t\frac{\sigma^2}{2}\right)^{-1}\left(1+\alpha t(1-t)\frac{\sigma^2}{2}\right)^{-1}$$

The two factors in the last product are the Laplace transforms of the density functions of U and V. This proves the theorem. □

The above result suggests looking at $\{(Z_{tn}/n)|Z_n>0\}$ for fixed n and variable $t\in[0,1]$ as a stochastic process with parameter t, say $X_n(t)$. One can then study the sequence of processes $X_1(t)$, $X_2(t),\ldots$. This was done by Lamperti and Ney (1968), who showed that the processes converge (in the sense of finite dimensional distributions) to a limiting diffusion process.

A similar situation exists for the Q-process $\{\hat{Z}_n\}$ constructed in section 14. There one considers the process $\{\hat{Z}_{nt}/n\}$ on $0\leqslant t<\infty$ and again proves convergence to a diffusion. The details are in the above reference.

Another type of limiting diffusion is obtained by Lamperti in his study of continuous state space branching processes, a brief account of which appears in chapter VI.

Complements and Problems I

1. If $p_0>0$, $p_1>0$ and $p_0+p_1<1$, then all the non-zero states of $\{Z_n; n\geqslant 0\}$ communicate.

2. Let $\{\pi_k; n\geqslant 1\}$ be as defined in lemma 7.2. Then
(i) $\pi_k>0$ if and only if $k\in \mathsf{S}=\{j; j\geqslant 1, P_n(1,j)>0$ for some $n\geqslant 1\}$.
(ii) $\pi_k<\infty$ for all $k\geqslant 1$.
Hints:
(i) If $k\in \mathsf{S}$ then

$$\gamma^n\pi_k=\sum_i \pi_i P_n(i,k)\geqslant \pi_1 P_n(1,k)>0\,.$$

(ii) $\infty>\gamma=\gamma\pi_1=\sum_k \pi_k P(k,1)$.

If $p_0>0$, then

$$P_1(k,1)=k p_0^{k-1} p_1>0,$$

and hence $\pi_k<\infty$ for all k.

If $p_0=0$, let $a_n^{(k)}=P_n(1,k)/P_n(1,1)$, and observe that for each $k\geqslant 1$

$$a_{n+1}^{(k)}\leqslant A_k+p_1^{k-1}a_n^{(k)},\qquad n\geqslant 1,$$

where $A_k=$constant. This implies that $\{a_n^{(k)}; n\geqslant 1\}$ is bounded.

3. Show that $\varepsilon(t)$ as defined in (9.7) can be written in the form

$$\varepsilon(t)=\sum_{k=3}^{\infty} p_k \sum_{j=2}^{k-1} \sum_{i-1}^{j-1} (1-t^i), \qquad 0\leqslant t<1 .$$

This shows that $\varepsilon(t)$ is $\geqslant 0$, is decreasing, and $\downarrow 0$ as $t\uparrow 1$.

4. Prove corollary 9.1.
Note that

$$\frac{f(t)-t}{(1-t)^2}\rightarrow\frac{\sigma^2}{2} \quad \text{as } t\rightarrow 1 ,$$

and that $n[1-f_n(s)]\rightarrow 2/\sigma^2$ as $n\rightarrow\infty$.

5. As in section 8, let T be the extinction time of $\{Z_n\}$. If $m<1$, then $P\{T>n\} =1-f_n(0)\sim -m^n Q(0)$. Thus if $\theta<-\log m$, then $E\{\exp(\theta T)\}<\infty$, and in particular T has all moments.

6. Show that if $m<1$ and $f''(1)<\infty$, then $\sum j^2 b_j<\infty$, where b_j is defined in (8.4).
Hint: Show that $m^{-n} f_n(s)$ is bounded for $n\geqslant 0, 0\leqslant s\leqslant 1$, and hence that $\lim_{s\rightarrow 1} Q''(s)$ exists; apply (11.30).

7. *An open problem* is to determine the rate of decay of $f_n(s)-1$ when $m<1$ and $EZ_1 \log Z_1=\infty$. The methods used in the proof of theorem 10.3 may be applicable.

8. Prove the analog of theorem 13.1 for the case $EZ_1 \log Z_1=\infty$.

9. *Tail behavior of W* (for $m>1$),

Let $F(t)=P\{W\leqslant t\}$, where W is as in theorem 10.4.

Let $L(x)=\int_0^x [1-F(t)]\,dt$.

Prove that $L(x)$ is slowly varying by studying its Laplace transform; and deduce that $\int x^p dF(x)<\infty$ for $0\leqslant p<1$ (see Athreya (1971a)).

10. *Tail behavior of the conditional limit distribution* when $m<1$ (corollary 8.1).
(a) Show that $\{b_j\}$ is degenerate if and only if $f(s)=(1-m)+ms$.
Hint: Use the functional equation (8.9).
(b) Show that in the non-degenerate case, $b_j>0$ for infinitely many j's.
(c) If $\{p_j\}$ is aperiodic then so is $\{b_j\}$.

11. *Continuation:* Recall that $\sum jb_j<\infty$ if and only if $\sum p_j j \log j<\infty$. This suggests an analog of theorem 10.2 for the subcritical case; namely that $\sum j(\log j)^\alpha b_j<\infty$ if and only if $\sum j(\log j)^{\alpha+1} p_j<\infty$.

12. *Continuation:* By analogy with the supercritical case (problem 9), is $\sum_{k=1}^{i} \sum_{j=k}^{\infty} b_j$ slowly varying, and is $\sum j^\beta b_j<\infty$ for $0\leqslant\beta<1$?

13. *Conjecture:* If $m>1$, $\alpha>0$, and $\beta>0$, then

$$\int x^\beta |\log x|^\alpha dF(x)<\infty \quad (F \text{ as in problem 9})$$

if and only if

$$\sum j^\beta |\log j|^{\alpha+1} p_j<\infty .$$

14. *Relation between branching and queueing processes.* Let T denote the busy period in a single server queue with Poisson input and general service time distribution, initiated with an arrival at $t=0$. Thus T is the time until the server is first free; and if N units are processed in this time, and X_i denotes the service time of the ith then $T=\sum_{i=1}^{N} X_i$. Now think of the first customer in the queue as the initial ancestor in a Galton-Watson process; and let the arrivals during his service period be his "offspring". Continuing in this way, the arrivals to the queue during any customer's service period are to be considered as his offspring in the corresponding Galton-Watson process. The offspring distribution in the G-W process is given by

$$p_j=\int_0^\infty e^{-\lambda t}\frac{(\lambda t)^j}{j!}\,dF(t),$$

where F is the service time distribution, and λ is the parameter of the Poisson input process.

From known results about the total number of particles in a G-W process (see Harris (1963), p. 32), it follows that the busy period T has a defective distribution (i.e. $P\{T<\infty\}<1$) if and only if $m=\sum j p_j=\lambda\int t\,dF(t)>1$. In the non-defective case, $ET<\infty$ if and only if $m<1$.

For further exploitation of these ideas see M. Neuts (1969) and D.G. Kendall (1951).

15. Investigate the recurrence property of the Q-process (section 14) in the case $m<1$ and $EZ_1\log Z_1<\infty$.

16. *An alternate approach to the conditional limit theorem in the subcritical case (J. Williamson).*

The p.g.f. of Z_n, given $Z_n>0$, is $P_n(s)\equiv(f_n(s)-f_n(0))(1-f_n(0))^{-1}$. Verify that for $n\geqslant 1$, $R_n(s)=P_n(s)\,(P_{n-1}(s))^{-1}$ is a p.g.f., where we define $P_0(s)\equiv 1$. (In fact, $R_n(s)$ is a power series in $f_{n-1}(s)$ with non-negative coefficients and $R_n(1)=1$.) Thus, $P_n(s)=\prod_{j=1}^{n} R_j(s)$. Let $\{Y_j\}$ be a sequence of independent random variables, with Y_j having p.g.f. $R_j(s)$. Then $P_n(s)$ is the p.g.f. of $S_n\equiv Y_1+Y_2+\cdots+Y_n$. Hence, the existence of a conditional limit distribution of Z_n, given $Z_n>0$, is reduced to that of a limit distribution for the sequence $\{S_n\}$. The random variables Y_j being non-negative, integer-valued, and independent, it is enough to show that $\sum_j P(Y_j\neq 0)<\infty$ (by the Borel-Cantelli lemma).

Now,

$$\begin{aligned} P(Y_n\neq 0) &= 1-R_n(0)\\ &= 1-\frac{(1-f_{n-1}(0))}{(1-f_n(0))}f'(f_{n-1}(0)) \quad \text{(by L'Hospital's rule)}\\ &= \frac{1-f_{n-1}(0)}{1-f_n(0)}H(1-f_{n-1}(0)), \end{aligned}$$

where $H(x)=x^{-1}(1-f(1-x))-f'(1-x)$ for $0<x\leqslant 1$. Since $f_n(0)\to 1$ as $n\to\infty$, and $((1-f_{n-1}(0))/(1-f_n(0)))\to 1/f'(1)<\infty$, it suffices to prove that

$\sum_n H(1-f_n(0))<\infty$. But $H(1-s)$ can be written as $(1-s)r'(s)$, where

$$r(s)=\frac{1-f(s)}{1-s}=\sum r_k s^k.$$

Interchanging the summations and using the fact that

$$1=\int_0^1 k y^{k-1}\,dy=\sum_{n=0}^{\infty}\int_{f_n(0)}^{f_{n+1}(0)} k y^{k-1}\,dy \geqslant c(1-m)\sum_{n=0}^{\infty} k f_n^{k-1}(0)(1-f_n(0))$$

gives

$$\sum H(1-f_n(0))=\sum_{n=0}^{\infty}\sum_{k=0}^{\infty} k r_k f_n^{k-1}(0)(1-f_n(0))\leqslant c(1-m)^{-1}\sum r_k.$$

We know from section 11 that $\sum j \lim_{n\to\infty} P(Z_n=j\,|\,Z_n>0)<\infty$ iff $\sum j(\log j)p_j<\infty$. Hence, $\sum_j E(Y_j)<\infty$ iff $\sum j(\log j)p_j<\infty$. Establish this directly.

17. Use the geometric convergence of $f_n(s)$ to q to show that when $m>1$, $P\{Z_n\to\infty\}=1-q$.

Hint: Assume $\gamma\equiv f'(q)>0$. Then, $\lim \gamma^{-n}(f_n(s)-q)=Q(s)$ exists for $|s|<1$. We may conclude that for any fixed $j\neq 0$, $\gamma^{-n} f_n^{(j)}(0)\to Q^{(j)}(0)$, and hence $\sum_n f_n^{(j)}(0)<\infty$, implying $\sum_n P(Z_n=j)<\infty$. By the Borel-Cantelli lemma we conclude that for each $j\neq 0$ the process $\{Z_n\}$ visits j only finitely many times. The case $\gamma=0$ is simpler still since the convergence rate is dominated by $\text{const}\cdot\delta^n$ for any $\delta>0$.

18. The result of Darling for case $m=\infty$ (see remark 5 of section 10) has recently been extended by Seneta (as yet unpublished technical report) as follows. Let $g(t)=-\log(1-f(1-e^{-t}))$. If $g(t)$ is convex or concave on $[0,\infty)$, and $0\leqslant \lim_{t\to\infty}(g(t)/t)=c\leqslant 1$, then there exists a sequence of positive constants $\{\rho_n\}$ such that $\rho_n\log(Z_n+1)$ converges in law to a proper distribution, which is continuous and strictly increasing on $(0,\infty)$, and has a jump q at zero. Necessary and sufficient conditions for $\rho_n\sim c^n$ are given.

Chapter II

Potential Theory

1. Introduction

If $P=\{P(i,j);\, i,j=0,1,2,\ldots,\}$ is the transition function of any stationary Markov chain, then $\{\eta_i;\, i=0,1,\ldots\}$ is called a *stationary* or *invariant measure* for P if $\eta_i \geqslant 0$ and

$$\eta_j = \sum_{i=0}^{\infty} \eta_i P(i,j), \qquad j \geqslant 0. \tag{1}$$

If in addition $\sum \eta_i < \infty$ (or without loss of generality, if $\sum \eta_i = 1$), then $\{\eta_i\}$ is a *stationary distribution.*

The sequence $\{\tau_i;\, i=0,1,\ldots\}$ is an *invariant function* or *harmonic function* for P if $\tau_i \geqslant 0$ and

$$\tau_i = \sum_{j=0}^{\infty} P(i,j)\,\tau_j, \qquad i \geqslant 0. \tag{2}$$

If $P_n(i,j)$ is the n-step transition function constructed from $P(i,j)$, then

$$G(i,j) = \sum_{n=0}^{\infty} P_n(i,j) \tag{3}$$

is the *Green function* or *renewal function* of the Markov chain.

The relevance of these quantities in the study of Markov processes has been the subject of much work in recent years. Pertinent background reading can be found, for example in Kemeny, Snell and Knapp (1966), Neveu (1964), and Spitzer (1964).

In the present chapter we shall study the existence, uniqueness, and asymptotic properties of the functions η, τ and G for the Galton-Watson process. Except for the use of a few results from the general theory, our treatment well be self-contained. However, to properly understand the motivation the reader will need a little background in Markov chains at about the level of the above references.

Stationary measures are constructed in section 2. In sections 3 and 4 we develop some refined results on the asymptotic behavior of $P_n(i,j)$.

In I.7 and theorem I.11.3 we considered $P_n(i,j)$ as $n\to\infty$ for fixed i and j; and we now need to study the situation when j and n both $\to\infty$. These local limit theorems, in addition to being of some interest in their own right, are used in the description of the asymptotics of the stationary measures in section 6, and of the Green function in sections 7 and 8. They also yield subsidiary results on the limiting behavior of Z_n.

Section 5 develops some more refined properties of W; including a sharp global limit law, and properties of the limiting density function. Harmonic functions and some boundary theory are discussed in the last two sections.

Some of the proofs in this chapter involve rather lengthy calculations, and can be omitted at a first reading. The subsequent chapters are entirely independent of the present one.

Remark. Some of the results about generating functions derived in chapter I for real s will here be needed for complex s. In most cases the proofs for the complex case are identical to the real one, and for these we will feel free to use the results in the complex setting without further comment.

2. Stationary Measures: Existence, Uniqueness, and Representation

We note at the outset that if we stay strictly with the definition (1.1) then the only stationary measure is the trivial one

$$\eta_0^*=1, \qquad \eta_i^*=0 \quad \text{for } i\geqslant 1 .$$

(Of course any stationary measure is only determined up to a multiplicative constant, unless boundary conditions are added.) That $\{\eta_i^*\}$ is such a measure is obvious. That there are no others is shown by Harris (1963) as follows. Set $j=0$ in (1.1) and note that $P(0,0)=1$. Then (1.1) becomes $\sum_{i=1}^{\infty} \eta_i^* P(i,0)=0$. But this is impossible if $p_0>0$ (unless $\eta_i^*=0$ for $i\geqslant 1$), since in that case $P(i,0)>0$, $i\geqslant 1$. On the other hand if $p_0=0$ then $P(i,j)=0$ for $i>j$ and hence (1.1) becomes $\eta_j^*=\sum_{i=0}^{j} \eta_i^* P(i,j)$. If k is the smallest positive integer for which $\eta_k^*>0$, then $\eta_k^*=\eta_k^* P(k,k)$, which is impossible since the assumptions of section 1.1 imply that $P(k,k)<1$ for $k>0$. Thus $\eta_k^*=0$ for $k\geqslant 1$.

We will find, however, that if we exclude the state 0, and call $\{\eta_i\}$ a stationary measure if $\eta_i\geqslant 0$ and

$$\eta_j=\sum_{i=1}^{\infty} \eta_i P(i,j), \qquad j\geqslant 1 , \tag{1}$$

then there will be non-trivial solutions. We adopt this definition from now on.

The situation is again quite different in the critical and non-critical cases. In the critical case the existence and uniqueness question is already settled by Theorems I.7.1 and I.7.3. (Earlier but more complicated proofs under extra conditions appear in Karlin, McGregor (1967a, b) and Kesten, Ney, Spitzer (1966)). We restate the result here.

Theorem 1. *When* $m=1$ *the sequence* $\{\pi_j; j \geqslant 1\}$ *defined by Lemma I.7.2 is the unique (up to multiplicative constants) stationary measure for the Galton-Watson process.*

We will see another proof of this fact in section 7.

Theorem I.7.2 tells us that $\sum \pi_j = \infty$ for $m=1$. To get more precise estimates of the behavior of π_j for large j is a more delicate question which we defer to later in this chapter.

Theorem 2. *The generating function* $U(s) = \sum \eta_j s^j$ *of any stationary measure is analytic for* $|s| < q$ *and (if normalized so that* $U(p_0) = 1$*) satisfies*

$$U[f(s)] = 1 + U(s). \tag{2}$$

Conversely, if $U(s) = \sum_{j=1}^{\infty} \eta_j s^j$, $\eta_j \geqslant 0$, $|s| < q$, *satisfies (2), then* $\{\eta_j\}$ *is a stationary measure.*

Proof. That (2) is satisfied by a stationary generating function is trivially verified

$$\left(\sum_{j=1}^{\infty} \eta_j s^j = \sum_{i=1}^{\infty} \eta_i \sum_{j=1}^{\infty} P(i,j) s^j = \sum \eta_i [f^i(s) - f^i(0)] = U[f(s)] - U[p_0] \right).$$

The remainder of the argument is very much like the proof of Theorem I.7.2. We observe that $\infty > \eta_j = \sum \eta_i P(i,j) \geqslant \sum \eta_i p_0^{i-1} p_j$, and taking j so that $p_j > 0$ this implies that $U(s) < \infty$ for $|s| < p_0$. Using (2) this implies that $U(s) < \infty$ for $|s| \leqslant f_n(p_0)$, $n = 1, 2, \ldots$; *i.e.* for $|s| < q$ since $f_n(p_0) \to q$ as $n \to \infty$. The converse follows by comparing coefficients in (2). □

We turn next to a description of the stationary measures for *the subcritical process.*

Let

$$U(s,t) = \sum_{n=-\infty}^{\infty} [\exp\{Q(0)[1 - \mathscr{B}(s)] m^{n-t}\} - \exp\{Q(0) m^{n-t}\}], \tag{3}$$

where $\mathscr{B}$ and Q are as in sections I.8 and I.11. (It is left as an exercise to show that this series converges. See complements for hints.)

Theorem 3. *Spitzer* (1967). *If* $m<1$ *and* $EZ_1 \log Z_1 < \infty$ *then for every probability measure* ν *on* $[0,1)$

$$U(s) = \int_0^1 U(s,t)\,\nu(dt) \tag{4}$$

is the generating function of a stationary measure. Conversely every stationary measure has a representation (4) *for some probability measure* ν *on* $[0,1)$.

Remark. Stationary measures as we have defined them are only determined up to multiplicative constants, but $U(s,t)$ as defined in (3) is normalized so that $U(f(0),t)=1$; and thus $U[f(0)]=1$.

Outline of Proof. Let k_i be any sequence of integers tending to ∞. The construction rests on the fact all sequences

$$\eta_j = \lim_{i\to\infty} G(k_i, j), \qquad j \geqslant 1, \tag{5}$$

are stationary measures (provided the limits in (5) exist), where G is the Green function defined in (1.3); and all stationary measures are obtained as mixtures of these sequences. More precisely, let I denote the set of integers and let $\hat{x}=$ the fractional part of $-\log x/\log m$. Then the function

$$\rho(x,y) = \begin{cases} |\hat{x}-\hat{y}| + \left|\dfrac{1}{x} - \dfrac{1}{y}\right|, & x, y \in I, \\ |\hat{x}-y| + \left|\dfrac{1}{x}\right|, & x \in I,\ y \in [0,1), \\ |x-y|, & x, y \in [0,1), \end{cases} \tag{6}$$

is a metric on $I \cup [0,1)$.

We shall see that a sequence $\{x_\alpha;\ \alpha=1,2,\ldots\} \subset I$ is Cauchy in the ρ metric if an only if $\{G(x_\alpha,j)\}$ is Cauchy (in the ordinary metric on the real line). Furthermore we shall show that in the topology defined by ρ the functions

$$\eta_j(x) = \begin{cases} G(x,j), & x \in I, \\ \text{coefficient of } s^j \text{ in } U(s,x), & x \in [0,1), \end{cases} \tag{7}$$

are continuous on $I \cup [0,1)$. In other words if $k_i \to \infty$ in such a way that the fractional part of $-\log k_i/\log m \to t \in [0,1)$ then $G(k_i,j) \to \eta_j(t) =$ coef. of s^j in $U(s,t)$.

Thus, since $I \cup [0,1)$ is the completion of I, correspondingly $\{\eta_j(x);\ x \in I \cup [0,1)\}$ is the completion of $\{G(x,j);\ x \in I\}$; and hence $\{\eta_j(t) = \text{coef. of } s^j \text{ in } U(s,t);\ t \in [0,1)\}$ will be stationary measures. (We will also give an explicit proof of this.) The Martin boundary theory for

Markov chains (see e.g. Kemeny, Snell, Knapp (1966)) then assures us that all stationary measures have the (Poisson) representation (4).

Proof. Turning to the details let

$$
\begin{aligned}
V(s,k) &= \sum_{j=1}^{\infty} G(k,j)s^j = \sum_{n=0}^{\infty} [f_n^k(s) - f_n^k(0)] \\
&= \sum_{n=0}^{\infty} [\{1-(1-f_n(0))(1-\mathscr{B}_n(s)\}^k - \{1-(1-f_n(0))\}^k],
\end{aligned} \tag{8}
$$

where

$$\mathscr{B}_n(s) = \frac{f_n(s)-f_n(0)}{1-f_n(0)}; \tag{9}$$

and recall that by Theorem I.8.1 $\mathscr{B}_n(s) \to \mathscr{B}(s)$ (note that this is also true for complex s), and $[1-f_n(0)]m^{-n} \to -Q(0)$. Using these two facts (see the complements for further details) one can show that

$$
\begin{aligned}
\lim_{k_i\to\infty} V(s,k_i) &= \lim_{k_i\to\infty} \sum_{n=0}^{\infty} [\{1+Q(0)m^n[1-\mathscr{B}(s)]\}^{k_i} - \{1+Q(0)m^n\}^{k_i}] \\
&= \lim_{k_i\to\infty} \sum_{n=0}^{\infty} [\exp\{Q(0)m^n[1-\mathscr{B}(s)]k_i\} - \exp\{Q(0)m^n k_i\}],
\end{aligned} \tag{10}
$$

where $k_i \to \infty$ is a subsequence of $\{k\}$. Now define a_i, h_i, f_i and α_i by

$$
\begin{aligned}
&a_i = -\log k_i/\log m, \\
&h_i = \text{integral part of } a_i, \\
&f_i = \text{fractional part of } a_i, \\
&\alpha_i = n - h_i \ (=\text{integer}).
\end{aligned} \tag{11}
$$

Then by (10)

$$
\begin{aligned}
\lim_{k_i\to\infty} V(s,k_i) &= \lim_{k_i\to\infty} \sum_{\alpha_i=-h_i}^{\infty} [\exp\{Q(0)[1-\mathscr{B}(s)]m^{\alpha_i-f_i}\} - \exp\{Q(0)m^{\alpha_i-f_i}\}] \\
&= \lim_{k_i\to\infty} \sum_{j=-\infty}^{\infty} [\exp\{Q(0)[1-\mathscr{B}(s)]m^{j-f_i}\} - \exp\{Q(0)m^{j-f_i}\}].
\end{aligned} \tag{12}
$$

Thus $\{V(s,k_i)\}$ is a Cauchy sequence if and only if $\{k_i\}$ is Cauchy in the ρ-metric; and the space $\{V(s,k); k\in I\}$ is completed by adjoining the limit points $\{U(s,t); t\in[0,1)\}$. Thus as $k_i\to\infty$, $f_i\to t$,

$$\sum_j G(k_i,j)s^j \to \sum_j \eta_j(t)s^j, \qquad 0 \leq s < 1. \tag{13}$$

Hence

$$G(k_i,j) \to \eta_j(t), \qquad j \geq 1. \tag{14}$$

This completes the proof. □

Although (14) assures us that the coefficients in the expansion of $U(s)$ are stationary measures, we can also verify this directly by showing that

$$U[f(s)] = U(s) + 1$$

(see Theorem 2). Namely, using the formula $\mathscr{B}[f(s)] = m\mathscr{B}(s) + 1 - m$ (see I.8.5), we get

$$\begin{aligned}
U[f(s)] &= \int_0^1 \sum_{n=-\infty}^{\infty} [\exp\{Q(0)[1 - m\mathscr{B}(s) - 1 + m] m^{n-t}\} \\
&\qquad\qquad - \exp\{Q(0) m^{n-t}\}] \, v(dt) \\
&= \int_0^1 \sum_{n=-\infty}^{\infty} [\exp\{Q(0)[1 - \mathscr{B}(s)] m^{1+n-t}\} \\
&\qquad\qquad - \exp\{Q(0) m^{1+n-t}\}] \, v(dt) \\
&\quad + \int_0^1 \sum_{n=-\infty}^{\infty} [\exp\{Q(0) m^{1+n-t}\} - \exp\{Q(0) m^{n-t}\}] \, v(dt) \, .
\end{aligned} \tag{15}$$

The last sum $= \lim_{n\to\infty} \exp\{Q(0) m^{n-t}\} - \lim_{n\to-\infty} \exp\{Q(0) m^{n-t}\} = 1 - 0 = 1$. (Recall that $Q(0) < 0$.) Thus the first integral on the right is $U(s)$ and the second is 1.

Examples. The linear generating function $f(s) = 1 - m + ms$ was excluded by assumption from our general development to avoid trivialities in certain cases. In the present context, however, it provides a non-trivial but easily computable example which already suggests the result of Theorem 2. In this case

$$f_n(s) = 1 - m^n + m^n s \, ,$$
$$\mathscr{B}_n(s) = s = \mathscr{B}(s) \, ,$$
$$Q(0) = -1 \, ,$$

$$P(i,j) = \begin{cases} \binom{i}{j} m^j (1-m)^{i-j} & \text{if } i \geqslant j \, , \\ 0 & \text{if } i < j \, . \end{cases}$$

Thus substituting into (3) and using the definition (7)

$$U(s,t) = \sum_{j=1}^{\infty} \eta_j(t) s^j$$

$$= \sum_{n=-\infty}^{\infty} \{e^{-(1-s)m^{n-t}} - e^{-m^{n-t}}\}$$

$$= \sum_{n=-\infty}^{\infty} e^{-m^{n-t}} \left[\sum_{j=1}^{\infty} \frac{m^{(n-t)j}}{j!} s^j \right].$$

Equating coefficients of s^j we get

$$\eta_j(t) = \sum_{n=-\infty}^{\infty} e^{-m^{n-t}} \frac{m^{(n-t)j}}{j!}, \tag{16}$$

which can be directly verified to be a stationary measure.

To produce another stationary measure from (16) let us apply (4) with v equal to Lebesgue measure. Then some calculation shows that

$$\int_0^1 \eta_j(t)\,dt = \frac{-1}{j \log m} \equiv \eta_j \qquad \text{(say)}. \tag{17}$$

Its generating function is

$$U(s) = \sum \eta_j s^j = \frac{\log(1-s)}{\log m}.$$

For general f it is not hard to verify that

$$U(s) = \frac{\log[1-\mathscr{B}(s)]}{\log m}$$

is the generating function of a stationary measure by checking (2). This example was first used by Harris ((1963), p. 25) to prove the existence of stationary measures for the subcritical process.

The supercritical case. (i) If $q=0$ there are no stationary measures, since then $p_0=0$, $P(i,j)=0$ for $i>j$, and $P_n(i,j)\to 0$ as $n\to\infty$ for all $i,j \geq 1$. But if μ_j is stationary then $\mu_j = \sum_{i=1}^{\infty} \mu_i P(i,j) = \sum_{i=1}^{j} \mu_i P(i,j) = \sum_{i=1}^{j} \mu_i P_n(i,j)$ for all n. Thus, necessarily $\mu_j=0$ for all j.

(ii) If $q>0$ then the construction can be handled by reduction to the subcritical case. If $f'(1)=m$, $f(q)=q$, then define $f^*(s)=q^{-1} f(qs)$. We have seen that this is the p.g.f of a subcritical process with mean $f'(q)<1$. Let P^* denote the transition function of the $*$-process. Then

$P^*(i,j)=P(i,j)q^{j-i}$, and hence $\{\eta_j\}$ is a stationary measure for P if and only if $\{q^j\eta_j\}\equiv\{\eta_j^*\}$ is a stationary measure for P^*.

Note that also $P_n^*(i,j)=P_n(i,j)q^{j-i}$, and hence $G(i,j)=G^*(i,j)q^{i-j}$.

Examples. (i) Let $\mathscr{B}^*(s)$ = the generating function of the conditioned limit law for the subcritical $*$-process. Then

$$V(s)=\frac{\log\left[1-\mathscr{B}^*\left(\frac{s}{q}\right)\right]}{\log\gamma}, \qquad \gamma=f'(q),$$

is the generating function of a stationary measure for the supercritical process determined by $P(i,j)$.

(ii) Kingman (1965) first demonstrated the non-uniqueness of stationary measures for the supercritical case. See his paper for further examples.

3. The Local Limit Theorem for the Critical Case

In order to obtain further information about stationary measures (such as their asymptotic properties), and about the other functions defined in section 1, we will need to know a lot more about the asymptotic behavior of the transition functions $P_n(i,j)$. We have already seen examples of such results (which are called local limit theorems) in sections I.7, I.8, I.9, I.11; but we will need sharper ones. The critical case will be treated in this section, and the supercritical case in the next.

For fixed i and j, we saw in corollary I.9.2 that (when $m=1$) $P_n(i,j) \sim c i\pi_j n^{-2}$ as $n\to\infty$, but we now want to let $j\to\infty$ also. In view of the global exponential limit law of section I.9, it comes as no surprise that there is also a local exponential law.

Theorem 1. *If P is aperiodic*[4], *$m=1$, and $EZ_1^2\log Z_1<\infty$, and if $k,n\to\infty$ in such a way that k/n remains bounded, then, for fixed $i>1$,*

$$\lim\frac{n^2}{i}\left(\frac{\sigma^2}{2}\right)^2 P_n(i,k)e^{\frac{2k}{\sigma^2 n}}=1\,. \tag{1}$$

The existence of EZ_1^2 is clearly necessary for this result, and one might guess that it is sufficient, but there is no published proof. H. Kesten, P. Ney, and F. Spitzer (1966) (hereafter referred to as KNS) have given a proof under the slightly stronger assumption $EZ_1^2\log Z_1<\infty$, but even in this case the proof is long and involved. Under the further assumption

$$EZ_1^{3+\delta}<\infty \quad \text{for some } \delta>0 \tag{2}$$

[4] P is called aperiodic if g.c.d. $\{k: P(1,k)>0\}=1$; it has period d if the g.c.d. is $d>1$.

a relatively straight forward proof is possible, and this is the one we will give here.

The assumption of aperiodicity is also not necessary. If the period is d, and $k \to \infty$ in the semigroup generated by $\{k: P(1,k)>0\}$, then (1) holds with 1 replaced by d on the right side (see KNS).

Finally, we will prove (1) only for $i=1$, and refer the reader to KNS for the extension to general i. We proceed via several lemmas.

Lemma 1.

$$|1-f_n(s)| \leqslant 2(1-f_n(0)). \tag{3}$$

Proof.

$$|1-f_n(s)| = \left|1-P_n(1,0) - \sum_{j=1}^{\infty} P_n(1,j)s^j\right|$$

$$\leqslant |1-P_n(1,0)| + \sum_{j=1}^{\infty} P_n(1,j) = 2[1-P_n(1,0)]. \qquad \square$$

We will need a sharper estimate on $f_n(s)$ than so far obtained, and this is given in the next lemma.

Lemma 2. *If* $m=1$ *and* $EZ_1^{3+\delta}<\infty$ *for some* $\delta>0$, *then for* $|s| \leqslant 1$, $s \neq 1$,

$$\frac{1}{1-f_n(s)} = \frac{1}{1-s} + na + (a^2-b)D_n(s) + U_0(s) + U_n(s), \tag{4}$$

where $a=f''(1)/2$, $b=f'''(1)/6$; $D_n(s)=\sum_{j=0}^{n-1} [(1-s)^{-1}+aj]^{-1}$, $|U_n(s)| \leqslant K_1 n^{-\varepsilon}$; $|U_0(s)| \leqslant K_2$; $\varepsilon>0$; $K_1, K_2 < \infty$.

This estimate was given by Harris (1963, page 20) under the hypothesis $EZ_1^4<\infty$.

Proof. Take $|s| \leqslant 1$, $s \neq 1$. Observe that

$$\frac{1}{1-f(s)} = \frac{1}{1-s} \sum_{k=0}^{2} \left(\frac{f(s)-s}{1-s}\right)^k + \frac{(f(s)-s)^3}{(1-s)^4} \cdot \frac{1-s}{1-f(s)}. \tag{5}$$

From the Taylor expansion of f we obtain

$$1-f(s) = (1-s) - a(1-s)^2 + b(1-s)^3 + o(1-s)^{3+\delta}. \tag{6}$$

(Note that $(1-s)/(1-f(s))$ is bounded due to the aperiodicity of f). From (5) and (6) we see that

$$\frac{1}{1-f(s)} = \frac{1}{1-s} + a + (a^2-b)(1-s) + \Gamma(s),$$

where $\Gamma(s) \leqslant K|1-s|^{1+\delta}$, and iterating this relation we get

$$\frac{1}{1-f_n(s)} = \frac{1}{1-s} + na + (a^2-b)\sum_{j=0}^{n-1}[1-f_j(s)] + \sum_{j=0}^{n-1}\Gamma[f_j(s)]. \tag{7}$$

Now, lemma 1 and theorem I.9.1 imply that

$$|1-f_n(s)| \leqslant \frac{\text{constant}}{n}. \tag{8}$$

Hence for all $n \geqslant 1$

$$\left|\sum_{j=0}^{n-1} [1-f_j(s)]\right| \leqslant c_1 \log n, \qquad |s| \leqslant 1, \tag{9}$$

and

$$\Gamma[f_n(s)] \leqslant \frac{\text{constant}}{n^{\delta+1}}. \tag{10}$$

Let

$$v_n(s) = 1 - f_n(s) - \frac{1}{\dfrac{1}{1-s} + na}.$$

Using (7) we see that

$$v_n(s) = \frac{-(a^2-b)\sum_{j=0}^{n-1} [1-f_j(s)] - \sum_{j=0}^{n-1} \Gamma[f_j(s)]}{\left[\dfrac{1}{1-s} + na\right]\left[\dfrac{1}{1-f_n(s)}\right]} \leqslant c \frac{\log n}{n^2}. \tag{11}$$

We now rewrite (7) in the form

$$\begin{aligned} \frac{1}{1-f_n(s)} = \frac{1}{1-s} + na + (a^2-b)\left[\sum_{j=0}^{\infty} v_j(s) - \sum_{j=n}^{\infty} v_j(s)\right] \\ + (a^2-b)\sum_{j=0}^{n-1} \frac{1}{\dfrac{1}{1-s} + ja} + \sum_{j=0}^{\infty} \Gamma[f_j(s)] - \sum_{j=n}^{\infty} \Gamma[f_j(s)]. \end{aligned} \tag{12}$$

Letting

$$U_0(s) = \sum_{j=0}^{\infty} \{(a^2-b) v_j(s) + \Gamma[f_j(s)]\}, \quad U_n(s) = \sum_{j=n}^{\infty} \{(a^2-b) v_j(s) + \Gamma[f_j(s)]\},$$

and applying (10) and (11), we get lemma 2. □

Let

$$\lambda_n(s) = E(e^{isZ_n} | Z_n > 0).$$

Lemma 3. *If* $m=1$ *and* $EZ_1^{3+\delta} < \infty$, *then there exists an* $\varepsilon_0 > 0$ *and constants* $K_1, K_2 < \infty$ *such that*

$$\lambda_n\left(\frac{x}{n}\right) = \frac{1}{1-iax} + r_1(x,n) + r_2(x,n), \tag{13}$$

where

$$|r_1(x,n)| \leqslant \frac{K_1}{n}\left|\log \frac{x}{n}\right|, \qquad |r_2(x,n)| \leqslant K_2 \frac{\log n}{xn}$$

for $|x/n| < \varepsilon_0$.

Proof.

$$\lambda_n\left(\frac{x}{n}\right)=\frac{f_n\left(e^{\frac{ix}{n}}\right)-f_n(0)}{1-f_n(0)}=\frac{\dfrac{1}{1-f_n\left(e^{\frac{ix}{n}}\right)}-\dfrac{1}{1-f_n(0)}}{\dfrac{1}{1-f_n\left(e^{\frac{ix}{n}}\right)}},$$

which by Lemma 2

$$=\frac{\dfrac{e^{\frac{ix}{n}}}{1-e^{\frac{ix}{n}}}+\left[U_0\left(e^{\frac{ix}{n}}\right)-U_0(0)\right]+\left[U_n\left(e^{\frac{ix}{n}}\right)-U_n(0)\right]+(a^2-b)\left[D_n\left(e^{\frac{ix}{n}}\right)-D_n(0)\right]}{\dfrac{1}{1-e^{\frac{ix}{n}}}+na+U_0\left(e^{\frac{ix}{n}}\right)+U_n\left(e^{\frac{ix}{n}}\right)+(a^2-b)D_n\left(e^{\frac{ix}{n}}\right)}. \tag{14}$$

But

$$|D_n(s)-D_n(0)|\leqslant\sum_{j=0}^{n-1}\frac{1}{|1+ja|\cdot|1+ja(1-s)|}\leqslant c_1+c_2\log\frac{1}{|1-s|},\qquad |s|<1,$$

where the last inequality can be seen by writing

$$\sum_{j=0}^{n-1}\leqslant\sum_{j=0}^{\infty}\leqslant\sum_{j=0}^{|1-s|^{-1}}+\sum_{j=|1-s|^{-1}}^{\infty},$$

and noting that the first sum on the right is dominated by a constant $\cdot\log 1/|1-s|$, while the second is bounded.

We also observe that

$$|D_n(s)|\leqslant C_3\log n\,.$$

Using the above estimates and the bounds on U and U_n given in Lemma 2, one can deduce (13) from (14). □

Lemma 4. *If* $m=1$ *and* $EZ_1^{3+\delta}<\infty$, *then given any* $\varepsilon>0$ *there is a constant* c *and a sequence* $c_n\to 0$ *as* $n\to\infty$ *such that*

$$n\lambda_n(u)=\psi_n(u)+\psi(u),\qquad \varepsilon\leqslant|u|\leqslant\pi, \tag{15}$$

where $|\psi(u)|\leqslant c$ *and* $|\psi_n(u)|\leqslant c_n$.

Proof. If we set $s=e^{iu}$ then analogously to (14) we can write

$$n\lambda_n(u)=n\{1-f_n(s)\}\cdot$$
$$\cdot\left\{\frac{s}{1-s}+[U_0(s)-U_0(0)]+[U_n(s)-U_n(0)]+(a^2-b)\,[D_n(s)-D_n(0)]\right\}. \tag{16}$$

Using Lemma 2 we can see that

$$n\{1-f_n(s)\}=\frac{1}{a}+O\left(\frac{\log n}{n}\right). \tag{17}$$

Now

$$D_n(s)-D_n(0)=\sum_{j=0}^{n-1}\frac{-s}{[1+ja][1+ja(1-s)]}$$
$$=\sum_{j=0}^{\infty}-\sum_{j=n}^{\infty}\frac{-s}{[1+ja][1+ja(1-s)]}. \tag{18}$$

Denote the first sum by $D(s)$ and the second by $\bar{D}_n(s)$. Since $|u|>\varepsilon$ implies $|1-s|>\delta$ for some $\delta>0$, it follows that

$$|D(s)|\leqslant \text{constant}, \qquad |\bar{D}_n(s)|\leqslant \frac{\text{constant}}{n}, \tag{19}$$

and also of course $|1-s|^{-1}<\delta^{-1}$.

Using (17), (18), (19) in (16), and applying the estimates of Lemma 2, we see that $n\lambda_n(u)$ can be expressed in the form (15). □

Proof of Theorem 1 (under the condition $EZ_1^{3+\delta}<\infty$ and with $i=1$). Let

$$p_n(k)=P\{Z_n=k\,|\,Z_n>0,\,Z_0=1\}. \tag{20}$$

Then

$$p_n(k)=\frac{P_n(1,k)}{P\{Z_n>0\,|\,Z_0=1\}},$$

and hence by Theorem I. 9.1

$$\frac{P_n(1,k)}{p_n(k)}\sim\frac{1}{an}, \qquad a=\frac{\sigma^2}{2}.$$

Thus to prove (1) we must show that

$$n\,p_n(k)=\frac{1}{a}e^{-\frac{1}{a}\frac{k}{n}}+J(k,n),$$

where

$$\lim_{\substack{k\to\infty,\,n\to\infty\\ \frac{k}{n}\leqslant c}} J(k,n)=0.$$

By definition

$$\lambda_n(t)=\sum_{k=0}^{\infty}e^{itk}p_n(k),$$

and taking the inverse Fourier transform of both sides we have

$$p_n(k)=\frac{1}{2\pi}\int_{-\pi}^{\pi}e^{-itk}\lambda_n(t)\,dt=\frac{1}{2\pi n}\int_{-\pi n}^{\pi n}e^{-ix\frac{k}{n}}\lambda_n\!\left(\frac{x}{n}\right)dx.$$

We can then write

$$n p_n(k) = \frac{1}{a} e^{-\frac{1}{a}\frac{k}{n}} + \sum_{i=1}^{4} J_i, \tag{21}$$

where

$$J_1(k,n,A) = \frac{1}{2\pi}\int_{-A}^{A} e^{-i\frac{k}{n}x}\left\{\lambda_n\left(\frac{x}{n}\right) - \frac{1}{1-iax}\right\}dx,$$

$$J_2(k,n,A,\varepsilon) = \frac{1}{2\pi}\int_{A<|x|\leqslant \varepsilon n} e^{-i\frac{k}{n}x}\left\{\lambda_n\left(\frac{x}{n}\right) - \frac{1}{1-iax}\right\}dx,$$

$$J_3(k,n,\varepsilon) = \frac{1}{2\pi}\int_{|x|\leqslant \varepsilon n} e^{-i\frac{k}{n}x}\frac{dx}{1-iax} - \frac{1}{a}e^{\frac{1}{a}\frac{k}{n}},$$

$$J_4(k,n,\varepsilon) = \frac{1}{2\pi}\int_{\varepsilon n<|x|\leqslant \pi n} e^{-i\frac{k}{n}x}\lambda_n\left(\frac{x}{n}\right)dx = \frac{n}{2\pi}\int_{\varepsilon<|u|\leqslant\pi} e^{-iku}\lambda_n(u)\,du\,.$$

We must show that given any $\delta>0$, we can produce an $\varepsilon>0$ and $A<\infty$ such that

$$\lim_{\substack{(k,n)\to\infty \\ \frac{k}{n}<c}} J_i(k,n)\leqslant\delta, \qquad i=1,2,3,4. \tag{22}$$

From Theorem I.9.2 or formula (I.9.5) we see that $\lambda_n(x/n)\to[1-iax]^{-1}$ uniformly on $[-A,A]$ and hence (22) holds for $i=1$. For J_2 we use Lemma 3, taking $\varepsilon<\varepsilon_0$ (the number specified in that lemma). Then

$$|J_2| \leqslant \frac{K_1}{2\pi}\int_{A<|x|\leqslant\varepsilon n}\left|\log\frac{x}{n}\right|\frac{dx}{n} + \frac{K_2}{2\pi}\int_{A<|x|\leqslant\varepsilon n}\frac{\log n}{nx}dx\,.$$

By taking ε small, and then n big, the right side can be made arbitrarily small. This proves (22) for $i=2$.

We leave as an exercise the proof of the fact that (22) holds for $i=3$. Some hints are indicated in the complements.

Finally, we use Lemma 4 to write J_4 as

$$J_4 = \frac{1}{2\pi}\int_{\varepsilon\leqslant|u|\leqslant\pi} e^{-iku}\psi_n(u)\,du + \frac{1}{2\pi}\int_{\varepsilon\leqslant|u|\leqslant\pi} e^{-iku}\psi(u)\,du\,.$$

The first integral goes to zero (uniformly in k) as $n \to \infty$, and the second goes to zero as $k \to \infty$ by the Riemann-Lebesgue lemma. This completes the proof. □

Remark. Inspection of the proof shows that we have in fact also proved that

$$\sup_{k \geq 1} \frac{n^2}{i} P_n(i,k) \leqslant K < \infty \tag{23}$$

when $i=1$, and this inequality holds for all i.

While we are on the subject of local limit theorems, we should remark on the spectral representations of $P_n(i,j)$, obtained by S. Karlin and J. McGregor (1966 b), which provide another powerful tool for the study of local behavior. In the critical case this result takes the following form.

Theorem 2. *Let $f(s)=\sum_j p_j s^j$ be a nondegenerate p.g.f with $\sum j p_j = 1$. Assume $f(s)$ is analytic at $s=1$. Let $f^{-1}(s)$, defined on $[f(0),1]$, admit the expansion*

$$1 - f^{-1}(1-s) = \sum_{1}^{\infty} c_k s^k,$$

with $c_k \geqslant 0$. Then, $P_n(i,j)$ admits the representation

$$P_n(i,j) = \int_0^{\infty} e^{-n\xi} Q_j(\xi)\, d\psi_i(\xi),$$

where $Q_j(\xi)$ is a polynomial of degree j, and $\psi_i(\xi)$ is a signed measure of bounded variation.

It is shown in the above reference that the p.g.f.'s in the family H described in III.12 satisfy the hypothesis of theorem 2. The proof of the above theorem uses some elegant techniques of compact operators on Hilbert spaces and is beyond the scope of the present work.

For a refinement of the above result, and a discussion of the continuous time case, see also Karlin and McGregor (1968 c).

In the supercritical case an analgous result has been discussed in section I.11.

4. The Local Limit Theorem for the Supercritical Case

We turn now to the behavior of $P_n(i,j)$ as $(j,n) \to \infty$ in the supercritical case. We have already seen that if $EZ_1 \log Z_1 < \infty$ then $W_n \equiv Z_n/m^n \to W$, the latter random variable being non-degenerate, and in fact having an absolutely continuous distribution away from zero. Thus, there is a function $w(x) \geqslant 0$, such that

$$\lim_{n\to\infty} P\{x_1 < W_n \leqslant x_2 \mid Z_0 = i\} = \int_{x_1}^{x_2} w^{*i}(x)\,dx\,, \qquad 0 < x_1 < x_2 < \infty\,, \tag{1}$$

where $w^{*i}(x)$ is the i-fold convolution of $w(x)$. Expressing this "global" limit law in terms of P_n yields

$$\lim_{n\to\infty} \sum_{j=x_1 m^n}^{x_2 m^n} P_n(i,j) = \int_{x_1}^{x_2} w^{*i}(x)\,dx = \int_{x_1 m^n}^{x_2 m^n} m^{-n} w^{*i}(y m^{-n})\,dy\,, \tag{2}$$

which suggests that one look for a "local" law

$$m^n P_n(i,j) \sim w^{*i}(j m^{-n})\,. \tag{3}$$

In fact we can state a somewhat stronger result. Let

$$\gamma = f'(q)\,, \qquad \delta = -\frac{\log\gamma}{\log m}\,, \qquad \beta = m^{\frac{\delta}{3+\delta}}.$$

Theorem 1. *Let $\{Z_n; n \geqslant 1\}$ be a Galton-Watson process with $m > 1$ and $E(Z_1^2 \mid Z_0 = 1) < \infty$. Fix i. Then given any $\beta_0 < \beta$ there exists a constant $C = C(\beta_0)$ such that*

$$|m^n P_n(i,j) - w^{*i}(m^{-n} j)| \leqslant C\left[\frac{\beta^{-n}}{m^{-n} j}\right] + \beta_0^{-n} \tag{4}$$

for all $j \geqslant 1$.

This theorem is proved in Athreya, Ney (1970), and work leading up to it can be found in Chistyakov (1957) and Imai (1968).

The second moment assumption in this theorem is undesireable, and we shall prove instead a slightly weaker form of (4), but under the "right" hypothesis. This proof is by Dubuc (1970). It will make use of the following lemma (also due to Dubuc).

Lemma 1. *Assume f is the generating function of an aperiodic Galton-Watson process with $m > 1$ and $E Z_1 \log Z_1 < \infty$. Let*

$$\mathsf{S} = \bigcup_{n=0}^{\infty} \left\{ f_n\!\left(e^{\frac{i\omega}{m^n}}\right); \frac{\pi}{m} \leqslant \omega \leqslant \pi \right\},$$

and $\bar{\mathsf{S}}$ = closure of S. Let

$$\mu_k = \sup_{s \in \bar{\mathsf{S}}} |f_k'(s)|\,. \tag{5}$$

Then

$$\sum \mu_k < \infty\,.$$

Proof. Recall that $\gamma = f'(q) < 1$ and that by theorem I.11.1

$$\gamma^{-n} f_n'(s) \to Q'(s) < \infty \quad \text{for } 0 \leqslant s < 1\,. \tag{6}$$

Furthermore as a consequence of Theorem I.10.1

$$f_n\left(e^{\frac{i\omega}{m^n}}\right) \to \varphi(-i\omega)$$

uniformly on compact intervals. By lemma I.10.6 and continuity of φ

$$\sup_{\frac{\pi}{m} \leqslant |\omega| \leqslant \pi} |\varphi(-i\omega)| < 1 .$$

Combining these facts we see that there is a compact set C_0 interior to the unit disk, and an $N < \infty$ such that

$$f_n\left(e^{\frac{i\omega}{m^n}}\right) \in \mathsf{C}_0 \quad \text{for } n > N, \quad \frac{\pi}{m} \leqslant \omega \leqslant \pi . \tag{7}$$

But, due to the fact that $\{p_j\}$ is aperiodic, $f_n(e^{i\omega/m^n})$ is also in a compact set interior to the unit disk for $n \leqslant N$ and $\pi/m \leqslant \omega \leqslant \pi$. Hence there is a $p < 1$ such that

$$\bar{\mathsf{S}} \subset \{s: |s| \leqslant p\} \tag{8}$$

Thus $\mu_k \leqslant f_k'(p)$, which by (6) is $\leqslant$ constant $\cdot \gamma^k$. This proves the lemma. □

Theorem 2. *Let Z_n be an aperiodic Galton-Watson process with $m > 1$ and $E(Z_1 \log Z_1 | Z_0 = 1) < \infty$. Let $(j, n) \to \infty$ in such a way that $j m^{-n} \to c$. Then, for each fixed i,*

$$j P_n(i, j) \to c\, w^{*i}(c) . \tag{9}$$

Proof. As before, let

$$f_n(s) = \sum_{j=0}^{\infty} P\{Z_n = j\} s^j, \qquad |s| \leqslant 1 ,$$

s complex.

Then

$$P_n(k, j) = \frac{1}{2\pi} \int_{-\pi}^{\pi} f_n^k(e^{i\theta}) e^{-ij\theta} d\theta , \tag{10}$$

and integrating by parts

$$\begin{aligned} j P_n(k, j) &= \frac{1}{2\pi} \int_{-\pi}^{\pi} k f_n^{k-1}(e^{i\theta}) f_n'(e^{i\theta}) e^{i\theta(1-j)} d\theta \\ &= \frac{1}{2\pi} \int_{-\pi m^n}^{\pi m^n} k f_n^{k-1}\left(\exp\left\{\frac{i\omega}{m^n}\right\}\right) f_n'\left(\exp\left\{\frac{i\omega}{m^n}\right\}\right) \exp\left\{\frac{i\omega(1-j)}{m^n}\right\} m^{-n} d\omega . \end{aligned} \tag{11}$$

Now if we could take the limit as $n\to\infty$ under the integral on the right side we would have (recalling $jm^{-n}\to c$)

$$\lim_{n\to\infty} jP_n(k,j) = \frac{-1}{2\pi}\int_{-\infty}^{\infty} k\varphi^{k-1}(-i\omega)\varphi'(-i\omega)e^{-i\omega c}\,d\omega\,, \tag{12}$$

where $\varphi(s)=Ee^{-sW}=\lim_{n\to\infty} f_n(\exp\{-s/m^n\})$.

But the right side of (12)

$$= -\frac{1}{2\pi}\int_{-\infty}^{\infty} k\varphi^{k-1}(i\omega)\varphi'(i\omega)e^{i\omega c}\,d\omega = c\,w^{*k}(c). \tag{13}$$

It thus remains only to justify the passage of the limit through the integral in (11). To do this we will bound the integrand by an integrable function.

For $-\pi\leqslant\omega\leqslant\pi$ the integral is bounded by k. Suppose $\pi m^{\alpha-1}\leqslant\omega\leqslant\pi m^{\alpha}$, where α is a positive integer. Then the integrand is bounded by

$$\begin{aligned} km^{-n}\left|f_n'\left(e^{\frac{i\omega}{m^n}}\right)\right| &\leqslant k\left|m^{-\alpha}f_\alpha'\left(f_{n-\alpha}\left(e^{\frac{i\omega}{m^n}}\right)\right)\cdot m^{-(n-\alpha)}f_{n-\alpha}'\left(e^{\frac{i\omega}{m^n}}\right)\right| \\ &\leqslant km^{-\alpha}\left|f_\alpha'\left(f_{n-\alpha}\left(e^{\frac{i\omega}{m^n}}\right)\right)\right|, \end{aligned}$$

since $|m^{-n}f_n'(z)|\leqslant 1$ for $|z|\leqslant 1$. But for $\pi m^{\alpha-1}\leqslant\omega\leqslant\pi m^{\alpha}$, $f_{n-\alpha}(e^{i\omega/m^n})\in \mathsf{S}$ where S is defined in lemma 1. Hence the integrand in (11) is bounded by $km^{-\alpha}\mu_\alpha$, where $\sum\mu_\alpha<\infty$. The bound for $-\pi m^{\alpha-1}\geqslant\omega\geqslant-\pi m^{\alpha}$ is similar. This completes the proof. □

5. Further Properties of W; A Sharp Global Limit Law; Positivity of the Density

The local limit of the last section can be used to develop further results about W. After a lemma about the characteristic function of W, $\psi(u)\equiv Ee^{iuW}$, and one about the Lipschitz continuity of its density $w(\cdot)$, we go on to prove a sharper form of the global limit law (4.1), which will include an exponential convergence rate. We then prove the strict positivity of the density $w(\cdot)$. Without this result the local limit theorem would lose much of its strength, since it would then not be clear that m^n is always the right norming sequence.

For simplicity we shall assume throughout this section that $f(0)=0$, *and hence that* $q=0$. The following arguments have their appropriate analogs when $q>0$.

The parameters γ and δ are as defined in section 4. We remind the reader (section I.11) that for complex s

$$\lim_{n\to\infty} \gamma^{-n} f_n(s) = Q(s), \qquad |s|<1, \tag{1}$$

Q being the unique solution of

$$Q[f(s)] = \gamma Q(s), \qquad |s| \leqslant 1,$$

with

$$Q(0)=0, \qquad Q'(0)=1. \tag{2}$$

Lemma 1. *If* $EZ_1 \log Z_1 < \infty$, *then for real* u,

(i) $\lim_{k\to\infty} m^{\delta k} \psi(m^k u) = Q[\psi(u)], \quad u \neq 0;$

(ii) $\sup_{-\infty<u<\infty} |u|^{\delta} |\psi(u)| < \infty;$

(iii) $\lim_{k\to\infty} m^{(1+\delta)k} \psi'(m^k u) = \psi'(u) Q'[\psi(u)], \quad u\neq 0;$

(iv) $\sup_{-\infty<u<\infty} |u|^{1+\delta} |\psi'(u)| < \infty.$

Proof. (i) Iteration of the relation $\psi(mu) = f(\psi(u))$, yields

$$\psi(m^k u) = f_k[\psi(u)], \qquad k=1,2,\ldots. \tag{3}$$

The non-degeneracy of W implies that $|\psi(u)|<1$ for $u\neq 0$ (see lemma I.10.6), and hence by (1)

$$m^{\delta k} \psi(m^k u) = \frac{f_k[\psi(u)]}{\gamma^k} \to Q[\psi(u)] \quad \text{as } k\to\infty.$$

(ii) Since $\gamma^{-k} f_k(x) \uparrow Q(x)$ for $0<x<1$,

$$(m^k |u|)^{\delta} |\psi(m^k u)| \leqslant |u|^{\delta} \frac{f_k(|\psi(u)|)}{\gamma^k} \leqslant |u|^{\delta} Q(|\psi(u)|).$$

Since ψ is continuous

$$\beta \equiv \sup_{1\leqslant |u| \leqslant m} |\psi(u)| < 1,$$

and hence

$$\sup_{\substack{1\leqslant |u| \leqslant m \\ 0 \leqslant k}} (m^k |u|)^{\delta} |\psi(m^k u)| \leqslant \sup_{1\leqslant |u|\leqslant m} |u|^{\delta} Q(|\psi(u)|) \leqslant m^{\delta} Q(\beta) < \infty.$$

Thus

$$\sup_{1\leqslant |u|} |u|^{\delta} \psi(|u|) < \infty,$$

while trivially

$$\sup_{|u|\leqslant 1} |u|^{\delta} \psi(|u|) < \infty.$$

(iii) Differentiation of (3) (note that since $EZ_1 \log Z_1 < \infty$, $EW<\infty$, and hence ψ' exists) yields

$$m^k \psi'(m^k u) = \psi'(u) f_k'(\psi(u)),$$

and hence

$$m^{k(1+\delta)}\psi'(m^k u)=\gamma^{-k}\psi'(u)\,f_k'(\psi(u))\to Q'[\psi(u)]\,|\psi'(u)| \quad \text{as } k\to\infty. \tag{4}$$

(iv) From (4)

$$(m^k|u|)^{1+\delta}|\psi'(m^k u)|\leqslant |u|^{1+\delta}|\psi'(u)|\frac{f_k'(|\psi(u)|)}{\gamma^k},$$

and since f_k'/γ^k is monotone increasing, the above is

$$\leqslant |u|^{1+\delta}|\psi'(u)|\,Q'(|\psi(u)|).$$

Hence

$$\sup_{\substack{1\leqslant|u|\leqslant m\\ 0\leqslant k}}\{(m^k|u|)^{1+\delta}|\psi'(m^k u)|\}\leqslant m^{1+\delta}Q'(\beta)<\infty,$$

and since

$$\sup_{|u|\leqslant 1}|u|^{1+\delta}|\psi'(u)|<\infty$$

the proof is complete. □

Lemma 2. *If* $EZ_1\log Z_1<\infty$, *then for each* $\varepsilon>0$ $w(x)$ *is Lipschitz continuous of order* $\delta'=\min(\delta,1)$ *in* $[\varepsilon,\infty)$.

Proof. Since $\psi'(u)=i\int_0^\infty e^{iux}xw(x)dx$ is integrable (by lemma 1), and $xw(x)$ is a density, we can invert the Fourier transform and get

$$xw(x)=\frac{1}{2\pi i}\int_{-\infty}^{\infty}e^{-iux}\psi'(u)du.$$

Hence for any $y_1, y_2>0$

$$\begin{aligned} w(y_1)-w(y_2)&=\frac{i}{2\pi}\frac{y_1-y_2}{y_1y_2}\int_{-\infty}^{\infty}e^{-iuy_2}\psi'(u)du\\ &\quad+\frac{i}{2\pi}\frac{1}{y_1}\int_{-\infty}^{\infty}(e^{-iuy_2}-e^{-iuy_1})\psi'(u)du. \end{aligned} \tag{5}$$

Since ψ' is integrable, the absolute value of the first integral is

$$\leqslant c\frac{|y_1-y_2|}{y_1y_2}. \tag{6}$$

To estimate the second integral we decompose its range into

$$\mathsf{A}=\{u\colon |u|\,|y_2-y_1|\leqslant 1\}$$

and

$$\bar{\mathsf{A}} = \text{complement of } \mathsf{A}.$$

Since

$$\frac{|e^{-iuy_2} - e^{-iuy_1}|}{|u|\,|y_2 - y_1|} \leqslant c \quad \text{for } u \in \mathsf{A},$$

we have

$$\int_{\mathsf{A}} |e^{-iuy_2} - e^{-iuy_1}|\,|\psi'(u)|\,du \leqslant c\,|y_2 - y_1| \int_{\mathsf{A}} |u|\,|\psi'(u)|\,du$$
$$= c'\,|y_2 - y_1| \quad \text{if } \delta \geqslant 1;$$

and by lemma 1 (iv)

$$\begin{aligned} &= c\,|y_2 - y_1| \int_{\mathsf{A}} |u|^{1+\delta} |\psi'(u)| \frac{du}{|u|^\delta} \\ &\leqslant c''\,|y_2 - y_1| \int_{\mathsf{A}} |u|^{-\delta} du \leqslant c_1 |y_2 - y_1|^\delta \quad \text{if } \delta < 1. \end{aligned} \tag{7}$$

On the other hand

$$\int_{\bar{\mathsf{A}}} |e^{-iuy_2} - e^{-iuy_1}|\,|\psi'(u)|\,du \leqslant 2 \int_{\bar{\mathsf{A}}} |\psi'(u)|\,du,$$

and again applying lemma 1 (iv), $\leqslant c_2 |y_2 - y_1|^\delta$.

Combining this with (5), (6) and (7) yields

$$|w(y_1) - w(y_2)| \leqslant c_0' \frac{|y_1 - y_2|}{y_1 y_2} + c_0'' \frac{|y_1 - y_2|^{\delta'}}{y_1},$$

which implies lemma 2. □

We can now easily deduce the following sharper form of the global limit law (first published in Athreya, Ney (1970)).

Theorem 1. *Let* $\{Z_n; n = 0, 1, 2, \ldots\}$ *be a supercritical Galton-Watson process with* $EZ_1^2 < \infty$ *and* $P\{Z_0 = r\} = 1$ *for some* r. *Then for any* $\beta_0 < \beta = m^{\delta/(3+\delta)}$, *and fixed* $0 < x_1 < x_2 < \infty$,

$$|P\{x_1 < W_n \leqslant x_2\} - P\{x_1 < W \leqslant x_2\}| = o(\beta_0^{-n}).$$

Proof. Without loss of generality we can take $r = 1$. We write

$$\begin{aligned} |P\{x_1 < W_n \leqslant x_2\} - P\{x_1 < W \leqslant x_2\}| &\leqslant \left| m^{-n} \sum_{j = x_1 m^n}^{x_2 m^n} \{m^n P_n(1, j) - w(m^{-n} j)\} \right| \\ &+ \left| \sum_{j = x_1 m^n}^{x_2 m^n} m^{-n} w(m^{-n} j) - \int_{x_1}^{x_2} w(x)\,dx \right| \equiv |J_1^{(n)}| + |J_2^{(n)}|. \end{aligned} \tag{8}$$

Due to theorem 4.1 each term in the first sum is $o(\beta_0^{-n})$ (uniformly for j in the range of summation). Since there are $(x_2 - x_1) m^n$ terms in the sum,

$$J_1^{(n)} = o(\beta_0^{-n}).$$

For the second term in (8) we write

$$|J_2^{(n)}| \leqslant \left| \sum_{x_1 \leqslant j m^{-n} \leqslant x_2} \int_{j m^{-n}}^{(j+1) m^{-n}} \{w(x) - w(m^{-n} j)\} \, dx \right|,$$

which by lemma 2

$$\leqslant K \sum_{j = x_1 m^n}^{x_2 m^n} \int_{j m^{-n}}^{(j+1) m^{-n}} |x - m^{-n} j|^{\delta'} dx = K \sum_{j = x_1 m^n}^{x_2 m^n} \int_0^{m^{-n}} y^{\delta'} dy$$

$$= K'(x_2 - x_1) m^{-\delta' n}.$$

Since $\beta_0 < \beta < m^{\delta'}$, this proves the theorem. □

We turn to the positivity of $w(x)$. As remarked before, if $w(c_0) = 0$ for some $c_0 > 0$ then $m^n P_n(i, c_0 m^n) \to 0$, and we would thus only have an upper bound for the growth rate of P_n rather than its asymptotic behavior.

Theorem 2. *If* $E Z_1 \log Z_1 < \infty$, *then the density* $w(x)$ *is strictly positive for all* $x > 0$.

Proof. The details of the proof are given in Athreya-Ney (1970), so we only outline the idea.

Since $|\psi'|$ is integrable (lemma 1), $x w(x)$ and hence $w(x)$ is continuous for $x > 0$. Since there must be at least one point where $w(\cdot) > 0$, there must also be an open interval $I = (a, b)$ on which $w(\cdot) > 0$.

From the identity $\psi(m u) = f[\psi(u)]$ we see that

$$w^{*i}(x) = \sum_j P(i, j) m w^{*j}(m x); \tag{9}$$

where $w^{*i}(x) = i$-fold convolution of w. But since $m > 1$ there must be integers k_1, k_2 such that $1 \leqslant k_1 < m < k_2$ and $P(1, k_1) > 0$, $P(1, k_2) > 0$, and thus $c_i = \min[P(i, i k_1), P(i, i k_2)] > 0$. By (9)

$$w^{*i}(x) \geqslant c\, w^{*i k_1}(m x), \qquad w^{*i}(x) \geqslant c\, w^{*i k_2}(x), \qquad i = 1, 2, \ldots. \tag{10}$$

If $w^{*i}(x) > 0$ for $x \in I$, then $w^{*ik}(x) > 0$ for $x \in kI$, and hence by repeated application of (10)

$$w^{*i}(x) > 0 \quad \text{for } x \in \bigcup_{n, N \geqslant 0} \left(\frac{k_1}{m}\right)^N \left(\frac{k_2}{m}\right)^n I \equiv \mathsf{S} \quad \text{(say)}. \tag{11}$$

Now note that if $u(x)$ and $v(x)$ are any densities positive on the sets A and B respectively then

$$(u * v)(x) > 0 \quad \text{for } x \in \{y + z : y \in \mathsf{A}, z \in \mathsf{B}\}. \tag{12}$$

But by (11) there is a sequence of intervals $I_n = (a_n, b_n) \subset \mathsf{S}$ such that $a_n, b_n \to 0$, and hence by (12) and the fact that $w(x) > 0$ for $x \in I$,

$$w^{*2}(x) > 0 \quad \text{for } x \in I + I_n,$$

implying

$$w^{*2}(x)>0 \quad \text{for } x\in I=(a,b). \tag{13}$$

But we can choose an n_0 such that $a+a_{n_0}<b$, and then by (12) and (13)

$$w^{*2}(x)>0 \quad \text{for } x\in(a,b+b_{n_0}).$$

Continuing in this way we show that $w^{*3}(x)>0$ on $(a,b+2b_{n_0})$; and given any $0<d<\infty$ there is a K_0 such that

$$w^{*k_0}(x)>0 \quad \text{on } (a,d) \quad \text{for all } k_0\geqslant K_0.$$

Now by choosing d so that $k_2/m<d/a$, and then applying (11) with $i=k_0$, $N=0$ and I replaced by (a,d) we see that $w^{*k_0}(x)>0$ on (a,∞); and then re-applying it with $n=0$ and I replaced by (a,∞) we get $w^{*k_0}(x)>0$ for $x\in(k_1/m)^N(a,\infty)$, $N=0,1,2,\ldots$. Thus

$$w^{*k_0}(x)>0 \quad \text{for } x>0 \quad \text{and } k_0\geqslant K_0. \tag{14}$$

By the second equation in (10)

$$w(x)\geqslant c\,w^{*k_2}(m\,x)\geqslant\cdots\geqslant c^r\,w^{*k_2^r}(m^r x), \qquad r\geqslant 1. \tag{15}$$

Choose $k_0\geqslant K_0$ so that $k_0=k_2^r$ for some positive integer r. (This is possible since $k_2>1$.) Then (14) and (15) imply

$$w\left(\frac{x}{m^r}\right)=w^{*\left(\frac{k_0}{k_2^r}\right)}\left(\frac{x}{m^r}\right)\geqslant c^r\,w^{*k_0}(x)>0 \quad \text{for } x>0.$$

Hence $w(x)>0$ for $x>0$, proving the theorem. □

Remark. When $EZ_1\log Z_1=\infty$, the above argument can be used to establish the positivity of the density of $\lim(Z_n/C_n)$, where the $C_n's$ are as in section 10.

6. Asymptotic Properties of Stationary Measures

In the critical case $(m=1)$ we saw (Theorem 2.1) that there is a unique stationary measure (under the normalization $\sum\eta_j p_0^j=1$). In this case much is known about $\{\eta_j;\, j=1,2,\ldots\}$. The results depend on moment assumptions about Z_n: the more stringent moment conditions leading to more refined theorems.

Theorem 1 (Kesten, Ney, Spitzer). *If* $m=1$ *and* $\sigma^2<\infty$ *then*

$$\lim_{n\to\infty}\frac{1}{n}\sum_{j=1}^{n}\eta_i=\frac{2}{\sigma^2}. \tag{1}$$

Proof. Let $U(s)=\sum_{j=1}^{\infty}\eta_j s^j$. Then $U(0)=0$ and iterating (2.2) we see that

$$U[f_n(0)]=n. \tag{2}$$

Now given any $t\in(0,1)$ we can find an $n\geqslant 0$ such that

$$f_n(0)\leqslant t\leqslant f_{n+1}(0), \tag{3}$$

and hence (due to the monotonicity of U) such that

$$[1-f_{n+1}(0)]\,U[f_n(0)]\leqslant(1-t)\,U(t)\leqslant[1-f_n(0)]\,U[f_{n+1}(0)]. \tag{4}$$

Thus by Theorem I.9.1

$$U(t)\sim\frac{1}{1-t}\,\frac{2}{\sigma^2}\quad \text{as } t<1. \tag{5}$$

Hence (1) follows from the Hardy-Littlewood Tauberian theorem. □

The next result strengthens the Cesaro limit in (1) to an ordinary limit and makes use of the local limit Theorem 3.1; hence it requires the slightly stronger moment condition used in that result.

Theorem 2. (KNS) *If* $m=1$ *and* $EZ_1^2\log Z_1<\infty$ then

$$\lim\eta_j=\frac{2}{\sigma^2}. \tag{6}$$

Proof.

$$\left|\eta_j-\frac{2}{\sigma^2}\right|\leqslant\left|\eta_j-\left(\frac{\sigma^2}{2}\right)n^2P_n(1,j)\right|+\left|\left(\frac{\sigma^2}{2}\right)n^2P_n(1,j)-\left(\frac{2}{\sigma^2}\right)e^{-\frac{2j}{n\sigma^2}}\right|$$
$$+\left(\frac{2}{\sigma^2}\right)\left|e^{-\frac{2j}{n\sigma^2}}-1\right|. \tag{7}$$

Pick any $\varepsilon>0$. First choose $\eta>0$ so that $|e^{-2j/n\sigma^2}-1|<\varepsilon/3$ for $j/n<\eta$. Then choose j_0 so that $|(\sigma^2/2)n^2P_n(1,j)-(2/\sigma^2)\exp\{-2j/n\sigma^2\}|<\varepsilon/3$ for $j\geqslant j_0$ and $j/n\leqslant\eta$. (This can be done by theorem 3.1). Finally, by corollary I.9.2 choose n (possibly depending on j), such that $|\eta_j-(\sigma^2/2)n^2P_n(1,j)|<\varepsilon/3$. Thus $|\eta_j-2/\sigma^2|<\varepsilon$ for $j\geqslant j_0$. □

A stronger moment assumption can be parleyed into a much sharper result.

Theorem 3 (Karlin-McGregor (1967 a, b)). *If* $m=1$ *and* $EZ_1^4<\infty$, *then*

$$\eta_i=\frac{2}{\sigma^2}-\left(1-\frac{4b}{\sigma^4}\right)\frac{1}{i}+\frac{e_i}{i} \tag{8}$$

where $b=f'''(1)/6$ *and* $\sum_{i=1}^{\infty}e_i^2<\infty$.

When $m \neq 1$, we saw in section 2 that there are many stationary measures associated with each f, and gave a representation for them. Very little is known about their asymptotic properties.

In (2.17) we had an example of a stationary measure η_j such that

$$\eta_j \sim \frac{c}{j}. \tag{9}$$

In the problem section at the end of the chapter we will see that there is an $\{\eta_j\}$ and a subsequence j_k such that $\eta_{j_k} > \text{constant}/\sqrt{j_k}$, and thus (9) does not always hold.

However, a weaker (or average) version of (9) does hold.

Theorem 4 (C. Lipow (1971)). *If* $1 < m < \infty$ *and* $p_0 > 0$, *or* $m < 1$ *and* $\sum p_j j \log j < \infty$; *and if* $\{\eta_j; j \geq 1\}$ *is a stationary measure, then*

$$\sum_{i=1}^{n} \eta_i q^i \sim \frac{\log n}{-\log \gamma} \quad \text{as } n \to \infty. \tag{10}$$

(As before, $\gamma = f'(q)$).

Proof. We have seen in section 2 that $U(s) = \sum_1^\infty \eta_i s^i$ satisfies

$$U(f(s)) = U(s) + 1, \tag{11}$$

and by iteration

$$U(f_n(s)) = U(s) + n. \tag{12}$$

Recalling the definition (I.11.1)

$$Q_n(s) = \frac{f_n(s) - q}{\gamma^n}.$$

and using (12), we can obtain the identity

$$\frac{\log \gamma}{\log(q - f_n(0))} U[f_n(0)] = 1 + \frac{\log \gamma}{\log(q - f_n(0))} U(0) - \frac{\log(-Q_n(0))}{\log(q - f_n(0))}. \tag{13}$$

But by hypothesis and theorem I.11.1, $0 < c_1 \leq -Q_n(0) \leq c_2 < \infty$, and hence the 2nd and 3rd terms on the right side of (13) go to zero as $n \to \infty$. Hence

$$\lim_{n \to \infty} \frac{\log \gamma}{\log(q - f_n(0))} U[f_n(0)] = 1. \tag{14}$$

Since for any $t \in [0, q)$ we can find an $n \geq 0$ such that $f_n(0) \leq t \leq f_{n+1}(0)$, we can use an inequality like (3) to conclude that

$$\lim_{t \uparrow q} \log \gamma \frac{U(t)}{\log(q - t)} = 1.$$

A Tauberian theorem (p. 423, Feller, vol. II) now implies (10). □

7. Green Function Behavior

In this section we study the limits (when they exist) of

$$G(i,j) = \sum_{n=0}^{\infty} P_n(i,j). \tag{1}$$

Theorem 1 (KNS). *If* $m=1$, $EZ_1^2 \log Z_1 < \infty$ *and* $(i,j) \to \infty$ *in such a way that* $i/j \to \beta$, *then*

$$\lim [G(i,j) - \delta_{ij}] = \frac{2}{\sigma^2} \min(1,\beta). \tag{2}$$

If $(i,j) \to \infty$ *in such a way that* $i/j \to 0$ *then we have the stronger result*

$$\frac{G(i,j) - \left(\frac{2}{\sigma^2}\right)\left(\frac{i}{j}\right)}{\left(\frac{i}{j}\right)} \to 0 \tag{3}$$

(whether or not $i \to \infty$*). If* $i=j$, *then*

$$\lim_{i \to \infty} G(i,i) = 1 + \frac{2}{\sigma^2}. \tag{4}$$

Idea of proof. Setting $\beta=1$ in (2) yields (4). To prove (2), one uses the definition (1), the local limit theorem of section 3, and several other estimates on $P_n(i,j)$.

Pick any $\varepsilon > 0$, and then decompose the sum (1) into

$$G(i,j) - \delta_{ij} = \sum_{n \leq \varepsilon i} P_n + \sum_{n > \varepsilon i} P_n \equiv S_1(\varepsilon) + S_2(\varepsilon). \tag{5}$$

One can show (KNS) that

$$\sup_{k \geq 1} \sqrt{in}\, P_n(i,k) \leq K < \infty,$$

and this implies that

$$S_1(\varepsilon) \leq \text{constant} \sqrt{\varepsilon}.$$

The main term in $\lim G(i,j)$ comes from S_2, and we will indicate how it is obtained. Recall the definition

$$p_n(j) = P\{Z_n = j | Z_n > 0, Z_0 = 1\},$$

and let $p_n^{*k}(j)$ be the k-fold convolution of p_n with itself. Then using theorem 3.1

$$\begin{aligned} P_n(i,j) &= \sum_{k=1}^{i} \binom{i}{k} [1-P_n(1,0)]^k P_n^{i-k}(1,0)\, p_n^{*k}(j) \\ &= \sum_{k=1}^{N} \binom{i}{k} [1-P_n(1,0)]^k P_n^{i-k}(1,0) \frac{1}{(an)^k} \frac{j^{k-1}}{(k-1)!} e^{-\frac{j}{an}} + R, \end{aligned} \tag{6}$$

where N is a fixed integer, and R is a remainder term depending on i,j,n,N. Summing this expression over $n > \varepsilon i$, one shows that the main term in $S_2(\varepsilon)$ is

$$\sum_{k=1}^{N} \sum_{n>\varepsilon i} \frac{i^k}{(an)^{2k}} e^{-\frac{i+j}{an}} \frac{j^{k-1}}{k!(k-1)!}. \tag{7}$$

Writing

$$\frac{i^k j^{k-1}}{(an)^{2k}} = \frac{1}{a} \left(\frac{i}{i+j}\right)^k \left(\frac{j}{i+j}\right)^{k-1} \left(\frac{i+j}{an}\right)^{2k} \left(\frac{a}{i+j}\right),$$

and using the fact that $i/(i+j) \to \beta/(1+\beta)$, $j/(i+j) \to 1/(1+\beta)$, one can approximate the sum in (7) by an integral, and get

$$\lim_{\substack{i\to\infty \\ N\to\infty \\ \varepsilon\to 0}} (S_1+S_2) = \frac{1}{a} \sum_{k=1}^{\infty} \frac{\theta^k(1-\theta)^{k-1}}{k!(k-1)!} \int_0^{\infty} x^{-2k} e^{-\frac{1}{x}} dx, \tag{8}$$

where the limits are taken successively in the order indicated, and $\theta = \beta/(1+\beta)$. (Of course the main difficulty in the above calculations is to keep track of the remainder terms.) Thus

$$\lim [G(i,j) - \delta_{ij}] = \frac{1}{a} \sum_{k=1}^{\infty} \frac{\theta^k(1-\theta)^{k-1}}{k!(k-1)!} (2k-2)!,$$

and evaluation of the right side yields (2).

The details of the above calculations, as well as those in the proof of (3) are in KNS. □

If $i \to \infty$ and j is fixed then the Martin boundary theory tells us that limits of the relativized Green function are stationary measures. We thus get another proof of existence, and also a proof uniqueness of the stationary measure in the critical case.

Theorem 2 (KNS). *If $m=1$ and $\sigma^2 < \infty$ then*

$$\lim_{i\to\infty} G(i,j) = \eta_j, \tag{9}$$

where η_j is the unique stationary measure of Theorems I.7.1 and II.2.1, normalized so that $\sum_{j=1}^{\infty} \eta_j p_0^j = 1$.

Remark. In section I.7 we had to assume that $p_1>0$ to assure the existence of π_j (which $=\eta_j$ when $m=1$). The above theorem does not require this assumption.

Proof. Let

$$V(s,i)=\sum_{j=1}^{\infty} G(i,j)s^j=\sum_{n=0}^{\infty}\sum_{j=1}^{\infty} P_n(i,j)s^j=\sum_{n=0}^{\infty}[f_n^i(s)-f_n^i(0)]$$

$$=\sum_{n=0}^{\infty} i f_n^{i-1}(0)[f_n(s)-f_n(0)]+\sum_{n=0}^{\infty}\sum_{\alpha=2}^{i}\binom{i}{\alpha} f_n^{i-\alpha}(0)[f_n(s)-f_n(0)]^{\alpha}$$

$$\equiv I_1(i,s)+I_2(i,s)\,. \tag{10}$$

We shall prove that

$$\lim_{i\to\infty} I_2(i,s)=0 \tag{11}$$

and

$$\lim_{i\to\infty} I_1(i,s)=U(s)\equiv\sum_{j=1}^{\infty}\eta_j s^j\,. \tag{12}$$

For (11) we use the following estimates: Given a fixed k, there exist constants c_1 and c_2 (c_2 depending on s) such that for all i, n, and all $0\leqslant\alpha\leqslant k$

$$f_n^{i-\alpha}(0)\leqslant c_1\exp\left\{\frac{-2i}{\sigma^2 n}\right\} \tag{13}$$

and

$$|f_n(s)-f_n(0)|\leqslant\frac{c_2}{n^2},\qquad |s|<1\,. \tag{14}$$

To obtain the first inequality use the fact that $(1-\varepsilon)^k\leqslant e^{-\varepsilon k}$ for small ε, and apply $1-f_n(0)\sim 2/\sigma^2 n$ to $f_n^{i-\alpha}(0)\leqslant\text{constant}\cdot[1-(1-f_n(0))]^i$, $\alpha=0,1,\ldots,k$. The second inequality follows from Corollary I.9.2.

Noting that $\binom{i}{\alpha}\leqslant i^{\alpha}$ we can use (13) and (14) to conclude that

$$\begin{aligned} I_2(i,s)&\leqslant c\sum_{\alpha=2}^{i}\sum_{n=1}^{\infty}\left(\frac{i}{n^2}\right)^{\alpha}\exp\left\{\frac{-2i}{\sigma^2 n}\right\}\\ &=\sum_{\alpha=2}^{\infty}\frac{c}{i^{\alpha-1}}\sum_{n=1}^{\infty}\left(\frac{i}{n}\right)^{2\alpha}\exp\left\{\frac{-2i}{\sigma^2 n}\right\}\cdot\frac{1}{i}\,.\end{aligned} \tag{15}$$

The sum on the right hand side converges, as can be seen by comparing it with the integral

$$\int_0^{\infty} x^{-2\alpha}\exp\left\{\frac{-2}{\sigma^2 x}\right\}dx\,.$$

This proves (11). Turning to (12), we note first that due to corollary I.9.2

$$\frac{\sigma^2}{2}\cdot \lim_{n\to\infty} n^2[f_n(s)-f_n(0)]=U(s). \tag{16}$$

Also

$$\sum_{n=1}^{\infty} i f_n^{i-1}(0)[f_n(s)-f_n(0)]=\sum_{n=1}^{\infty}\left(\frac{i f_n^i(0)}{n^2}\right)\left(\frac{n^2[f_n(s)-f_n(0)]}{f_n(0)}\right). \tag{17}$$

But $\lim_{i\to\infty} i f_n^i(0)=0$ for each n, and hence by (16) and a truncation argument,

$$\begin{aligned}\lim_{i\to\infty} I_1(i,s)&=\lim_{i\to\infty}\sum_{n=1}^{\infty}\left(\frac{i f_n^i(0)}{n^2}\right)\left(\frac{2}{\sigma^2}U(s)\right)\\ &=\frac{2}{\sigma^2}U(s)\lim_{i\to\infty}\sum_{n=1}^{\infty}\left(\frac{i}{n}\right)^2\exp\left\{\frac{-2i}{\sigma^2 n}\right\}\frac{1}{i}\\ &=\frac{2}{\sigma^2}U(s)\int_0^{\infty}\frac{1}{x^2}\exp\left\{\frac{-2}{\sigma^2 x}\right\}dx=U(s).\end{aligned}$$

This proves (11) and hence the theorem. □

In the supercritical case S. Dubuc (1970) has studied the asymptotic behavior of the Green function $G(i,j)$. We state his result without proof.

Theorem 3. *Let* $f(s)=\sum_j p_j s^j$ *be aperiodic (in the sense that* $f(s)=1$, $|s|\leqslant 1$, *implies* $s=1$), *and assume that* $m>1$, *and* $\sum j(\log j)^2 p_j<\infty$. *Let* $j_n\to\infty$ *in such a way that* $j_n m^{-n}\to y>0$. *Then,*

$$\lim_{n\to\infty} j_n G(i,j_n)=h_y(i),$$

where $h_y(i)=\sum_{-\infty}^{\infty} y m^n w^{*i}(y m^n)$.

8. Harmonic Functions

Recall that $\{h_i; i=0,1,2,\ldots\}$ is a harmonic function for the Galton-Watson process if $h_i\geqslant 0$ and $h_i=\sum_{j=0}^{\infty} P(i,j)h_j$, $i\geqslant 1$.

Theorem 1. *If* $m=1$ *then* $\{h_i=i; i\geqslant 0\}$ *is the unique (up to multiplicative constants) harmonic function for the Galton-Watson process.*

Proof. Clearly $\sum_j P(i,j)j=i$, and hence $h_i=i$ is a harmonic function. To prove uniqueness it is sufficient (see Kemeny, Snell, Knapp (1966)) to verify that if

$$\lim_{\nu\to\infty}\frac{G(i,k_\nu)}{G(1,k_\nu)}=h_i^* \tag{1}$$

exists for some sequence of integers $k_\nu \to \infty$, then $h_i^* = i$. If we were willing to assume that $EZ_1^2 \log Z_1 < \infty$, then (1) would follow at once from (7.3). However, if we want to assume nothing beyond $m=1$, then further argument is needed.

We first show that if (1) is satisfied, then for any fixed N,

$$\lim_{\nu\to\infty} \frac{G_N(i,k_\nu)}{G(1,k_\nu)} = 0\,, \tag{2}$$

where G_N is the truncated Green function $\left(G_N = \sum_{n=0}^{N} P_n\right)$. This part of the argument is due to J. Sims (unpublished work). If h^* exists as the limit in (1) then it must be a harmonic function and we write the fact that it satisfies (1.2) in the notation

$$P h^* = h^*\,. \tag{3}$$

Iterating this we get

$$P_{N+1} h^* = h^*\,.$$

Thus

$$\begin{aligned} h_i^* &= \sum_j P_{N+1}(i,j) h_j^* = \sum_j P_{N+1}(i,j) \liminf_{\nu\to\infty} \frac{G(j,k_\nu)}{G(1,k_\nu)} \\ &\leqslant \liminf_{\nu\to\infty} \sum_j P_{N+1}(i,j) \frac{G(j,k_\nu)}{G(1,k_\nu)} \qquad \text{(by Fatou's lemma)} \\ &= \liminf_{\nu\to\infty} \left[\frac{G(i,k_\nu)}{G(1,k_\nu)} - \frac{G_N(i,k_\nu)}{G(1,k_\nu)}\right] \\ &= h_i^* - \limsup_{\nu\to\infty} \frac{G_N(i,k_\nu)}{G(1,k_\nu)}\,. \end{aligned}$$

Hence

$$0 \leqslant -\limsup_{\nu\to\infty} \frac{G_N(i,k_\nu)}{G(1,k_\nu)} \leqslant 0\,,$$

and this implies (2).

We complete the proof by showing that if $k_\nu \to \infty$ and (2) is satisfied then

$$\lim_{\nu\to\infty} \frac{G(i,k_\nu)}{G(1,k_\nu)} = i\,. \tag{4}$$

This part of the proof is taken from KNS. Pick any $\varepsilon > 0$ and then choose N so that $P_n^{i-1}(1,0) \geqslant 1-\varepsilon$ for all $n > N$. Then we have

$$\begin{aligned}\frac{G(i,k_\nu)}{G(1,k_\nu)} &= \frac{\sum_{n=0}^{\infty} P_n(i,k_\nu)}{G(1,k_\nu)} \geqslant \frac{\sum_{n=0}^{\infty} i P_n(1,k_\nu) P_n^{i-1}(1,0)}{G(1,k_\nu)} \\ &\geqslant (1-\varepsilon)\frac{\sum_{n=N+1}^{\infty} i P_n(1,k_\nu)}{G(1,k_\nu)} = (1-\varepsilon) i \left[\frac{G(1,k_\nu)-G_N(1,k_\nu)}{G(1,k_\nu)}\right].\end{aligned} \tag{5}$$

Applying (2) and the fact that ε is arbitrary we conclude that

$$\liminf_{\nu\to\infty} \frac{G(i,k_\nu)}{G(1,k_\nu)} \geqslant i\,. \tag{6}$$

Conversely, we note that

$$\begin{aligned} G(i,k_\nu) &= \sum_{n=0}^{\infty} P_n(i,k_\nu) = \sum_{n=1}^{\infty}\sum_{j=0}^{\infty} P(i,j)P_{n-1}(j,k_\nu) + P_0(i,k_\nu) \\ &= \sum_{j=0}^{\infty} P(i,j)G(j,k_\nu) + P_0(i,k_\nu)\,,\end{aligned}$$

and by iteration

$$= \sum_{j=0}^{\infty} P_n(i,j)G(j,k_\nu) + \sum_{\alpha=0}^{n-1} P_\alpha(i,k_\nu)\,. \tag{7}$$

Thus, setting $i=1$,

$$\begin{aligned} 1 &\geqslant \limsup_{\nu\to\infty} \sum_{j=0}^{\infty} P_{n_0}(1,j)\frac{G(j,k_\nu)}{G(1,k_\nu)} \\ &\geqslant P_{n_0}(1,l)\limsup_{\nu\to\infty}\frac{G(l,k_\nu)}{G(1,k_\nu)} + \liminf_{\nu\to\infty}\sum_{j\neq l} P_{n_0}(1,j)\frac{G(j,k_\nu)}{G(1,k_\nu)}\,,\end{aligned} \tag{8}$$

where l is any (fixed) positive integer, and n_0 is chosen so that $P_{n_0}(1,l)>0$.

Applying (6) to (8) we see that

$$1 \geqslant P_{n_0}(1,i)\limsup_{\nu\to\infty}\frac{G(i,k_\nu)}{G(1,k_\nu)} + \sum_{j\neq i} P_{n_0}(1,j)j\,. \tag{9}$$

But $\sum_{j\neq i} P_{n_0}(1,j)j = 1 - iP_{n_0}(1,i)$, and substituting this in (9) we conclude that

$$\limsup_{\nu\to\infty}\frac{G(i,k_\nu)}{G(1,k_\nu)} \leqslant i\,. \tag{10}$$

Now (6) and (10) imply (4) and the theorem. □

In the supercritical case the situation is more complicated, there being many harmonic functions.

Recall that if $\{X_n; n\geqslant 0\}$ is a Markov chain, and if $f(X_n)\to Y$ a.s., then $H_A(i)\equiv P\{Y\in A|X_0=i\}$ is a harmonic function for each Borel set A.

Also if the above measures are absolutely continuous, then the associated densities $h_x(i)$ are harmonic functions. (Note that $\sum P(i,j)H_A(j) = \sum P(i,j)P\{Y \in A | X_0 = j\} = \sum P(i,j)P\{Y \in A | X_1 = j\} = P\{Y \in A | X_0 = i\} = H_A(i)$). We use this fact to construct a family of harmonic functions.

We cannot directly use the fact that $Z_n m^{-n} \to W$ a.s., since the left side is not of the form $f(Z_n)$. However, if we take logarithms to the base m, then

$$\log_m Z_n - n \to \log_m W,$$

and letting $\langle \alpha \rangle =$ the fractional part of α,

$$\langle \log_m Z_n - n \rangle = \langle \log_m Z_n \rangle \to \langle \log_m W \rangle \equiv Y.$$

Now we can write $f(Z_n) = \langle \log_m Z_n \rangle \to Y$ a.s., and apply the agrument in the previous paragraph to conclude that

$$P\{\langle \log_m W \rangle \leqslant x | Z_0 = i\} \equiv H_x(i)$$

is a harmonic function.

Now

$$\begin{aligned} H_x(i) &= P\{\langle \log_m W \rangle \leqslant x | Z_0 = i\} \\ &= P\{n \leqslant \log_m W \leqslant x + n \text{ for some } -\infty < n < \infty | Z_0 = i\} \\ &= P\left\{ \bigcup_{n=-\infty}^{\infty} [\omega : m^n \leqslant W(\omega) \leqslant m^{x+n}] \,\middle|\, Z_0 = i \right\}. \end{aligned}$$

But the distribution function

$$G(x) = P\{\log_m W \leqslant x | Z_0 = i\}$$

has associated density

$$g(x) = (\log_e m) m^x w^{*i}(m^x),$$

where $w(\cdot)$ is the density of the random variable W. Hence the density associated with $H_x(i)$ is

$$\sum_{n=-\infty}^{\infty} (\log_e m) m^{x+n} w^{*i}(m^{x+n}).$$

Multiplication by a constant still leaves h a harmonic function, and letting $m^x = y$ we conclude

Theorem 2. *If* $E Z_1 \log Z_1 < \infty$, *and* $m > 1$, *then for any* $y \geqslant 0$

$$h_y(i) \equiv \sum_{n=-\infty}^{\infty} y m^n w^{*i}(y m^n) \tag{11}$$

is a harmonic function for the Galton-Watson process.

That the $h_y(i)$ are harmonic can also be easily verified directly. With $\psi(u)=E\,e^{iuW}$, we saw in chapter I that

$$\psi(mu)=f[\psi(u)],$$

and this implies that

$$w^{*i}(x)=\sum_k P(i,k)\,m\,w^{*k}(mx). \tag{12}$$

Now (11) and (12) imply that $h_y(\cdot)$ is harmonic.

For any non-negative measure v on $[0,\infty)$ it is also clear that

$$\int h_y(i)\,dv(y)\equiv h_v(i) \tag{13}$$

are harmonic. The difficult question is whether (13) gives us all the harmonic functions.

The answer is affirmative, as has been shown by S. Dubuc (1970). Under a slightly stronger hypothesis than in Theorem 2 he establishes that $\{h_y(i);\, y\geqslant 0\}$ includes all the extremal harmonic functions, and hence (13) yields all the harmonic functions.

Theorem 3. *Let* $f(s)=\sum_0^\infty p_j s^j$ *be aperiodic and* $m>1$. *Assume that* $\sum_2^\infty j(\log j)^2 p_j<\infty$. *Then, if* $h(\cdot)$ *is any extremal harmonic function*

$$h(0)\neq 0 \quad \textit{implies} \quad h(i)=h(0)\,q^i,$$

and

$$h(0)=0 \quad \textit{implies there exists a} \quad y>0 \quad \textit{such that}$$

$$h(i)=h(1)\,\frac{h_y(i)}{h_y(1)},$$

where $h_y(\cdot)$ *is as in* (13).

The proof is nontrivial and we refer the reader to Dubuc's paper. In another deep work on the harmonic functions for supercritical branching processes, Dubuc (1971) has obtained the following estimates on their asymptotic behavior.

Theorem 4. *Under the hypothesis of Theorem* 3 *and the extra moment condition* $\sum_j j^2 p_j<\infty$, *we have:*

(i) $h_y(\cdot)$ *is an extremal harmonic function for each* $y>0$;

(ii) *every harmonic function* $h(x)$ *is such that* $h(i)=O(\sqrt{i})$ *as* $i\to\infty$, *and there exist positive, finite constants* A *and* B *such that for every* n

$$A\sqrt{n}\leqslant \sum_{x=1}^{n}\frac{h(x)}{\sqrt{x}}\leqslant B\sqrt{n}.$$

In the same paper Dubuc gives several examples of harmonic functions, two of which are described below. (His definition of harmonic function drops non-negativity.)

1. Let $z_1, z_2, \ldots, z_n$ be n complex numbers such that $f(z_i)=z_{i+1}$, $1 \leqslant i \leqslant n-1$, and $f(z_n)=z_1$. Then, $h(i)=\sum_{r=1}^{n} z_r^i$ is a harmonic function.

2. More generally, if $\sum p_j R^j < \infty$ for some $R>1$, and if f maps $K=\{z: |z| \leqslant R\}$ into K, and μ is a measure on K such that for every continuous function φ on K

$$\int_K \varphi(z)\, d\mu(z) = \int_K \varphi(f(z))\, \mu(dz),$$

then $h(x)=\int_K z^x\, d\mu(z)$ is harmonic.

9. The Space-Time Boundary

With any Markov chain there is associated a so-called space-time process. Let S denote the state-space of the original Markov chain, and I be the set of non-negative integers. The space-time process is a M. C. with state space $\mathsf{S} \times I$, and transition function

$$\mathscr{P}[(i_1, n_1), (i_2, n_2)] = \begin{cases} P(i_1, i_2) & \text{if } n_2 = n_1 + 1, \\ 0 & \text{otherwise}, \end{cases} \tag{1}$$

where $i_1, i_2 \in \mathsf{S}, n_1, n_2 \in I$, and $P(\cdot,\cdot)$ is the transition function for the original chain. The n-step transition function is

$$\mathscr{P}_n[(i_1, n_1), (i_2, n_2)] = \begin{cases} P_n(i_1, i_2) & \text{if } n_2 = n_1 + n, \\ 0 & \text{otherwise}. \end{cases} \tag{2}$$

The Green function for the space-time chain is then

$$G[(i_1,n_1),(i_2,n_2)] = \sum_{n=0}^{\infty} \mathscr{P}_n[(i_1,n_1),(i_2,n_2)] = P_{n_2-n_1}(i_1,i_2). \tag{3}$$

Now fix a point (i_0, n_0), and consider all functions

$$\lim_{(j,N)\to\infty} \frac{G[(i,n),(j,N)]}{G[(i_0,n_0),(j,N)]} = \lim_{(j,N)\to\infty} \frac{P_{N-n}(i,j)}{P_{N-n_0}(i_0,j)} \equiv h(i,n) \tag{4}$$

for all possible sequences $(j,N)\to\infty$. All of these are harmonic functions for the $\mathscr{P}$-process; and any harmonic function for this process can be expressed as a mixture of the limits in (4) with respect to some measure. (See e.g. Kemeny, Snell, Knapp (1966) for a discussion of these matters.)

For the Galton-Watson process with $m>1$, the local limit theorem 4.2 tells us that if $(j,n)\to\infty$ so that $j m^{-n} \to x$, then

$$\lim m^n P_n(i,j) = w^{*i}(x). \tag{5}$$

Thus if $(j,N)\to\infty$ in (4) so that $jm^{-N}\to x$, then

$$\lim \frac{P_{N-n}(i,j)}{P_{N-n_0}(i_0,j)} = \lim m^{n-n_0} \frac{m^{N-n} P_{N-n}(i,j)}{m^{N-n_0} P_{N-n_0}(i_0,j)}$$

$$= m^{n-n_0} \frac{w^{*i}(xm^n)}{w^{*i_0}(xm^{n_0})} = c\,m^n w^{*i}(xm^n),$$

where the constant c depends on x, i_0, n_0. Let

$$h_x(i,n) = c\,m^n w^{*i}(xm^n).$$

Referring again to the by now familiar relation (see e.g. (5.9))

$$w^{*i}(x) = \sum_j P(i,j)\,m\,w^{*j}(mx),$$

we see that

$$h_x(i,n) = \sum_j P(i,j)\,h_x(j,n+1) = \sum_{j,N} \mathscr{P}[(i,n),(j,N)]\,h_x(j,N),$$

and thus h_x are harmonic functions.

There may be other such functions. For some further remarks on this matter see Athreya and Ney (1970) and Karlin (1967).

Complements and Problems II

1. Show that the series for $U(s,t)$ in (2.3) converges.

Hint: Note that U is of the form

$$\sum_{n=-\infty}^{0} + \sum_{n=1}^{\infty} [\exp\{-am^n\} - \exp\{-bm^n\}],$$

where $0<a<b<\infty$, and $0<m<1$. The first sum converges trivially, and in the second, each term is bounded by

$$1 - e^{-bm^n} \leqslant bm^n.$$

2. For another example of stationary measures see Harris (1963), p. 28.

3. In the subcritical case, not much is known about the harmonic functions for the Galton-Watson process. However, when $f(s)$ has a fixed point $s_0>1$ (which of course implies the existence of all moments), then the transformation

$$\hat{f}(s) = \frac{f(1+s(s_0-1))-1}{s_0-1}$$

determines a supercritical process $\{\hat{Z}_n\}$, whose transition function can be determined from that of $\{Z_n\}$. It should then be possible to determine the harmonic functions of $\{\hat{Z}_n\}$ from those of $\{Z_n\}$.

4. *Open problem:* The local limit theorem (analog of theorems 4.1 and 4.2) when $EZ_1 \log Z_1 = \infty$.

5. *Also open:* Local limit theorems in the critical case when $\sigma^2=\infty$. When $\{p_k\}$ is in the domain of a stable law, see Slack (1968) for corresponding global limit laws.

6. Show that if $h(\cdot)$ is a harmonic function for a supercritical G-W process and if $h(i)\to 0$ as $i\to\infty$, then $h(i)=h(0)q^i$.

Hint: If $h(\cdot)$ is harmonic for a Markov chain Z_n, then $h(Z_n)=Y_n$ is a martingale. If $h\geqslant 0$ then $\lim Y_n=Y$ exists almost surely. If h is bounded then $\lim EY_n=EY$, and since Y_n is a martingale, $EY=EY_0$. Furthermore $h(i)=E(Y_0|Z_0=i)$. These facts are true for general Markov chains.

Now if $\{Z_n\}$ is a Galton-Watson process then we may write

$$h(i)=E(Y;A|Z_0=i)+E(Y;B|Z_0=i),$$

where $A=\{\omega: Z_n(\omega)\to\infty\}$ and $B=\{\omega: Z_n(\omega)\to 0\}$. If $h(i)\to 0$ then $h(i)=0+h(0)q^i$.

7. If $m\leqslant 1$, then the bounded harmonic functions are constant.

Hint: Use the ideas of problem 6 and the facts that $P(A)=0$ and $q=1$.

8. Regarding the asymptotic behavior of stationary measures discussed in section 6 in relation to expression (6.9) and theorem 6.4, the following example is in Lipow (1971).

Let $f(s)=1-m+ms$, and consider the stationary measure

$$\eta_j(t)=(j!)^{-1}\sum_{\alpha=-\infty}^{\infty}\exp\{-m^{\alpha-t}\}m^{(\alpha-t)j},\qquad 0\leqslant t<1.$$

Now $\exp\{-m^{\alpha-t}\}m^{(\alpha-t)j}$ has a maximum value of $e^{-j}j^j$ (when $\alpha-t=\log j/\log m$). Thus

$$\eta_j(t)\leqslant\int_{-\infty}^{\infty}(j!)^{-1}\exp\{-m^{\alpha-t}\}m^{(\alpha-t)j}\,d\alpha+\frac{e^{-j}j^j}{j!}=[-j\log m]^{-1}+\frac{e^{-j}j^j}{j!}\sim\frac{1}{\sqrt{2\pi j}}.$$

If $m=\frac{1}{2}$ and $j_k=2^k$ then

$$\eta_{j_k}(0)\geqslant(j_k!)^{-1}\exp\{-m^{-k}\}m^{-kj_k}=\frac{e^{-j_k}j_k^{j_k}}{j_k!}\sim\frac{1}{\sqrt{2\pi j_k}}.$$

Thus $\eta_{j_k}(0)\sim 1/\sqrt{2\pi j_k}$, and (6.9) is false on the subsequence j_k.

9. Here are a few more steps in the proof of (2.10). Let

$$a_n(k_i)=f_n^{k_i}(s)-f_n^{k_i}(0),$$
$$b_n(k_i)=\exp\{Q(0)m^n[1-\mathscr{B}(s)]k_i\}-\exp\{Q(0)m^n k_i\}.$$

Note that $\sum a_n$ and $\sum b_n$ converge. We must show that

$$\lim_{k_i\to\infty}\sum_{n=0}^{\infty}[a_n(k_i)-b_n(k_i)]=0,$$

and hence it is sufficient to prove that

$$\lim_{k\to\infty}\sum_{n=0}^{\infty}(f_n^k(s)-\exp\{Q(0)m^n[1-\mathscr{B}(s)]k\})=0. \qquad (*)$$

To do this, use $|x^k - y^k| = |x-y|(x^{k-1} + x^{k-2}y + \cdots + y^{k-1}) \leqslant |x-y|/(1-y)$ for $x, y \in [0,1]$ to bound the summand in (*) by

$$\frac{|f_n(s) - \exp\{Q(0)m^n[1-\mathscr{B}(s)]\}|}{1-\exp\{Q(0)m^n[1-\mathscr{B}(s)]\}}.$$

Since $|e^x - 1 - x| \leqslant x^2/2$ for small x,

$$\sum_{n=0}^{\infty} \frac{|\exp\{Q(0)m^n[1-\mathscr{B}(s)]\} - 1 - Q(0)m^n[1-\mathscr{B}(s)]|}{1-\exp\{Q(0)m^n[1-\mathscr{B}(s)]} < \infty,$$

and hence if we can show that

$$\sum_n \frac{|f_n(s) - 1 - Q(0)m^n[1-\mathscr{B}(s)]|}{Q(0)m^n[1-\mathscr{B}(s)]} < \infty,$$

then one can use the Lebesgue dominated convergence theorem to take the limit $(k \to \infty)$ through the sum in $(*)$, and draw the desired conclusion. Thus is is sufficient to show that

$$\sum_{n=1}^{\infty} \left| \frac{f_n(s) - 1}{m^n} - Q(s) \right| < \infty.$$

We leave it to the reader to show that this will be so if $\sum p_k k \log k < \infty$. (Use the methods of section I.11.)

10. Some hints for the proof of (3.22) for $i=3$. First note that

$$2\pi\, g_n(t) \equiv \int_{-\varepsilon n}^{\varepsilon n} \frac{e^{-itx}dx}{1-ix} = \int_{-\varepsilon n}^{\varepsilon n} \frac{\cos tx}{1+x^2} dx + \int_{-\varepsilon n}^{\varepsilon n} \frac{x \sin tx}{1+x^2} dx.$$

Next show that each term on the right converges (as $n \to \infty$) to a function of t, uniformly for $0 \leqslant t \leqslant c < \infty$. This is easy for the cos terms, and can be done for the sin term by decomposing the integral into two parts, namely

$$\int_{-\varepsilon n}^{\varepsilon n} \frac{x \sin tx}{1+x^2} dx = \int_{|x| < t^{-1}} + \int_{t^{-1} \leqslant |x| \leqslant \varepsilon n},$$

and integrating the second of these integrals by parts. One thus obtains that $g_n(t)$ converges uniformly (for $0 \leqslant t \leqslant c$) to some function $h(t)$ which must be continuous since it is a uniform limit of continuous functions.

But, letting $g(t) = e^{-t}$ for $t \geqslant 0$, $=0$ for $t < 0$, it follows from the Plancherel theorem that $\int_{-\infty}^{\infty} |g_n(t) - g(t)|^2 dt \to 0$; hence that $\int_0^c |g_n(t) - e^{-t}|^2 dt \to 0$; hence $g_n(t) \to e^{-t}$ almost surely on $[0, c]$. Hence $h(t) = e^{-t}$ a.s., and thus due to the continuity of h, equality holds everywhere on $[0, c]$, for any $c < \infty$.

Chapter III

One Dimensional Continuous Time Markov Branching Processes

1. Definition

In the Galton-Watson process the lifetime of each particle was one unit of time. A natural generalization is to allow these lifetimes to be random variables. Instead of the discrete time Markov chain $\{Z_n; n=0,1,2,\ldots\}$ of Chapter I, we must consider a process $\{Z(t); t \geqslant 0\}$, where $Z(t)$ = the number of particles at time t. This process will in general not be Markovian, unless the lifetimes are independent, exponentially distributed random variables. It is the latter process which we will study in this chapter. The general non-Markovian case will be considered in Chapter IV.

Much of the early work on the continuous time process, including the derivation of the Kolmogorov and functional equations of sections 2 and 3, and some results on moments and limit laws, was done by Kolmogorov and Dmitriev (1947), and by B. A. Sevastyanov (1951). For many other references see Harris (1963).

Definition. A stochastic process $\{Z(t,\omega); t \geqslant 0\}$ on a probability space $(\Omega, \mathbb{F}, P)$ is called a *one dimensional continuous time Markov branching process* if:

(i) its state space is the set of non-negative integers;

(ii) it is a stationary Markov chain with respect to the fields $\mathbb{F}_t = \sigma\{Z(s,\omega); s \leqslant t\}$, (for any collection D of real-valued, Borel measurable random variables on $(\Omega, \mathbb{F}, P)$, $\sigma(D)$ denotes the sub-σ-algebra of $\mathbb{F}$ generated by D);

(iii) the transition probabilities $P_{ij}(t)$ satisfy

$$\sum_{j=0}^{\infty} P_{ij}(t)\, s^j = \left[\sum_{j=0}^{\infty} P_{1j}(t)\, s^j\right]^i \tag{1}$$

for all $i \geqslant 0$ and $|s| \leqslant 1$.

Properties (i) and (ii) say that $Z(t)$ is a continuous time Markov process on the integers; while (iii) characterizes the basic branching property.

Throughout this chapter we will refer to the above process as the "continuous" (branching) process.

As might be expected, most of the theory of the Galton-Watson process carries over to the continuous case, and could be developed by techniques very close to those of Chapter I. It is interesting, however, that the continuous process has much *additional* structure.

Rather than dwell on a catalog of results paralleling those of Chapter I, our objective in this chapter will be threefold:

(i) To develop and illustrate some simple techniques by which one can pass directly from the discrete to the continuous case; and thus make it unnecessary to rederive all the Galton-Watson limit theorems in their present setting.

(ii) Study some of the special structures of the continuous process.

(iii) Look at some of the finer relations between discrete and continuous processes (the embeddability problem).

Before turning to these matters, we consider briefly the construction of the continuous process.

2. Construction

One possible approach to the construction of the continuous branching process is to start with the transition probabilities

$$P_{ij}(\tau,\tau+t)=P\{Z(\tau+t)=j|Z(\tau)=i\},$$

which, due to the assumption of time homogeneity, satisfy

$$P_{ij}(\tau,\tau+t)=P_{ij}(t), \qquad \tau\geqslant 0. \tag{1}$$

These transition functions are determined by the infinitesimal probabilities

$$\begin{aligned} &0<a<\infty, \\ &0\leqslant p_i, \qquad \sum_{i=0}^{\infty} p_i=1, \end{aligned} \tag{2}$$

as solutions of the Kolmogorov forward and backward equations

$$\frac{d}{dt}P_{ij}(t)=-jaP_{ij}(t)+a\sum_{k=1}^{j+1}kP_{ik}(t)\,p_{j-k+1} \quad \text{(forward)} \tag{3}$$

and

$$\frac{d}{dt}P_{ij}(t)=-iaP_{ij}(t)+ia\sum_{k=i-1}^{\infty}p_{k-i+1}P_{kj}(t) \quad \text{(backward)}, \tag{4}$$

with boundary conditions

$$P_{ij}(0+)=\delta_{ij}=\begin{cases}1 & \text{for } i=j,\\ 0 & \text{for } i\neq j.\end{cases} \tag{5}$$

The infinitesimal probabilities are to have probabilistic interpretation in terms of a branching process constructed as follows. It starts with a specified number of particles. If a given particle is alive at a certain time, its additional life length is a random variable which is exponentially distributed with parameter a. Upon death it leaves k offspring with probability $p_k, k=0,1,2,\ldots$. As usual, particles act independently of other particles, and of the history of the process. Consequently

$$P_{ij}(\tau)=iap_{j-i+1}\tau+o(\tau) \quad \text{as } \tau\to 0 \quad \text{if } j\geqslant i-1, \quad j\neq i, \tag{6}$$

$$P_{ii}(\tau)=1-ia\tau+o(\tau) \quad \text{as } \tau\to 0, \tag{7}$$

and

$$P_{ij}(\tau)=o(\tau) \quad \text{as } \tau\to 0 \quad \text{if } j<i-1. \tag{8}$$

Thus, if the process is in state i at a given time, it continues there for an amount of time which is exponentially distributed with parameter ia, and then jumps into state $j\geqslant i-1$ with probability p_{j-i+1}. It then stays in j for an exponentially distributed amount of time (parameter ja) and jumps to $k\geqslant j-1$ with probability p_{k-j+1}, etc. The process so constructed is called the minimal process (see e.g. Chung (1967)). It is easy to argue by standard methods (see e.g. Feller, Vol. I, Chapter 17) that the transition function of this process must satisfy (3), (4), (5). We will exploit its probabilistic properties in section 7.

Unfortunately, although (3), (4), (5) always have solutions satisfying $\sum_j P_{ij}(t)\leqslant 1$, these solutions may not be unique. Related to this difficulty is the fact that the probabilistic interpretation of the previous paragraph may not always be valid. These problems arise when it is possible for the process to produce infinitely many particles in a finite time (by having infinitely many transitions in $(0,t)$), i.e., to explode; a situation which exists when there are solutions of (3), (4), (5) such that $\sum_j P_{ij}(t)<1$.

However, under an additional hypothesis to be given below, it is known that:

(i) There can be no explosions.

(ii) Equations (3), (4), (5) have a unique solution $P_{ij}(t)$ satisfying $\sum_j P_{ij}(t)=1$.

Let

$$f(s)=\sum_{j=0}^{\infty} p_j s^j.$$

Non-Explosion Hypothesis: For every $\varepsilon>0$

$$\int_{1-\varepsilon}^{1} \frac{ds}{f(s)-s}=\infty.$$

Then we have the following basic result:

Theorem 1. *The non-explosion hypothesis is necessary and sufficient for* $P\{Z(t)<\infty\}\equiv 1$, *where* $Z(t)$ *is the minimal process.*

The proof may be found in T. Harris (1963). Branching processes more general than the present one, and matters related to the explosion phenomenon, have been investigated by Savits (1969). We will not consider such processes and will *assume that* $f'(1)<\infty$ *throughout this chapter*, which ensures that the non-explosion hypothesis holds.

The Kolmogorov theorem assures us that there exists a probability space $(\Omega, \mathbb{F}, P)$, and a Markov process $\{Z(t); t \geqslant 0\}$ on this space having the transition function described above. However, it does not tell us enough about the sample space to assure us that we will be able to properly define certain random variables which we will need to consider (see e.g. (9) below).

We have already mentioned (section I.1) the Harris construction of a probability space for a branching process (chapter VI of his book (1963)). In this construction each point in the sample space represents a complete "family tree", specifying the time of birth, life length, ancestors and decendants of each particle; and an appropriate probability measure is constructed on the Borel extension of the cylinder sets of this space. The construction is in fact carried our for processes with arbitrary lifetime distributions (which we take up in the next chapter).

When the lifetime distribution is exponential (as in the present case), Harris proves that the process is Markovian, and has a transition function which satisfies the Kolmogorov equations above. (See section 11 of chapter VI of his book. Actually his statement is in terms of generating functions.)

He thus obtains a Markov process with the same transition mechanism as that described above, but with an underlying sample space *specifically identified as the space of family trees.*

This identification will at times be useful to us. For example, it enables us to define such random variables as

$$Z_t^{(i)}(u,\omega) = \text{the number of offspring of the } i\text{th of the } Z(t,\omega) \text{ particles existing at } t, \text{ which are alive at time } t+u. \tag{9}$$

In terms of these we have the following representation:

$$Z(t+u,\omega)=\begin{cases}\sum_{i=1}^{Z(t,\omega)} Z_t^{(i)}(u,\omega) & \text{for } Z(t,\omega)>0,\\ 0 & \text{for } Z(t,\omega)=0\end{cases} \tag{10}$$

where, conditioned on $Z(t,\omega)$, $\{Z_t^{(i)}(u,\omega); u \geqslant 0\}$ are independent processes equivalent to $\{Z(u,\omega); u \geqslant 0\}$. (We adopt the usual convention that, unless specifically indicated otherwise, $Z(u,\omega)$ stands for the process with $Z(0,\omega)=1$).

3. Generating Functions

Generating functions again play an important role in the analysis of the continuous process. Recall that

$$f(s)=\sum_{j=0}^{\infty} p_j s^j, \quad |s| \leqslant 1, \quad \text{and define} \tag{1}$$

$$u(s)=a[f(s)-s]. \tag{2}$$

Let $m=f'(1)$ and $\lambda=u'(1)=a(m-1)$. The number $0<a<\infty$, and the probability function $\{p_j; j=0,2,3,\ldots\}$ are the total datum of the process $\{Z(t); t \geqslant 0\}$. Define

$$\begin{aligned} F(s,t) &= \sum_{k=0}^{\infty} P\{Z(t)=k|Z(0)=1\} s^k \\ &= \sum_{k=0}^{\infty} P_{1k}(t) s^k = E s^{Z(t)}. \end{aligned} \tag{3}$$

From the Kolmogorov equations (2.3), (2.4), (2.5) we see at once that $F(s,t)$ satisfies

$$\frac{\partial}{\partial t} F(s,t)=u(s) \frac{\partial}{\partial s} F(s,t) \quad \text{(forward equation)}, \tag{4}$$

and

$$\frac{\partial}{\partial t} F(s,t)=u[F(s,t)] \qquad \text{(backward equation)}, \tag{5}$$

with the boundary condition

$$F(s,0)=s. \tag{6}$$

By theorem 2.1 there is a unique solution of (4), (5), (6), which is a probability generating function, and there is a unique Markov process with this generating function.

We note that (5) is equivalent to the equation

$$\int_s^{F(s,t)} [u(x)]^{-1} dx=t, \tag{7}$$

provided $u(x) \neq 0$ for $s \leqslant x \leqslant F(s,t)$, and that thus, in principle, F can also be obtained as the solution of (7). This relation was used by Harris (1963) in deriving the non-explosion condition, and we will also have occasion to use it later in the chapter.

Now using (1.1) and the Chapman-Kolmogorov equation

$$F(s,t+u) = \sum_{k=0}^{\infty} P_{1k}(t+u)s^k = \sum_{k=0}^{\infty} \sum_{j=0}^{\infty} P_{1j}(u) P_{jk}(t) s^k,$$

and interchanging order of summation

$$= \sum_j P_{1j}(u) F^j(s,t).$$

Thus

$$F(s,t+u) = F(F(s,t), u), \quad |s| \leqslant 1, \quad \text{for } u \geqslant 0, t \geqslant 0. \tag{8}$$

This is the analog of the functional iteration formula (I.1.4) for $f_n(s)$ in the Galton-Watson process.

4. Extinction Probability and Moments

Some elementary properties can immediately be derived from the Kolmogorov equations. We illustrate with a direct computation of the extinction probability.

First one can show exactly as in the discrete case (we do not repeat it here) that all states other than the zero state are transient. Thus, if we set

$$\mathsf{A} = \{\omega;\ Z(t,\omega) \to \infty \text{ as } t \to \infty\},$$

and

$$\mathsf{B} = \{\omega;\ Z(t,\omega) \to 0 \text{ as } t \to \infty\},$$

then

$$P(\mathsf{A} \cup \mathsf{B}) = 1.$$

The set B above is called the extinction set, and its probability the extinction probability. From the basic branching property it is clear that

$$P(\mathsf{B}|Z(0)=k) = [P(\mathsf{B}|Z(0)=1)]^k,$$

and hence it suffices to find $q \equiv P(\mathsf{B}|Z(0)=1)$. From the definition of $F(s,t)$ we note that

$$q(t) \equiv P\{Z(t)=0|Z(0)=1\} = F(0,t) \tag{1}$$

is a nondecreasing function of t. Furthermore, from the backward equation (3.5) we see that $q(t)$ is differentiable in t and satisfies

$$\frac{d}{dt} q(t) = u(q(t)). \tag{2}$$

Integrating, and noting that $q(0)=0$ we get

$$q(t)=\int_0^t u(q(y))\,dy\,. \tag{3}$$

Let q^* be the smallest root in $[0,1]$ of the equation $f(s)=s$, or equivalently $u(s)=0$. Since $q(t)$ is nondecreasing and (2) holds, we have $u(q(t))\geqslant 0$, and since $u(0)\geqslant 0$ it follows that $q(t)\leqslant q^*$ for all $t\geqslant 0$. But $q=\lim_{t\to\infty} q(t)$. Thus $q\leqslant q^*$.

Now suppose $u(q)>0$. Then by continuity of $u(x)$ we have

$$\lim_{t\to\infty}\int_0^t u(q(y))\,dy=\infty, \quad \text{while } q(t) \text{ lies in } [0,1].$$

This contradicts (3). Hence we must have $u(q)=0$, and by the definition of q^*, get $q=q^*$. Summarizing, we have

Theorem 1. *The extinction probability q is the smallest root in $[0,1]$, of the equation $u(s)=0$.*

Remark. Of course, probabilistically the above result is rather obvious. If one denotes the size of the nth generation by $\zeta_n(\omega)$, then the sequence $\{\zeta_n(\omega);\, n=0,1,2,\ldots\}$ forms a discrete time Galton-Watson process with offspring p.g.f. $f(s)$. The event $\{\omega;\, Z(t,\omega)\to 0 \text{ as } t\to\infty\}$ is the same as the event $\{\omega;\, \zeta_n(\omega)\to 0 \text{ as } n\to\infty\}$ and hence the probability of two events are the same. (The same argument will apply to the age-dependent branching process to be discussed in Chapter IV.) This argument can be made rigorous. (See Harris (1963).) In section 5 we will give yet another proof of the same result.

As usual, the moments of the process can also be obtained from the Kolmogorov equations. We shall see later (Corollary 6.1) that $m<\infty$ implies $m_1(t)=EZ(t)<\infty$ for all t. Differentiating both sides of the backward equation (3.5) with respect to s and then letting s go to 1 from below we get $(d/dt)m_1(t)=\lambda m_1(t)$, with boundary condition $m_1(0)=1$. This yields

$$m_1(t)=e^{\lambda t}. \tag{4}$$

Now assume $\sum_{j=1}^{\infty} j^2 p_j<\infty$. Again (as we shall see later by Corollary 6.1) $m_2(t)\equiv EZ^2(t)$ is finite for all t. By using the backward equation we see that $m_2(t)$ satisfies the differential equation

$$\frac{d}{dt}m_2(t)=u''(1)\,e^{2\lambda t}+\lambda m_2(t)\,,$$

with boundary condition $m_2(0)=1$.

It is easily verfied that the solution of the above is

$$m_2(t)=\begin{cases} u''(1)\lambda^{-1}(e^{2\lambda t}-e^{\lambda t}) & \text{if } \lambda\neq 0, \\ u''(1)t & \text{if } \lambda=0. \end{cases} \tag{5}$$

5. Examples

The Kolmogorov equations are usually quite hard to solve, and hence the distribution of $Z(t)$ can rarely be determined explicitly.

However, there are two important cases where they can be solved, namely the Yule process and the birth-death process.

Yule process or binary fission. Here $f(s)=s^2$. Thus each particle lives an exponential length of time and then splits into two particles. The backward equation becomes:

$$\frac{\partial}{\partial t}F(s,t)=a[F^2(s,t)-F(s,t)] \quad \text{with } F(s,0)=s.$$

The solution is

$$F(s,t)=se^{-at}[1-(1-e^{-at})s]^{-1}.$$

Birth and death process. Consider the simple birth and death process with rates $n\lambda$ and $n\mu$ for birth and death respectively, when the process is in state n. It is not hard to see that such a process is a branching process with parameters.

$$a=\lambda+\mu,$$

and

$$f(s)=(\mu+\lambda s^2)a^{-1}.$$

The backward equation becomes

$$\frac{\partial}{\partial t}F(s,t)=[\lambda F^2-(\lambda+\mu)F+\mu],$$

with $F(s,0)=s$. The solution is

$$F(s,t)=\frac{\mu(s-1)-e^{(\mu-\lambda)t}(\lambda s-\mu)}{\lambda(s-1)-e^{(\mu-\lambda)t}(\lambda s-\mu)}.$$

Remarks. 1. In the Yule case if instead of splitting into two particles each particle splits into exactly k particles $(k>2)$, then the generating function is given by

$$F(s,t)=se^{-at}[1-(1-e^{-a(k-1)t})s^{k-1}]^{-1/(k-1)}.$$

2. Both the Yule and birth-death processes are special cases of the linear fractional case; the infinitesimal generating functions in either case being of the form $u(s)=a[f(s)-s]$, where $f(s)$ is a quadratic p.g.f.; i.e. $f(s)\equiv a_0+(1-a_0)s^2$, $0\leqslant a_0\leqslant 1$.

6. The Embedded Galton-Watson Process and Applications to Moments

A useful device for studying continuous time Markov processes is to reduce questions about them to the discrete case—and we now proceed to do this (sections 6 and 7) in our special setting. The key tool is the following elementary result.

Theorem 1. *For every* $\delta > 0$ *the sequence*

$$\{Z_n(\omega)\} \equiv \{Z(n\delta, \omega)\}$$

is a Galton-Watson process with offspring p.g.f. $g(s) \equiv F(s, \delta)$.

Proof. That $\{Z_n\}$ is a Markov chain with stationary transition probabilities follows at once from the corresponding properties of $\{Z(t)\}$. Let

$$g_{(n)}(s) = E[s^{Z_n} | Z_0 = 1].$$

Then by the iteration formula (3.8)

$$g_{(n)}(s) = g[g_{(n-1)}(s)].$$

And hence

$$g_{(n)}(s) = g_n(s) = \text{the } n\text{th iterate of } g(s). \quad \square$$

The idea in applying this theorem is to use our knowledge about the Galton-Watson process to obtain a result about $\{Z(n\delta);\, n = 0, 1, 2, \ldots\}$ for each δ, and then to go from this to the whole process $\{Z(t)\}$.

As a simple example, note that the extinction of the process $\{Z(t);\, t \geqslant 0\}$ is equivalent to that of $\{Z(n\delta);\, n = 0, 1, 2, \ldots\}$ for each $\delta > 0$, and thus q satisfies $q = F(q, \delta)$. In other words $F(q, t)$ is independent of t and thus using (3.5) we see that $u(q) = 0$.

In using the above technique we have to verify, for each $\delta > 0$, that certain hypotheses hold for $Z(\delta)$, before we can apply a result on Galton-Watson processes to $\{Z(n\delta);\, n = 0, 1, 2, \ldots]$. Our initial data are just the offspring p.g.f. $f(s)$ and the lifetime parameter a. Thus, the problem of relating conditions on $Z(\delta)$ to $f(s)$ and a becomes very relevant. The following result asserts an equivalence principle of conditions on $f(s)$ and $F(s, \delta)$.

Theorem 2. *Let* $\{Z(t, \omega);\, t \geqslant 0\}$ *be a Markov branching process. Let* $\phi(x)$ *be a function from* $(0, \infty)$ *to* $(0, \infty)$ *such that there exists* $c \geqslant 0$ *satisfying*

(i) $\phi(x)$ *is bounded on* $(0, c)$ *and convex on* (c, ∞), *and*

(ii) $\phi(xy) \leqslant K\phi(x)\phi(y)$ *for some* K *and all* x, y *in* (c, ∞) *(*K *independent of* x *and* y*).*

Then $E\phi(Z(t)) < \infty$ *for any* $t > 0$ *if and only if*

$$\sum_{j=0}^{\infty} \phi(j) p_j < \infty.$$

We defer the proof of this theorem until Section IV.5, where we prove it for the more general age dependent branching process. There are two important corollaries.

Corollary 1. *Let* $r \geqslant 1$. *Then* $EZ^r(t) < \infty$ *for any* $t > 0$ *if any only if* $\sum_{j=0}^{\infty} j^r p_j < \infty$.

Harris has shown the "if" part of the corollary for the case when r is an integer. He used a differential equations method that would break down if r were not an integer.

Corollary 2. *For any* $t > 0$, $\alpha \geqslant 1$, $\beta \geqslant 0$,

$$E\{Z^\alpha(t)|\log Z(t)|^\beta\} < \infty \quad \textit{if and only if} \quad \sum_{j=2}^{\infty} j^\alpha (\log j)^\beta p_j < \infty .$$

Of course, the proofs of these corollaries are trivial since the functions $\phi(x) = |x|^r$ $(r \geqslant 1)$ and $\phi(x) = |x|^\alpha |\log x|^\beta$ $(\alpha \geqslant 1, \beta \geqslant 0)$ satisfy the hypothesis of theorem 2. We shall use the corollaries crucially later. Recall that we already have made use of corollary 1 in Section 4.

7. Limit Theorems

The asymptotic behavior of $Z(t)$ for large t is very similar to that of the discrete time Galton-Watson process and we can use our knowledge of the embedded discrete process to prove limit laws for the continuous case.

We will indicate how this is done for the three main limit laws in the various criticality classes. In section 8 we give the main convergence result for generating functions. Many of the more refined results of the previous two chapters can similarly be carried over to the continuous process.

To begin with, we know from section 4 that $E(Z(t)|Z(0)=1) = e^{\lambda t}$. This suggests that the population grows at an exponential rate and indeed we have a martingale convergence attesting to this Malthusian law of growth.

Theorem 1. *The family* $\{Z(t)e^{-\lambda t}, \mathbb{F}_t; t \geqslant 0\}$ *is a non-negative martingale and hence*

$$\lim_{t \to \infty} Z(t) e^{-\lambda t} = W \quad \textit{exists a.s.} \tag{1}$$

Proof. Immediate from the Markov property, and the representation in (2.10). □

The Supercritical Case. Once again we are faced with the question of the nontriviality of W. From section 4 we know that if $f'(1)=m\leqslant 1$, i.e. if $u'(1)=\lambda\leqslant 0$, then the process becomes extinct with probability 1. Thus $P\{W=0\}=1$ if $\lambda\leqslant 0$. However, W could be zero even if $\lambda>0$. In fact, we know from Theorem I.10.1 that if

$$W_\delta \equiv \lim_{n\to\infty} Z(n\delta)e^{-n\lambda\delta}, \qquad \delta>0, \tag{2}$$

then either

$$P\{W_\delta=0\}=1, \quad \text{or } EW_\delta=1, \tag{3}$$

with the latter holding if and only if

$$E\{Z(\delta)\log Z(\delta)\}<\infty. \tag{4}$$

But from Theorem 1 we know that $W_\delta=W$ a.s. Applying Theorem I.10.4 in addition to the above facts we get

Theorem 2. *Let W be as defined in Theorem 1 and assume $\lambda>0$. Then*

$$P\{W=0\}=1 \quad \text{or } EW=1, \tag{5}$$

the latter holding if and only if

$$\sum p_j j\log j<\infty. \tag{6}$$

Furthermore, when (6) holds

(i)
$$P\{W=0\}=q, \tag{7}$$

and

(ii) *There exists a continuous density function $w(x)$ on $(0,\infty)$ such that for $0<x_1<x_2<\infty$*

$$P(x_1\leqslant W\leqslant x_2)=\int_{x_1}^{x_2} w(x)\,dx. \tag{8}$$

Remarks. 1. In section 9 we will obtain a different proof of the first part of Theorem 2 through the study of split times.

2. The additional structure of the continuous process enables us to get more information than in discrete time. In fact $\varphi(u)=E(e^{-uW})$ satisfies the functional equation

$$\varphi(u)=\int_0^\infty a f[\varphi(ue^{-\lambda y})]e^{-ay}\,dy, \tag{9}$$

from which one can show that

$$\varphi^{-1}(u)=(1-u)\exp\left\{\int_1^u\left[\frac{m-1}{f(x)-x}+\frac{1}{1-x}\right]dx\right\} \quad \text{for } 0\leqslant u\leqslant 1, \tag{10}$$

where φ^{-1} is the inverse function of φ. The proof is deferred to section 8, since it will make use of theorem 8.2.

3. The analogue of theorem I.10.3 for the present setting can be obtained along similar lines.

Before turning to the other limit laws we state a lemma which is useful in translating limit theorems into the continuous setting.

Lemma 1 (Kingman (1963)). *Let* $h(\cdot)$ *be a continuous function on* $(0, \infty)$. *If* $\lim_{n\to\infty} h(n\delta)=c(\delta)$ *exists for every* $\delta>0$, *then* $\lim_{t\to\infty} h(t)=c$ *exists, and thus necessarily* $c(\delta)\equiv c$ *for all* $\delta>0$.

The critical case. Now assume $\lambda=0$ and $j^2 p_j<\infty$.

Let

$$\sigma^2=a\left[\sum_{j=0}^{\infty} j^2 p_j-1\right] \tag{11}$$

and

$$H(x, t)=P\left\{Z(t)>\frac{\sigma^2}{2} t x \mid Z(t)\neq 0\right\}. \tag{12}$$

From theorems 6.1 and I.9.2 we see that

$$\lim_{n\to\infty} H(x, n\delta)=e^{-x} \quad \text{for each } \delta>0. \tag{13}$$

But since the process $Z(t)$ is continuous in probability, $H(x, t)$ is continuous in t for each fixed x. Hence applying lemma 1 we get the Kolmogorov-Yaglom exponential limit law.

Theorem 3. *If* $\lambda=0$ *and* $\sigma^2<\infty$ *then*

$$\lim_{t\to\infty} P\left\{Z(t)>\frac{\sigma^2}{2} t x \mid Z(t)>0\right\}=e^{-x}. \tag{14}$$

Again it is instructive to see that a direct proof of theorem 3 is also possible, without reference to theorem 6.1 or Kingman's lemma.

Lemma 2. *Assume that* $\lambda=0$ *and* $\sigma^2=a[\sum j^2 p_j-1]<\infty$. *Then*

$$\lim_{t\to\infty} \frac{1}{t}\left[\frac{1}{1-F(s, t)}-\frac{1}{1-s}\right]=\frac{\sigma^2}{2} \tag{15}$$

uniformly for $s\in[0, 1)$.

Proof. Since $\lambda=0$, $F(s, t)\geqslant s$ for $0\leqslant s<1$; also $u(s)>0$ for such s, unless $f(s)\equiv s$, in which case the lemma is trivial. Now since

$$\lim_{y\to 1-} \frac{u(y)}{(1-y)^2}=\frac{\sigma^2}{2} \tag{16}$$

we see that given any $\eta>0$, there exists an $\varepsilon>0$ such that

$$\left[\frac{\sigma^2}{2}+\eta\right]^{-1}\left[\frac{1}{1-F(s,t)}-\frac{1}{1-s}\right]\leqslant\int_s^{F(s,t)}\frac{dy}{u(y)}\leqslant\left[\frac{\sigma^2}{2}-\eta\right]^{-1}\left[\frac{1}{1-F(s,t)}-\frac{1}{1-s}\right] \tag{17}$$

uniformly for $0\leqslant t$, $1-\varepsilon\leqslant s<1$.

Hence by (3.7)

$$\frac{\sigma^2}{2}-\eta\leqslant\frac{1}{t}\left[\frac{1}{1-F(s,t)}-\frac{1}{1-s}\right]\leqslant\frac{\sigma^2}{2}+\eta \tag{18}$$

uniformly for $0\leqslant t$, $1-\varepsilon\leqslant s<1$.

To complete the proof of the lemma it is thus sufficient to show that

$$\lim_{t\to\infty}\frac{1}{t}\left[\frac{1}{1-F(s,t)}-\frac{1}{1-s}\right]=\frac{\sigma^2}{2} \tag{19}$$

uniformly for $0\leqslant s\leqslant 1-\varepsilon$, or equivalently that

$$\lim_{t\to\infty} t[1-F(s,t)]=\frac{2}{\sigma^2} \tag{20}$$

uniformly for $0\leqslant s\leqslant 1-\varepsilon$.

But by (16)

$$\lim_{\delta\to 0}\delta\int_\tau^{1-\delta}\frac{dy}{u(y)}=\frac{2}{\sigma^2}\quad\text{uniformly for } 0\leqslant\tau\leqslant 1-\varepsilon,$$

and hence

$$\lim_{t\to\infty}[1-F(s,t)]\int_s^{F(s,t)}\frac{dy}{u(y)}=\frac{2}{\sigma^2}\quad\text{uniformly for } 0\leqslant s\leqslant 1-\varepsilon.$$

But by (3.7) this implies (20). □

Proof of Theorem 3. This now follows along exactly the same lines as the proof of theorem I.9.2, by using lemma 2 to show that

$$\lim_{t\to\infty} E\left[\exp\left\{-\alpha\frac{Z(t)}{t}\right\}\middle| Z(t)>0\right]=\left[1+\frac{\alpha\sigma^2}{2}\right]^{-1}.$$

The Subcritical case. There is again an exact analog of the Yaglom theorem.

Theorem 4. *If* $\lambda<0$, *then*

$$\lim_{t\to\infty} P[Z(t)=j\,|\,Z(t)>0]=b_j,\quad j\geqslant 0 \tag{21}$$

exists, where $b_j\geqslant 0$ *and* $\sum_{j=0}^{\infty} b_j=1$.

Proof. Use theorem I.8.1 and its corollary, theorem 6.1, Kingman's lemma, and the fact that $Z(t)$ is continuous in probability. The idea is the same as in the first proof of theorem 3, and we leave out the details. □

For further applications of the above method, see also Conner (1967 a).

8. More on Generating Functions

Convergence rates for generating functions are also important in the present theory, and we therefore give the analog of the key result of section I.11. This is then used to get the limiting transform (7.10) for the supercritical case. We shall assume in this section that $u'(1)\neq 0$ and thus rule out the critical case. The proofs of theorems 1 and 2, and corollary 1 below follow from the results of section I.11, theorem 6.1, and the backward equation (3.5).

Theorem 1. *Let* $\beta=-u'(q)$. *Without any further assumptions*

$$\lim_{t\to\infty} e^{\beta t}\frac{\partial}{\partial s}F(s,t)\equiv A'(s) \tag{1}$$

exists for $0\leqslant s<1$. *The function* $A'(s)$ *has the following properties:*

(i) $A'(s)\equiv 0$ *in* $[0,1)$ *if and only if* $m<1$ *and*

$$\sum_{j=0}^{\infty} p_j j\log j=\infty\,;$$

(ii) *in all other cases* $A'(s)>0$ *in* $[0,1)$ *and*

$$\lim_{s\to q} A'(s)=1\,.$$

Now define

$$A(s)=\int_q^s A'(x)\,dx,\qquad 0\leqslant s<1\,. \tag{2}$$

Then we have

Corollary 1. *If* $u'(1)\neq 0$ *then*

$$\lim_{t\to\infty} e^{\beta t}[F(s,t)-q]=A(s),\qquad 0\leqslant s<1\,. \tag{3}$$

Theorem 2. *If* $m>1$, *or if* $m<1$ *and* $\sum_{j=0}^{\infty} p_j j\log j<\infty$, *then* $A(s)$ *is is the unique solution of*

$$A[F(s,t)]=e^{-\beta t}A(s) \tag{4}$$

with

$$A(q)=0,\qquad A'(q)=1\,.$$

Since $A'(s)>0$, the map $s\to w=A(s)$ takes $[0,1)$ monotonically into $[A(0),\infty)$. Thus there is an inverse map $w\to s=B(w)$ having a power series expansion valid near $w=0$. From (4) we see that

$$F(s,t)=B[e^{-\beta t}A(s)],$$

and hence from the backward equation (3.5) evaluated at $t=0$ we see that (5) below holds for s near q. Since both sides are analytic, we get

Corollary 2 (Karlin-McGregor (1968 a)).

$$u(s)=-\beta\frac{A(s)}{A'(s)},\qquad 0\leqslant s<1. \tag{5}$$

In theorem 3 below we will solve this equation, and obtain the Laplace transform of the limit distribution in the supercritical case (thus establishing (7.10)). Though we will not exploit it here, it can be used to obtain a number of finer results about the limit variable. It is a good illustration of information available in the continuous process (and a special class of discrete processes called embeddable, to be introduced in section 12), which we do not have in the general discrete case.

Theorem 3 (Harris (1951), Sevastyanov (1951), Karlin, McGregor (1968 a).) *Let* $W=\lim_{t\to\infty}e^{-\lambda t}Z(t)$ *for a supercritical continuous time branching process with infinitesimal generating function* $u(s)=a[f(s)-s]$. *Let* $\varphi(u)=Ee^{-uW}$ *and* φ^{-1} *be the inverse function of* φ. *Assume that* $\sum p_j j\log j<\infty$. *Then*

$$\varphi^{-1}(x)=(1-x)\exp\left\{\int_1^x\left(\frac{f'(1)-1}{f(s)-s}+\frac{1}{1-s}\right)ds\right\},\qquad q<x\leqslant 1.$$

Proof. The following proof is due to Karlin and McGregor. Take $q<x<y<1$. Then by (5)

$$-\int_y^x\frac{A'(s)}{A(s)}ds=\int_y^x\frac{\beta ds}{u(s)}=\int_y^x\left(\frac{\beta}{u(s)}+\frac{\beta}{\lambda(1-s)}\right)ds-\frac{\beta}{\lambda}\int_y^x\frac{ds}{1-s}\qquad(\lambda=u'(1)),$$

$$=\frac{\beta}{\lambda}\int_y^x\left(\frac{m-1}{f(s)-s}+\frac{1}{1-s}\right)ds+\frac{\beta}{\lambda}\log\frac{1-x}{1-y}. \tag{6}$$

Hence, letting $\beta/\lambda=\alpha$, we see from (6) that

$$\frac{A(y)}{A(x)}=\left(\frac{1-x}{1-y}\right)^{\alpha}\exp\left\{\alpha\int_y^x f^*(s)ds\right\}, \tag{7}$$

where

$$f^*(s)=\frac{m-1}{f(s)-s}+\frac{1}{1-s};$$

or equivalently

$$(1-y)^\alpha A(y)=(1-x)^\alpha A(x)\exp\left\{\alpha\int_y^x f^*(s)\,ds\right\}. \tag{8}$$

Now note that

$$f^*(s)\sim(1-m)^{-1}(1-s)^{-1}\left(m-\frac{1-f(s)}{1-s}\right)\quad \text{as } s\uparrow 1.$$

But by corollary I.10.2

$$\frac{1}{s}\left(m-\frac{1-f(1-s)}{s}\right)$$

is integrable near zero if and only if $\sum(j\log j)p_j<\infty$. Hence under this hypothesis the integral in (7) and (8) converges (as $y\uparrow 1$). Hence

$$\lim_{y\uparrow 1}(1-y)^\alpha A(y)\equiv K \text{ exists.} \tag{9}$$

Thus

$$K=(1-x)^\alpha A(x)\exp\left\{\alpha\int_1^x f^*(s)\,ds\right\},$$

or

$$\left(\frac{K}{A(x)}\right)^{\frac{1}{\alpha}}=(1-x)\exp\left\{\int_1^x f^*(s)\,ds\right\}. \tag{10}$$

The proof is completed by showing that

$$\left(\frac{K}{A(x)}\right)^{\frac{1}{\alpha}}=\varphi^{-1}(x). \tag{11}$$

This is obtained from the relation

$$F(s,t)=B[e^{-\beta t}A(s)] \tag{12}$$

as follows. Set $s=\exp\{-\eta e^{-\lambda t}\}$.
Then

$$\begin{aligned}F(s,t)&=F(\exp\{-\eta e^{-\lambda t}\},t)\\&=E(\exp\{-\eta e^{-\lambda t}Z(t)\}) \qquad (13)\\&\to\varphi(\eta)\quad\text{as } t\to\infty \text{ by theorem 7.1.}\end{aligned}$$

On the other hand

$$\lim_{t\to\infty} e^{-\beta t}A(e^{-\eta e^{-\lambda t}})$$

$$=\lim_{t\to\infty}[e^{-\beta t}(\eta e^{-\lambda t})^{-\alpha}]\left[\frac{1-e^{-\eta e^{-\lambda t}}}{\eta e^{-\lambda t}}\right]^{-\alpha}[1-e^{-\eta e^{-\lambda t}}]^\alpha A(e^{-\eta e^{-\lambda t}}). \tag{14}$$

But $\beta=\lambda\alpha$, and

$$\frac{1-e^{-\eta e^{-\lambda t}}}{\eta e^{-\lambda t}} \to 1 \quad \text{as } t\to\infty .$$

Also by (9)

$$[1-e^{-\eta e^{-\lambda t}}]^{\alpha} A(e^{-\eta e^{-\lambda t}})\to K \quad \text{as } t\to\infty ,$$

Hence by (14)

$$\lim_{t\to\infty} e^{-\beta t} A(e^{-\eta e^{-\lambda t}})=K\eta^{-\alpha},$$

and by (12) and (13) we see that for large η

$$\varphi(\eta)=B(K\eta^{-\alpha}) .$$

Applying A to both sides

$$A[\varphi(\eta)]=K\eta^{-\alpha},$$

and letting $u=\varphi(\eta)$, $\eta=\varphi^{-1}(u)$

$$A(u)=K[\varphi^{-1}(u)]^{-\alpha}$$

for u small, and hence for all u. □

9. Split Times

We will here examine the continuous process more carefully by separating it into two processes: one the process of "jump" or "split" times, i.e. the times at which particles split; the other the process of jumps, i.e. the discrete time process representing the sequence of states into which the process goes at the split times. (See Athreya-Karlin (1967).)

Before making these notions precise we remark that it is no loss of generality to assume that $p_1=0$; since if $p_1>0$ then the process $\{Z(t);\, t\geqslant 0\}$ is equivalent to another Markov branching process $\{\tilde{Z}(t);\, t\geqslant 0\}$ with parameters given by

$$\tilde{a}=a(1-p_1),$$

$$\tilde{f}(s) = \frac{f(s)-p_1 s}{1-p_1} = \sum_{j=0}^{\infty} \tilde{p}_j s^j ,$$

so that $\tilde{p}_1=0$. It is easy to see from elementary properties of Poisson processes that the split time sequence τ_n of the $Z(t)$ process can be constructed from the sequence $\tilde{\tau}_n$ of the $\tilde{Z}(t)$ process by inserting between $\tilde{\tau}_{n-1}$ and $\tilde{\tau}_n$ a random number N_n of randomly distributed points; their locations being independently and uniformly distributed in $(\tilde{\tau}_{n-1},\tilde{\tau}_n)$, while $P(N_n=k)=p_1^k(1-p_1)$. Hence *we assume for the remainder of this section that* $p_1=0$. With this convention the discontinuities of the sample path $Z(t,\omega)$ coincide with the times at which the particles split. Accordingly we call these discontinuities the *split times*.

We recall the construction of the minimal process. Let $\{\xi_i; i=1,2,\ldots\}$ be a sequence of independent integer valued random variables with p.g.f. $f(s)$. Let $\eta_i=\xi_i-1$, $E\eta_i\equiv\mu=m-1$, $S_0=k$, $(k>0)$, $S_n=k+\sum_{i=1}^{n}\eta_i$, $n\geqslant 1$, and $N=\inf\{n; S_n=0\}$, $(=\infty$ *if* $S_n\neq 0$ for any n). Given the sequence $\{\xi_i\}$, let $T_1, T_2, \ldots, T_N$ be mutually independent exponential random variables with means

$$E(T_j|\{\xi_i\})=(aS_{j-1})^{-1} \quad \text{for } 1\leqslant j\leqslant N\,.$$

According to the discussion of section 2, $(\Omega, \mathbb{F}, P)$ is a probability space on which all these random variables are defined. Now set

$$\begin{aligned} \tau_0&=0, \qquad \tau_n=T_1+T_2+\cdots+T_n, && 1\leqslant n\leqslant N\,,\\ Z(t)&=S_{n-1} \quad \text{for } \tau_{n-1}\leqslant t<\tau_n, && 1\leqslant n\leqslant N\,,\\ &=0 \qquad\ \ \text{for } \tau_N\leqslant t\,. \end{aligned}$$

Then $\{Z(t,\omega); t\geqslant 0\}$ is a continuous time branching process with $P\{Z(0,\omega)=k\}=1$, $\{\tau_n(\omega); n\geqslant 0\}$ is the split time process, and $\{S_n(\omega); n\geqslant 0\}$ is the process of jumps.

If the process $Z(t)$ is not supercritical, then $E\eta_i\leqslant 0$, so that the random variable N is finite a.s. If the process is supercritical, then $E\eta_i>0$, and hence $P(N<\infty)<1$. But if $p_0=0$, then $P(N=\infty)=1$, and then the limit properties of the split times τ_n have a natural significance. It is this case which we shall study in detail, and we hence *assume that* $p_0=p_1=0$.

Thus $\tau_n(\omega)$ is well defined for any n, and is furthermore a stopping time[5] with respect to the family $\{\mathbb{F}_t\}$ of fields. Let $\mathscr{F}_n$ be the σ-algebra associated with stopping time $\tau_n(\omega)$ (i.e., containing all information about the process up to time τ_n), and let

$$Y_n(\omega)=\sum_{j=1}^{n}\left(T_j-\frac{1}{aS_{j-1}}\right), \qquad n\geqslant 1\,.$$

One can identify $\mathscr{F}_n$ with the σ-algebra generated by $\xi_1, \xi_2, \ldots, \xi_n$ and $T_1, T_2, \ldots, T_{n-1}$. We then have the following result

Theorem 1. *The family* $\{Y_n, \mathscr{F}_n; n=1,2,\ldots\}$ *is a martingale such that*

$$\lim_{n\to\infty} E\,Y_n^2<\infty\,, \tag{1}$$

[5] Note that our process can be assumed to be strong Markov without loss of generality, since the state space is discrete. Also it is known that τ_n, the time of the n'th jump is a stopping time for the process (see Chung (1967)).

and hence

$$\lim_{n\to\infty} Y_n \equiv Y \tag{2}$$

exists in mean square and a.s.

Corollary 1. $\lim_{n\to\infty} \tau_n/\log n = \lambda^{-1}$ a.s.

Theorem 2. $\lim_{n\to\infty} n e^{-\lambda\tau_n} = W/\mu$ a.s., *where W is as defined in* (7.4).

The strongest result about the τ_n's is

Theorem 3.

$$P\left\{\lim_{n\to\infty}\left(\tau_n - \frac{1}{\lambda}\log n\right) \text{exists and is finite}\right\} \quad \textit{is}$$

one or zero, as

$$\sum_{j=2}^{\infty} (j\log j)\, p_j < \infty, \quad \textit{or} \quad = \infty\,. \tag{3}$$

Corollary 2. $P\{W>0\}=1$ *or* 0 *according as* (3) *does or does not hold.*

Our proofs will exploit the facts that $Z(\tau_j)=S_j$, and that S_j is a sum of non-negative random variables. The key tool is the following general theorem on sums of non-negative independent and identically distributed random variables. We refer the reader to Athreya-Karlin (1967) for a proof, where he will also find more material on the split times.

Theorem 4. *Let $U_1, U_2, U_3, \ldots$ be i.i.d. non-negative (not necessarily integer valued) random variables. Let $S_n = c + U_1 + U_2 + \cdots + U_n$, where c is a non-negative constant. Assume that $0 < E(U_i) = \mu \leqslant \infty$, and that at least one of the following holds:*

a) $c>0$, b) $EU_i^{-1} < \infty$.

Then

(i) *For each $\gamma > 0$*

$$\lim_{n\to\infty} E\left(\frac{n}{S_n}\right)^{\gamma} = \frac{1}{\mu^{\gamma}},$$

(ii) $\lim_{n\to\infty} \sum_{i=1}^{n} [(1/S_i) - (1/i\mu)]$ *exists* a.s. *if and only if*

$$E U_1 |\log U_1| < \infty\,.$$

Proof of Theorem 1. Since T_i *and* S_i are $\mathscr{F}_i$ measurable,

$$E(Y_n|\mathscr{F}_{n-1}) = Y_{n-1} + E\left(T_n - \frac{1}{a S_{n-1}}\middle|\mathscr{F}_{n-1}\right\} \quad \text{a.s.}$$

But by the strong Markov property and the way the T_j's have been defined

$$E(T_n|\mathscr{F}_{n-1}) = \frac{1}{a S_{n-1}} \quad \text{a.s.}$$

Thus

$$E\left(T_n - \frac{1}{a S_{n-1}}\middle|\mathscr{F}_{n-1}\right) = 0 \quad \text{a.s.,}$$

and hence $\{Y_n\}$ is a martingale. Similarly

$$E(Y_n^2|\mathscr{F}_{n-1}) = Y_{n-1}^2 + 2\,Y_{n-1}\cdot 0 + E\left[\left(T_n - \frac{1}{a S_{n-1}}\right)^2\middle|\mathscr{F}_{n-1}\right] \quad \text{a.s.}$$

The expectation on the right is the variance of an exponential random variable with mean $(a S_{n-1})^{-1}$, and hence

$$E(Y_n^2|\mathscr{F}_{n-1}) = Y_{n-1}^2 + \left(\frac{1}{a S_{n-1}}\right)^2 \text{a.s.}$$

Taking expectation with respect to Y_{n-1} and iterating yields

$$E\,Y_n^2 = \frac{1}{a^2} \sum_{i=0}^{n-1} E\left(\frac{1}{S_i^2}\right). \tag{4}$$

But from Theorem 4 (i) we have

$$\sup_i E\left(\frac{i}{S_i}\right)^2 = k_2 < \infty. \tag{5}$$

and thus (4) and (5) imply

$$\sup_n E\,Y_n^2 < \infty. \tag{6}$$

To complete the proof we appeal to Doob's martingale convergence theorem. □

Proof of Corollary 1. Observe that

$$Y_n = \sum_{i=1}^{n} \left(T_i - \frac{1}{a S_{i-1}}\right) = \tau_n - a^{-1} \sum_{i=0}^{n-1} S_i^{-1}. \tag{7}$$

By Theorem 1, Y_n converges to a finite limit a.s. Also, by the strong law of large numbers, n/S_n converges to μ^{-1} a.s., and hence

$$(\log n)^{-1} \sum_{i=1}^{n} \frac{1}{a S_i} \to \frac{1}{a\mu} = \frac{1}{\lambda}. \quad \square$$

Remark. We can further assert that

$$\frac{E(\tau_n)}{\log n} \to \frac{1}{\lambda}. \tag{8}$$

In fact, since $E(T_i)=(1/a)E(1/S_{i-1})$, the limit relation (8) is a consequence of Theorem 4(i).

Proof of Theorem 2. From corollary 1 we see that

$$P\{\tau_n \to \infty\} = 1 .$$

We also have from theorem 7.1 that

$$P\{\lim_{t\to\infty} Z(t)e^{-\lambda t} = W\} = 1 .$$

Combining these two facts we get

$$P\{\lim_{n\to\infty} Z(\tau_n)e^{-\lambda\tau_n} = W\} = 1 .$$

Now note that $Z(\tau_n)=S_n$ and that by the strong law of large numbers $S_n/n \to \mu$. □

Theorem 3 is a refinement of theorem 2. To prove it we use the following

Lemma 1. *The two sets*

$\{(\lambda\tau_n - \log n)\text{ converges}\}$ *and* $\left\{\lim_{n\to\infty}\sum_{i=1}^{n}\left(\frac{1}{S_i}-\frac{1}{i\mu}\right)\text{exists}\right\}$ *coincide* a.s.

Proof. Consider $\lambda\tau_n - \log n$. After appropriate rearrangement it takes the form

$$\begin{aligned}\lambda\tau_n - \log n &= \lambda\sum_{i=1}^{n}\left(T_i - \frac{1}{aS_{i-1}}\right) + \sum_{i=1}^{n}\left(\frac{\mu}{S_{i-1}} - \frac{1}{i}\right) + \left(\sum_{i=1}^{n} i^{-1}\right) - \log n\\ &= \lambda Y_n + \left(\sum_{i=1}^{n}\frac{1}{i}\right) - \log n + \mu\sum_{i=1}^{n}\left(\frac{1}{S_{i-1}} - \frac{1}{i\mu}\right).\end{aligned}$$

According to theorem 1, $\lim Y_n = Y$ exists a.s. Also $\left(\sum_{i=1}^{n} i^{-1}\right) - \log n$ and $\sum_{i=2}^{n}((1/i-1)-(1/i))$ converge as $n\to\infty$. The assertion of the lemma now follows.

Proof of Theorem 3. By Theorem 4(ii) we know that $P\left\{\lim_{n\to\infty}\sum_{i=1}^{n}((1/S_i)-(1/i\mu))\text{ exists}\right\}>0$ if and only if $E\eta_1|\log\eta_1|<\infty$, which is equivalent to $\sum_{j=2}^{\infty}(j\log j)\,p_j<\infty$. Now use lemma 1 to complete the proof. □

Remarks. 1. This gives another proof of theorem 7.2, since theorem 3 implies that the two sets $\{\omega: \lambda\tau_n(\omega)-\log n \text{ converges}\}$ and $\{\omega: W(\omega)>0\}$ coincide. Note that in view of our assumption that $p_0=0$, the extinction probability q is zero.

2. The supercritical case with $p_0>0$ can be reduced to the case $p_0=0$ by the same kind of decomposition we discussed in section I.12.

10. Second Order Properties

The continuous time and split time processes also have decompositions and second order properties similar to those in sections I.12 and I.13 for the Galton-Watson case. We summarize some of the results here, not because anything surprisingly new happens, but because we will use them in the next section—where we will prove some qualitatively different theorems.

Since $Z(t,\omega)e^{-\lambda t}\to W(\omega)$ a.s., we would expect a decomposition for $Z(t,\omega)e^{-\lambda t}-W(\omega)$. Reference is again made to the same probability space $(\Omega, \mathbb{F}, P)$, with the functions $Z(t,\omega)$ taken so as to be right continuous.

Theorem 1. *Assume that* $p_0=0, \sum(j\log j)\,p_j<\infty$, *and* $P\{Z(0)=1\}=1$. *Then there is a set* A *in* $\mathbb{F}$ *such that* $P(\mathsf{A})=1$, *and a family of random variables* $\{W_t^{(j)}(\omega);\ j=1,2,\ldots,Z(t,\omega)\}$, *which when conditioned on* $Z(t,\omega)$ *are mutually independent and identically distributed as* W, *and such that*

$$Z(t,\omega)-e^{\lambda t}\,W(\omega)=\sum_{j=1}^{Z(t,\omega)}\left[1-W_t^{(j)}(\omega)\right] \tag{1}$$

for all $t\geqslant 0$, $\omega\in\mathsf{A}$.

Proof. That for each t there is an A_t such that the conclusion of the theorem holds for each $\omega\in\mathsf{A}_t$, follows exactly as in the discrete case. Let $\{t_\alpha;\ \alpha=1,2,\ldots\}$ denote the set of rationals and define $\mathsf{A}_r(\omega)=\bigcap_\alpha \mathsf{A}_{t_\alpha}(\omega)$. Then $P(\mathsf{A}_r)=1$, and the conclusion holds for all $\omega\in\mathsf{A}_r$. The right continuity of $Z(t,\omega)$ then implies the theorem.

By an entirely similar argument applied to the split times process $\{Z(\tau_k), \mathscr{F}_k;\ k=1,2,\ldots\}$ of section 9, we get:

Theorem 2. *Assume that the infinitesimal generating function* $u(s)=a[f(s)-s]$ *satisfies* $f(0)=f'(0)=0$ *and* $\sum(j\log j)p_j<\infty$. *Then there is a set* $\mathsf{A}\in\mathbb{F}$ *with* $P(\mathsf{A})=1$; *and a family of random variables*

$\{W_k^{(j)}(\omega); j=1,\ldots,Z(\tau_k,\omega)\}$, which conditioned on $\mathscr{F}_k$ are mutually independent and distributed as W, and are such that

$$Z(\tau_k,\omega)-e^{\lambda\tau_k}W(\omega)=\sum_{j=1}^{Z(\tau_k,\omega)}[1-W_k^{(j)}(\omega)] \tag{2}$$

for all $k\geq 0$ and $\omega\in\mathsf{A}$.

There are also the following versions of central-limit-type results.

Theorem 3. *Assume that $p_0=0, \sum j^2 p_j<\infty$ and $Z(0)=1$. Let $\sigma_w^2=$ variance W (which exists by arguments analogous to the discrete case). Then*

$$\frac{Z(t)-e^{\lambda t}W}{\sqrt{Z(t)}}\xrightarrow{d}N(0,\sigma_w^2). \tag{3}$$

"$\xrightarrow{d}$" means converges in distribution. $N(\mu,\sigma^2)$ is the normal distribution with mean μ and variance σ^2.

Proof. Use the representation in theorem 2, and argue exactly as in the discrete case.

The corresponding result for the split time process requires a little more proof.

Theorem 4. *Under the hypothesis of Theorem 2*

$$\frac{n\mu-e^{\lambda\tau_n}W}{\sqrt{n}}\xrightarrow{d}N(0,\sigma_w^2\mu^2+\theta^2), \tag{4}$$

where $\sigma_w^2=\operatorname{var}W$, and (μ,θ^2) are the (mean, var.) of a random variable with generating function $f(s)/s$. (As before, $\lambda=a\mu$.)

Proof. Let $\{\xi_j; j=1,2,\ldots\}$ be i.i.d. random variables with generating function $f(s)/s$. We have seen in section 7 that

$$\{Z(\tau_j)-Z(\tau_{j-1}); j=1,2,\ldots\} \tag{5}$$

where $\tau_j=$ the jth split time, is distributed as $\{\xi_j; j=1,2,\ldots\}$.

Let

$$S_n=\sum_{j=1}^{n}(\mu-\xi_j).$$

By theorem 2 we can write

$$\begin{aligned}n\mu-e^{\lambda\tau_n}W&=n\mu-Z(\tau_n)+Z(\tau_n)-e^{\lambda\tau_n}W\\&=S_n-Z(0)+\sum_{j=1}^{Z(\tau_n)}(1-W_n^{(j)}).\end{aligned}$$

Let $V_n^{(1)}=\{S_n-Z(0)\}/\sqrt{n}$, and $V_n^{(2)}=\sum_{j=1}^{Z(\tau_n)}(1-W_n^{(j)})/\sqrt{n}$; and we can thus write

$$\frac{n\mu-e^{\lambda\tau_n}W}{\sqrt{n}}\equiv V_n=V_n^{(1)}+V_n^{(2)}.$$

Letting $\psi(t)=E\,e^{it(1-W)}$ we see that

$$E\{e^{it\mathrm{V}_n^{(2)}}|\mathscr{F}_n\}=\left[\psi\left(\frac{t}{\sqrt{n}}\right)\right]^{Z(\tau_n)}. \tag{6}$$

But by the central limit theorem

$$\left[\psi\left(\frac{t}{\sqrt{n}}\right)\right]^n\to e^{-\frac{\sigma_W^2t^2}{2}}\quad \text{as } n\to\infty,$$

and since $Z(\tau_n)/n\to\mu$ a.s., we see from (6) that

$$E\{e^{it\mathrm{V}_n^{(2)}}\,|Z(\tau_n)\}\to e^{-\frac{\sigma_W^2\mu^2t^2}{2}}\quad \text{a.s.} \tag{7}$$

Thus

$$E(e^{it}V_n)=E\left\{e^{it\,\mathrm{V}_n^{(1)}}\left[E(e^{it\mathrm{V}_n^{(2)}}\,|\mathscr{F}_n)-e^{\frac{\sigma_W^2\mu^2t^2}{2}}\right]\right\} \tag{8}$$

$$+e^{-\frac{\sigma_W^2\mu^2t^2}{2}}E\,e^{it\,\mathrm{V}_n^{(1)}},$$

(where $\mathscr{F}_n$ is as in section 9).

Applying the central limit theorem again,

$$E\,e^{it\,\mathrm{V}_n^{(1)}}\to e^{-\frac{\theta^2t^2}{2}}.$$

and theorem 4 then follows from (7) and (8) by the bounded convergence theorem. □

11. Constructions Related to Poisson Processes

The construction of the minimal process from a "tree" whose branches have exponentially distributed lengths, suggests that there may be Poisson processes lurking in the background. In this section we give two results (one due to J. Lamperti (private communication from S. Watanabe), the other to D.G. Kendall (1966)) illustrating relations between Poisson and supercritical branching processes.

We have seen that the jumps of the branching process $\{Z(t);t\geqslant 0\}$ are i.i.d. random variables, while the times between jumps are independent and exponentially distributed. If these exponential distributions were all the same (i.e. had the same parameter) then $Z(t)$ would be a compound Poisson process. Of course they are *not* the same, the parameters depending on the state of the process. In fact, in a growing population the rate at which jumps occur increases with time. This suggests, however, that under a (random) time change, which appropriately slows the process down, we should get a compound Poisson process. This is the idea behind the following result of Lamperti.

Theorem 1. *Let* $\{Z(t,\omega); t \geqslant 0\}$ *be a supercritical branching process having infinitesimal generating function* $u(s)=a[f(s)-s]$, *with* $0<a<\infty$, $f(0)=f'(0)=0$, *and* $f'(1)<\infty$. *Define the cumulative process*

$$Y(t,\omega)=\int_0^t Z(s,\omega)\,ds\,, \tag{1}$$

its inverse

$$Y^{-1}(u,\omega)=\inf\{t;\, Y(t,\omega)=u\}\,, \tag{2}$$

and the modified time scale process

$$X(u,\omega)=Z[Y^{-1}(u,\omega),\omega]\,. \tag{3}$$

Then $\{X(u,\omega); u \geqslant 0\}$ *is a compound Poisson with*

$$E e^{-sX(t)}=e^{at[s^{-1}f(s)-1]}.$$

Remarks. 1. We shall give an elementary proof using split times. Lamperti's original proof used the general theory of time change via an additive functional of a Markov process. One notes that $Y(t)$ is a "nice" additive functional for $Z(t)$, and then deduces that the infinitesimal generator of the new Markov process $X(t)$ can be identified with that of a compound Poisson process.

2. Theorem 1 tells us that the $Z(t)$ process can be constructed from the compound Poisson process, thus yielding yet another construction of our original process.

Proof. Note that since $f(0)=p_0=0$, $Y(t,\omega)$ is strictly increasing and hence Y^{-1} is well defined. We take Z to be right continuous, and then X will be also. Let $\{\tau_i'(\omega);\ i=0,1,\ldots\}$ denote the sequence of discontinuities of the sample paths of $X(t,\omega)$. (As before, τ_i are the split times of $Z(t)$). Then

$$Y(\tau_i)=\tau_i', \qquad i=0,1,2,\ldots. \tag{4}$$

Let

$$\theta_i \equiv \tau_i' - \tau_{i-1}',$$

and

$$\xi_i \equiv X(\tau_i') - X(\tau_{i-1}')$$

denote the size of the ith inter-jump time and ith jump, respectively, for the X-process. From (4) and the above definitions we also see that

$$\theta_i = Z(\tau_{i-1}) \cdot [\tau_i - \tau_{i-1}],$$

and

$$\xi_i = Z(\tau_i) - Z(\tau_{i-1}).$$

From elementary properties of the split time process we then see that the θ_i are all exponentially distributed with mean a^{-1}, while the ξ_i are non-negative and integer valued with p.g.f. $f(s)/s$. Furthermore,

$$X(u,\omega) = \sum_{j=1}^{N(u,\omega)} \xi_j,$$

where $N(u,\omega)=r$ when $\theta_1+\cdots+\theta_r \leqslant u < \theta_1+\cdots+\theta_{r+1}$. Thus we can draw the conclusion of the theorem provided we can show that the random variables $\{\xi_i, \theta_j; i,j=1,2,\ldots\}$ are independent. This fact follows from the Markovian character of the $Z(t,\omega)$ process, from which we can infer that (for $\mathscr{F}_i$ as defined in section 9)

$$\sum_{k=0}^{\infty} P\{\xi_i = k \mid \mathscr{F}_{i-1}\} s^k = \frac{f(s)}{s},$$

$$P\{\theta_i > \alpha \mid \mathscr{F}_{i-1}\} = e^{-a\alpha},$$

and

$$P\{\xi_i = k, \theta_i > \alpha \mid \mathscr{F}_{i-1}\} = P\{\xi_i = k \mid \mathscr{F}_{i-1}\} P\{\theta_i > \alpha \mid \mathscr{F}_{i-1}\}.$$

By repeated application of these formulas the joint distribution of any collection of ξ_i's and θ_j's can be factored into their marginal distributions. □

Another interesting relation between branching and Poisson processes was established by Kendall (1966) in the case of binary fission (in which case the branching process is a Yule process).

Theorem 2. *Let $Z(t,\omega)$ have infinitesimal generating function $u(s)=\lambda[s^2-s]$, $0<\lambda<\infty$. Then, conditioned on W, the process*

$$\{Z[\lambda^{-1}\log(1+tW^{-1}),\omega];\ t\geqslant 0\}$$

is a Poisson process with unit rate. That is, for any $0<t_1<\cdots<t_k<\infty$, and integers $n_i \geqslant 0$, $i=1,\ldots,k$, and Borel subset B of $[0,\infty)$

$$\begin{aligned} P\{Z[\lambda^{-1}\log(1+t_iW^{-1}),\omega]=n_i,\ i=1,\ldots,k,\ W\in\mathsf{B}\} \\ = P\{W\in\mathsf{B}\}P\{N(t_i)=n_i,\ i=1,\ldots,k\}, \end{aligned} \tag{5}$$

where $\{N(t);\, t \geqslant 0\}$ *is a Poisson process with parameter 1 and* $P\{N(0)=Z(0)\}=1$.

Outline of Proof. Observe that the Yule process (see section 5) only takes jumps of one unit, and hence that it suffices to prove that the jump times of $Z[\lambda^{-1}\log(1+tW^{-1})]$ (when conditioned on W), form a Poisson process. This is the same as saying that when conditioned on W, the sequence

$$\{\tau_n''=W(e^{\lambda\tau_n}-1);\; n=0,1,2,\ldots\},$$

where τ_n = the nth split time of $Z(t)$, constitutes the epochs of a Poisson process. This is in turn equivalent to the assertion that (conditioned on W) the random variables

$$\{\tau_j''-\tau_{j-1}''=W(e^{\lambda\tau_j}-e^{\lambda\tau_{j-1}});\; j=1,2,\ldots\}$$

are mutually independent, independent of W, and exponentially distributed with unit parameter. To prove this we will establish the formula

$$E_k \equiv E\exp\left\{-W\left[\alpha+\sum_{j=1}^{k}\alpha_j(e^{\lambda\tau_j}-e^{\lambda\tau_{j-1}})\right]\right\}=\frac{1}{1+\alpha}\prod_{j=1}^{k}\frac{1}{1+\alpha_j} \tag{6}$$

$$\text{for } 0\leqslant\alpha, \quad 0\leqslant\alpha_j, \quad j=1,\ldots,k.$$

Let

$$g_k=e^{-\lambda\tau_k}\left[\alpha+\sum_{j=1}^{k}\alpha_j(e^{\lambda\tau_j}-e^{\lambda\tau_{j-1}})\right], \tag{7}$$

and recall from theorem 10.2 that

$$W=e^{-\lambda\tau_k}\sum_{j=1}^{Z(\tau_k)}W_k^{(j)}. \tag{8}$$

Thus combining (6), (7) and (8), and noting that $Z(\tau_k)=k+1$ for the Yule process, we have

$$E_k=E\left\{\exp\left(-g_k\sum_{j=1}^{k+1}W_k^{(j)}\right)\right\}. \tag{9}$$

Now in the special case of the Yule process one can show directly from the generating function in section 5 that W has unit exponential distribution (we leave demonstration of this point as an exercise).

Furthermore, conditioned on $\mathscr{F}_k$, the $W_k^{(j)}$ are i.i.d. as W. Hence by so conditioning—and then taking expectations—we can write (9) as

$$E_k=E\{(1+g_k)^{-(k+1)}\}. \tag{10}$$

Now observe the identity

$$g_k=\alpha_k+(g_{k-1}-\alpha_k)e^{-\lambda(\tau_k-\tau_{k-1})}, \tag{11}$$

and note that the distribution of $\lambda k(\tau_k - \tau_{k-1})$ conditioned on $\mathscr{F}_{k-1}$, is unit exponential. Then using the integration formula

$$\int_0^\infty \left(1+a+be^{-\frac{y}{k}}\right)^{-(k+1)} e^{-y} dy = (1+a)^{-1}(1+a+b)^{-k} \tag{12}$$

for any real numbers a, b, we obtain from (10) and (11) that

$$E_k = E\{E[(1+g_k)^{-(k+1)} | \mathscr{F}_{k-1}]\} = E\{(1+\alpha_k)^{-1}(1+g_{k-1})^{-k}\}.$$

This yields the recurrence relation

$$E_k = (1+\alpha_k)^{-1} E_{k-1},$$

and by iteration

$$E_k = E_1 \prod_{j=2}^{k} (1+\alpha_j)^{-1}. \tag{13}$$

It therefore remains only to prove that

$$E_1 = (1+\alpha_1)^{-1}(1+\alpha)^{-1}. \tag{14}$$

But this is obvious if we note by (10) that

$$\begin{aligned} E_1 &= E\{1+e^{-\lambda\tau_1}[\alpha+\alpha_1(e^{\lambda\tau_1}-1)]\}^{-2} \\ &= E\{1+\alpha_1+(\alpha-\alpha_1)e^{-\lambda\tau_1}\}^{-2}, \end{aligned}$$

which by (12)

$$= (1+\alpha_1)^{-1}(1+\alpha)^{-1}. \quad \square$$

Remarks. 3. Let $Z(t)$ be the Yule process. From the law of the iterated logarithm as applied to a Poisson process with unit rate we obtain the result that with probability one the set of limit points (as $t \to \infty$) of

$$\bar{Y}(t,\omega) \equiv \frac{X(t)-t}{\sqrt{2t \log\log t}}, \tag{15}$$

where $X(t) \equiv X(t,\omega) = Z(\lambda^{-1}\log(1+tW^{-1}), \omega)$, is the interval $[-1, +1]$.

Since the map $t \to \lambda^{-1}\log(1+tW^{-1})$ is monotone increasing in t for any $W > 0$, we conclude from (15) and theorem 2 that (with probability one) the limit points of

$$Y(\tau,\omega) = \frac{Z(\tau) - We^{\lambda\tau}}{\sqrt{2\,We^{\lambda\tau}\log\tau}} \quad (\text{as } \tau \to \infty) \tag{16}$$

coincide with $[-1, +1]$.

4. Kendall's theorem tells us how to construct realizations of the Yule process with a prescribed value for $\lim_{t\to\infty} Z(t,\omega)e^{-\lambda t}$. Indeed let

$\{X(t);\, t \geqslant 0\}$ be any Poisson process with unit rate, and W be an exponentially distributed random variable independent of the Poisson processes. Define $Z(t) = X(W(e^{\lambda t} - 1))$ for any $\lambda > 0$. Then $\{Z(t);\, t \geqslant 0\}$ is a Yule process with infinitesimal generating function $u(s) = \lambda(s^2 - s)$. Clearly this is only a reformulation of theorem 2.

5. As far as we know there has been no generalization of Kendall's result to a general continuous time Markov branching process. Any such attempt will have to consider the fact that the jumps of the process are still independent random variables, though no longer of unit size. Conditioning on W may, however, destroy this independence. The fact that W was exponential proved helpful in the proof of theorem 2. In this connection one can show that W is exponential if and only if the process is Yule. More generally, W has a Gamma distribution with parameters a and k ($a > 0$, k a non-negative integer) if and only if the infinitesimal generating function of the process is $u(s) = a[s^{k+1} - s]$. One can prove this by showing that given a and $f'(1)$, the functional equation $\varphi(\theta) = \int_0^\infty f(\varphi(\theta e^{-\lambda y}))e^{-ay}\,dy$ establishes a one-to-one correspondence between φ and f.

6. The proof of Kendall's theorem given here is different from his original proof in (1966).

12. The Embeddability Problem

We saw in section 6 (theorem 1) that for any $\delta > 0$ the sequence $\{Z(n\delta, \omega);\, n = 1, 2, 3, \ldots\}$ forms a Galton-Watson process with p.g.f. $g(s) = F(s, \delta)$. A natural converse question is the following: Given a Galton-Watson process $\{X_n(\omega);\, n = 0, 1, 2, \ldots\}$ with p.g.f. $g(s)$, does there exist a continuous time branching process $\{Z(t, \omega);\, t \geqslant 0\}$ with p.g.f. $\{F(s, t);\, t \geqslant 0\}$ such that for some $\delta > 0$, $\{Z(n\delta, \omega)\}$ is a stochastic process equivalent to $\{X_n(\omega)\}$? This is the embeddability problem. Analytically it is equivalent to asking whether we can find a semigroup $\{F(s, t);\, t \geqslant 0\}$ of p.g.f.'s satisfying $F(s, t + u) = F(F(s, t), u)$ for all $t, u \geqslant 0$, and $F(s, \delta) = g(s)$ for some $\delta > 0$. In his book (1963), Harris remarks that if $g(s)$ is an entire function and $g(0) = 0$ then the answer is negative.

A far more complete answer has been given by Karlin and McGregor (1968a, b). In this section, we shall give a brief account of their work.

In addition to its own intrinsic interest, the problem is important due to the fact that (as we have seen) the continuous time case lends itself more easily to analysis, and hence one can say more about embeddable Galton-Watson processes than non-embeddable ones. For example in the embeddable case one can assert that the density $w(x)$

of the limit random variable is an entire function away from origin and $\lim_{x\downarrow 0} x^{\alpha-1} w(x)=d,\ 0<d<\infty$ exists for some $\alpha>0$. There exist non-embeddable cases where $w(x)$ oscillates wildly near zero (Harris (1948)).

Unfortunately, large classes of commonly encountered p.g.f.'s are non-embeddable, the main exception being the linear fractional.

The more general question of embedding an analytic function $g(x)$ with a fixed point $x^*=g(x^*)$ in its domain of analyticity, into a one parameter semigroup $\{F(s,t);\ t\geqslant 0\}$, each analytic near x^* (suppressing the requirement that $F(s,t)$ be a probability generating function for each t), has a long history dating back to Fatou (see Harris (1963)). This problem is of interest not only in branching processes but also in the theory of conformal mapping, iterative processes of analysis, and elsewhere. Some principal contributors are Fatou, Koenigs, Baker, Jabotinsky, Szekeres, Karlin, McGregor, Dubuc and Kuczma, among others. We refer to the papers of Karlin and McGregor (1968 a, b) and the book of Kuczma (1967) for a complete bibliography.

To fix ideas we adopt the following

Definition. *A p.g.f.* $g(s)=\sum_{j=0}^{\infty} p_j s^j$ *is embeddable if there exists a one parameter family (indexed by* $t\geqslant 0$*) of p.g.f.'s* $F(s,t)$*, such that*

$$F(s,t+u)=F(F(s,t),u);\qquad t,u\geqslant 0,\qquad |s|\leqslant 1, \tag{1}$$

and

$$F(s,1)=g(s). \tag{2}$$

In the following discussion s will sometimes be taken as complex, and sometimes real. It will always be clear from the context which interpretation to make, and we will not bother to mention it explicitly. In the complex case $g(s)$ will represent the analytic function $\sum_{j=0}^{\infty} p_j s^j$, whenever the series converges absolutely. Being a p.g.f., $g(s)$ is analytic at least in the open unit disc. In some cases we will need analyticity in a larger domain. The following simplifications will be useful:

a) It is obvious that if $p_1=0$ then g is not embeddable, since for any Markov branching process $\{Z(t);\ t\geqslant 0\}$ one has

$$P\{Z(1)=1\}\geqslant P\{\text{parent particle does not die in } [0,1]\}=e^{-a}>0.$$

On the other hand the case $g(s)\equiv p_0+p_1 s$ with $0<p_1=1-p_0<1$ is easily seen to be embeddable in a process whose infinitesimal generating function is $u(s)\equiv a(1-\alpha)(1-s)$, where $0<a<\infty$, and $0<\alpha<1$ are chosen so that $a(1-\alpha)=-\log p_1$. To exclude these two trivial cases all the p.g.f.'s occurring henceforth will be taken to be of the form $g(s)=\sum_{i=0}^{\infty} p_i s^i$, with $p_1>0$, and $p_1+p_0<1$.

b) When $m=g'(1-)>1$, there exist two fixed points q and 1 of the transformation $y=g(s)$ with $0\leqslant q<1$. If we set

$$\hat{g}(s) = \frac{g((1-q)s+q)-q}{1-q},$$

then, as we saw in section I.12, $\hat{g}(s)$ is a p.g.f. with $\hat{g}'(1-)\equiv g'(1-)$ and $\hat{g}(0)=0$; hence with extinction probability $\hat{q}=0$. If g is embeddable in the semigroup $\{F(s,t);\, t\geqslant 0\}$ then $\hat{g}$ is embeddable in the semigroup defined by

$$\hat{F}(s,t) = \frac{F((1-q)s+q,t)-q}{1-q}.$$

Conversely, if $\hat{g}$ is embeddable in $\hat{F}(s,t)$ then g is embeddable in F. It is thus no loss of generality to assume in the supercritical case that $g(0)=0$.

c) If $m=g'(1-)<1$ then it may happen (for instance, if $g(s)$ is meromorphic) that a second real fixed point $\tilde{q}>1$ exists, such that g is analytic in $|s|<\tilde{q}+\varepsilon$ for $\varepsilon>0$. In this case the function $\tilde{g}$ defined by

$$\tilde{g}(s) = \frac{1}{\tilde{q}}\, g(\tilde{q}\, s)$$

is a supercritical p.g.f. It is easily checked that g is embeddable if and only if $\tilde{g}$ is, and in fact that the corresponding semigroups $F(s,t)$ and $\tilde{F}(s,t)$ are related by

$$\tilde{F}(s,t) = \frac{1}{\tilde{q}}\, F(\tilde{q}\, s,t).$$

The main results, then, are as follows. (We shall sketch the proof of theorem 2, and interpret the others. For complete proofs and further details see Karlin-McGregor (1968 a, b).)

Theorem 1 *(supercritical case). Let $g(s)$ be a p.g.f. analytic at $s=1$, with $g'(1)=m>1$, $g(0)=0$*[6]*, and $g'(0)=c>0$. Suppose that the singular points of $g(s)$ in the extended complex plane form a countable closed set S and that $g(s)$ is single valued in the complement of S. Then, g is embeddable only if it is a linear fractional function, i.e. if $g(s)=(1-p)s[1-ps]^{-1}$ for $0<p<1$.*

Thus if g is meromorphic then it is not embeddable unless it is linear fractional. Also, if g is entire, then it is not embeddable, this being Harris'

[6] By remark b) above, the assumption $g(0)=0$ can be dropped. Also by c), the conclusion of the theorem applies to some subcritical processes.

result as quoted earlier. Thus, Poisson, binomial, and negative binomial (except linear fractional) p.g.f.'s are non-embeddable. It turns out that if we don't insist that g be single valued in the complement of S, then we can have many embeddable p.g.f.'s, for example the following class:

$$\mathsf{G} \equiv \{g_{p,k}(s) = s(1-p)^{\frac{1}{k}}(1-ps^k)^{\frac{1}{k}}; 0 \leqslant k,\ 0<p<1\}.$$

To see this note first that the linear fractional p.g.f. $g_p(s)=(1-p)s[1-ps]^{-1}$ is embeddable in the semigroup

$$F_p(s,t) = e^{-dt}s[1-(1-e^{-dt})s]^{-1},$$

where $d=-\log(1-p)$. If $h_k(s)\equiv s^k$ then for each p and k the family

$$\{F_{p,k}(s,t) = h_k^{-1}[F_p(h_k(s),t)];\ t \geqslant 0\}$$

is a semigroup of p.g.f.'s in which $g_{p,k}$ can be embedded. Karlin and McGregor also give a characterization and an extremal property of the class G.

A corollary to their characterization is the fact that in the class of p.g.f.'s

$$g_p(s) \equiv s(1-p)^\alpha(1-ps)^{-\alpha}, \quad \text{where } \alpha>0, \quad 0<p<1,$$

only those with $\alpha=1$ are embeddable. These matters are somewhat too involved to go into here further.

A simple test which excludes certain p.g.f.'s from being embeddable is to check the sign of a quantity $e(g)$ given by $e(g)=3(g''(q))^2-2g'(q)g'''(q)$.

Theorem 2. *Let g be a p.g.f. and q the smallest positive root of the equation $g(s)=s$. Assume that $g^{(iv)}(q)<\infty$ (of course, this is always satisfied when $m>1$, since then $q<1$). Then the following hold:*

a) *If $e(g)>0$ then $g(s)$ is not embeddable.*

b) *If $e(g)=0$, $q>0$, and g is embeddable, then $g(s)$ is a linear fractional function.*

We shall prove this theorem after stating two results which are useful in checking whether $e(g)$ is $\geqslant 0$.

Theorem 3. *If $g(s)\equiv \sum_{k=0}^{\infty} p_k s^k$ is a p.g.f. such that*

$$p_k^2 - p_{k-1}p_{k+1} \geqslant 0 \quad \text{for } k=1,2,\ldots, \tag{3}$$

then

$$\left(\frac{g^{(r)}(s)}{r!}\right)^2 - \left(\frac{g^{(r-1)}(s)}{(r-1)!}\right)\left(\frac{g^{(r+1)}(s)}{(r+1)!}\right) \geqslant 0 \tag{4}$$

for $0\leqslant s<1$ and $r=1,2,\ldots$ (Note that when $r=2$, (4) says that $e(g)\geqslant 0$).

Theorem 4. *If g_1 and g_2 are p.g.f.'s for which* (3) *holds then the same is true of $g_1 g_2$.*

It is easily verified that (3) holds for the following g's:

$g(s)=e^{\lambda(s-1)}$, $\lambda \geqslant 0$ (Poisson);

$g(s)=(1-\beta)^{\alpha}(1-\beta s)^{-\alpha}$, $\alpha>1, 0<\beta<1$ (negative binomial);

and

$g(s)=(q+ps)^r$, $0<p<1$, $p+q=1$,

r a positive integer (binomial).

Using theorem 4 and limiting arguments, we see that (3) holds for all p.g.f.'s in

$$\mathsf{H} \equiv \left\{ h(s); h(s)=e^{\lambda(s-1)} \prod_{i=1}^{\infty} [(1-\beta_i)(1-\beta_i s)^{-1}]^{\alpha_i}(q_i+p_i s)^{r_i} \right\}, \tag{5}$$

where

$$0 \leqslant s \leqslant 1, \quad 0 \leqslant p_i < 1, \quad q_i+p_i=1, \; 0 \leqslant \beta_i < 1,$$

$$\alpha_i \geqslant 1, \text{ and } 0 \leqslant r_i = \text{an integer}, \; i=1,2,\ldots, \text{ with}$$

$$\sum_{i=1}^{\infty} (\alpha_i \beta_i + p_i r_i) < \infty.$$

(The latter condition is needed to make sure that the infinite product in the definition of $h(s)$ converges). Noting that $h(0)>0$ for every $h \in \mathsf{H}$, we see from part b of theorem 2 that the only element of H that is embeddable is the linear fractional function. The class H includes sums of independent Poisson, binomial, negative binomial variables, and their limits.

Sketch of proof of theorem 2

Case $m \neq 1$: If g is embeddable in $\{F(s,t); t \geqslant 0\}$ with infinitesimal generating function $u(s)$, then it can be shown (see complements) that

$$e^{-\beta} = F'(q,1) = g'(q) \equiv \gamma,$$

and (8.4) implies

$$A(g(s)) = \gamma A(s). \tag{6}$$

The hypothesis that $g^{(iv)}(q) < \infty$ implies that $A^{(iv)}(q) < \infty$. (This is not trivial to prove as can be seen by referring to Karlin-McGregor.) Successive differentiation of (6), and evaluation at q yields the formulas

$$A''(q) = (\gamma(1-\gamma))^{-1} g''(q), \tag{7}$$

and

$$A'''(q) = [\gamma(1+\gamma)(1-\gamma)^2]^{-1} [3(g''(q))^2 + (1-\gamma) g'''(q)],$$

which imply the identity

$$e(g)=\gamma^2(1-\gamma^2)\left[3(A''(q))^2-2A'''(q)A'(q)\right]. \tag{8}$$

If g is embeddable then (8.5) holds, and therefore using (8) we get

$$u'''(q)=-3\beta(\gamma^2(1-\gamma^2))^{-1}e(g). \tag{9}$$

But this is impossible when $e(g)>0$, since $u'''(q)\geqslant 0$, $\beta>0$ and $0<\gamma<1$. This proves part (a).

To prove (b) suppose $e(g)=0$, and hence conclude $u'''(q)=0$. But if $q>0$ this is possible only when $f(s)$ is a quadratic (recall $u(s)=a[f(s)-s]$), i.e. $f(s)=a_0+a_1s+a_2s^2$. Since we have shown (section 9) that it is no loss of generality to take $a_1=0$, we have $f(s)=a_0+(1-a_0)s^2$, $0<a_0<1$. This implies that $g(s)=F(s,1)$ is linear fractional.

Case $m=1$: Let $g(s)$ be embeddable in a semigroup $F(s,t)$ with infinitesimal generating function $u(s)$. Since $m=1$, $u(s)>0$ for $0\leqslant s<1$, and

$$A(s)\equiv\int_0^s\frac{1}{u(\xi)}\,d\xi,\qquad 0\leqslant s<1, \tag{10}$$

is well defined. From (3.7) it then follows that

$$A(F(s,t))=A(s)+t. \tag{11}$$

Under the hypothesis $g^{(iv)}(1)<\infty$ it has been shown by Karlin and McGregor that $A(s)$ admits the representation

$$A(s)=(1-s)^{-1}+e(g)a^{-1}\log(1-s)+\psi(s)+d \tag{12}$$

where $a=g''(1)/2$, $e(g)$ is as before, d is a constant, and $\psi''(s)$ is bounded on $0\leqslant s\leqslant 1$.

Differentiating (12) we obtain

$$(u(s))^{-1}=(1-s)^2-e(g)a^{-1}(1-s)^{-1}+\psi'(s),\qquad 0<s<1. \tag{13}$$

Since ψ' is bounded (13) implies

$$u'''(1-)=-2a^{-1}e(g).$$

Again, as before, if $u(s)$ is an infinitesimal p.g.f., then $u'''(1-)\geqslant 0$ with equality holding only if $g(s)$ is a linear fractional p.g.f. □

Remark. Note that the expression for the Laplace transform of W in theorem 8.3 is also valid for embeddable Galton-Watson processes.

Complements and Problems III

1. Some open problems when $m>1$ and $EZ_1 \log Z_1 = \infty$:

(a) Determine a function $\eta(t)$ such that $Z(t)/\eta(t)$ converges a.s. to a non-degenerate random variable.

(b) Determine the behavior of $A(s)$ (as defined in (8.3)) near 1. Recall that the fact that $(1-s)^{\alpha} A(s) \to \text{constant}$ when $EZ_1 \log Z_1 < \infty$ was crucially needed in the proof of theorem 8.3.

(c) Determine an analog of theorem 8.3, for the case $EZ_1 \log Z_1 = \infty$. Observe that in this case the integral in the previously derived expression for $\varphi^{-1}(x)$ diverges.

2. Show that when $f(s)=s^2$, the limit random variable W has an exponential distribution, and conversely. What about $f(s)=s^k$? Show also that in general, if $m>1$ and the parameter a are specified, then $f(s)$ is determined by the distribution of W.

3. Prove theorem 8.1 using the suggestion given. This result is in fact subsumed under a more general one in the next chapter. Also prove theorem 8.2.

4. *Extensions of 2nd order properties:*

(a) Discuss the analogs of theorems 10.1 and 10.2 when $EZ_1 \log Z_1 = \infty$.

(b) Determine the analogs of theorems 10.3 and 10.4 when $\{p_j\}$ is in the domain of attraction of a stable law.

5. Referring to the proof of theorem 12.2, show that

$$e^{-\beta} = F'(q,1) = g'(q) \equiv \gamma .$$

Hint: Let $a(t)=F'(q,t)$. Then by the backward equation $(d/dt)a(t)=u'(q)a(t)$.

Chapter IV

Age-Dependent Processes

1. Introduction

When particles have general (not necessarily exponential) lifetime distribution, then the process $\{Z(t); t \geqslant 0\}$, $Z(t)$ = the number of particles existing at t, is called an *age-dependent* or general time branching process. It is not Markovian, and the methods of the previous chapters do not apply. Such models were introduced and first studied by R. Bellman and T. Harris (1952).

The data of the process now consist of the same generating function $f(s) = \sum p_k s^k$ governing particle production; plus a distribution $G(\cdot)$ on $[0, \infty)$ describing particle life length. The process is initiated at time $t=0$ with a parent particle which lives for time T_0, and then produces k offspring with probability p_k. These live for times $T_{11}, \ldots, T_{1k}$, and then produce offspring according to $\{p_k\}$, etc. The T's are independent random variables with distribution $G(\cdot)$; particle production is independent of the present state or past history of the process; and the lifetime and particle production variables are independent.

This sketch of the process serves as the motivation for the rigorous construction of a probability space $(\Omega, \mathbb{F}, P)$ on which the process lives. The points ω of Ω are again to be thought of as "trees" whose branch lengths represent particle life lengths; the number of branches at a given vertex representing the number of offspring of a given particle. The σ-algebra $\mathbb{F}$ is to be taken large enough so that $Z(t, \omega)$, $t \geqslant 0$, (and other similar random variables which we will study) are measurable functions on $(\Omega, \mathbb{F})$. It is not difficult to imagine how the distributions $\{p_k\}$ and $G(t)$, and the previously described independence assumptions can be used to construct an appropriate probability measure P on $(\Omega, \mathbb{F})$; but carrying this out carefully is a somewhat lengthy task. As already observed in chapters I and III, such a construction has been carried out by T. Harris in the first half of chapter VI of his book (1963), and we will not repeat it here. Even though most of our work will consist of a purely analytical study of the generating function of $Z(t)$, we will also have occasion to talk about such matters as almost sure convergence, equivalence of various statements about the process, etc. In these instances we will not hesitate to refer to the space $(\Omega, \mathbb{F}, P)$.

As in the first three chapters, careful study of the generating function of the process will continue to be a key to our analysis. Whereas functional iteration was the main tool for the Galton-Watson process, and the Kolmogorov equations were heavily used in the continuous time Markov case, the present work will center around an integral equation satisfied by the generating function of $Z(t)$. This equation turns out to be a continuous time parameter analog of functional iteration. Wherever possible, we will reduce questions to analogs previously answered for the Galton-Watson process—but we will here be much less successful in achieving such reductions than we were for the continuous time Markov process.

We start with a heuristic derivation of the basic integral equation. A decomposition of the sample space Ω in accordance with the life-length and number of offspring of the initial particle suggests the relation:

$P\{Z(t)=k\}=\sum P$ {the initial particle dies at time y and produces j offspring; and in the remaining time $t-y$, these j particles give rise to a total of k offspring}, the sum being taken over $0\leqslant y\leqslant t$, $0\leqslant j$. Thus

$$P\{Z(t)=k\}=[1-G(t)]\,\delta_{1k}+\int_0^t dG(y)\sum_{j=0}^{\infty} p_j P^{*j}\{Z(t-y)=k\},$$

where $P^{*j}=$ the j-fold convolution of P, and δ_{ij} is the Kronecker delta. The first term on the right side takes care of the case when the parent lives longer than time t. Multiplying through by s^k, summing over k, and letting

$$F(s,t)=\sum_{k=0}^{\infty} P\{Z(t)=k\}\,s^k,$$

we get

$$F(s,t)=s[1-G(t)]+\int_0^t f[F(s,t-y)]\,dG(y), \qquad |s|\leqslant 1. \tag{1}$$

This is the basic equation.

Note that when G is the unit step function

$$G(t)=\begin{cases}0 & \text{for } t<1,\\ 1 & \text{for } t\geqslant 1,\end{cases}$$

then (1) reduces to a functional iteration formula for $f(s)$; while if

$$G(t)=\begin{cases}0 & \text{for } t<0,\\ 1-e^{-\lambda t} & \text{for } t\geqslant 0,\end{cases}$$

then (1), on differentiation, yields the Kolmogorov equation of chapter III. That the latter process is then also Markovian requires some further argument, which is given in section 11, chapter VI of Harris (1963).

The above derivation is of course purely suggestive, and tells us nothing about the properties of F, or even about the existence or uniqueness of solutions. By careful reference to the space $(\Omega, \mathbb{F}, P)$ one can make the above argument rigorous, thus assuring the existence of a solution—in fact of a solution which is a probability generating function. This is what is done by Harris. An alternative approach is to start with (1) as given, prove (purely analytically) that it has a unique solution, and that this solution is a generating function; and then go on to study its finer properties. This is the approach we will take here.

We assume throughout that $G(0+)=0$; i.e. that there is 0 probability of instantaneous death. (This is in fact no loss of generality, since the case $G(0+)>0$ can be reduced to $G(0+)=0$ by suitably modifying f.) Together with the assumption of finite mean particle production, which we also make throughout, this guarantees the a.s. finiteness of the process. Namely we have:

Theorem 1. *If* $G(0+)=0$ *and* $m<\infty$ *then* $P\{Z(t)<\infty\}=1$ *for each* $t\geqslant 0$.

Proof. See T. Harris (1963), pp. 138–139.

For a weaker sufficient condition see Savits (1969).

2. Existence and Uniqueness

Theorem 1. *Let f be a probability generating function and G a distribution on $[0,\infty)$ with $G(0+)=0$. Then (1.1) has a solution $F(s,t)$, which is a generating function for each t, and which is the unique bounded solution.*

We shall say that F is the generating function of the process determined by (f, G).

Proof (N. Levinson (1960), T. Harris (1963)). Let $F_0(s,t)\equiv s$, and define

$$F_{n+1}(s,t)\equiv s[1-G(t)]+\int_0^t f[F_n(s,t-y)]\,dG(y). \tag{1}$$

Since $F_1(s,t)=s[1-G(t)]+f(s)G(t)$ and $|s|\leqslant 1, 0\leqslant f(s)\leqslant 1$, we see that

$$|F_1(s,t)|\leqslant 1 \quad \text{and} \quad |F_1(s,t)-F_0(s,t)|\leqslant 1.$$

If $|F_n(s,t)| \leqslant 1$ then also clearly $|F_{n+1}(s,t)| \leqslant 1$ and hence by induction,

$$|F_n(s,t)| \leqslant 1, \qquad n \geqslant 0. \tag{2}$$

We claim further that

$$|F_{n+1}(s,t) - F_n(s,t)| \leqslant m^n G^{*n}(t), \qquad n \geqslant 0, \ |s| \leqslant 1, \tag{3}$$

where $G^{*n}(t)$ is the n-fold convolution of G with itself. We already have this fact for $n=0$. If it is true for n, then, by the mean value theorem,

$$\begin{aligned}|F_{n+2}(s,t) - F_{n+1}(s,t)| &\leqslant m \int_0^t |F_{n+1}(s,t-y) - F_n(s,t-y)| \, dG(y) \\ &\leqslant m^{n+1} G^{*(n+1)}(t),\end{aligned}$$

and (3) holds by induction. Consequently

$$|F_{n+k}(s,t) - F_n(s,t)| \leqslant \sum_{i=n}^{\infty} m^i G^{*i}(t), \qquad k=0,1,2,\ldots.$$

We shall show in lemma 1 of section 4 that $\sum_{i=0}^{\infty} m^i G^{*i}(t) < \infty$. Thus F_n is a Cauchy sequence and converges to a function F. Due to the monotonicity of $G^{*n}(t)$ the series $U(t) \equiv \sum_{n=1}^{\infty} m^n G^{*n}(t)$ converges uniformly for $t \leqslant t_0 < \infty$, and thus we have

$$\lim_{n\to\infty} F_n(s,t) = F(s,t) \quad \text{uniformly for } t \leqslant t_0, \ |s| \leqslant 1. \tag{4}$$

Due to the uniformity we may take limits through the integral in (1) to conclude that $F(s,t)$ is a solution of (1.1), and that it must be a probability generating function.

To prove uniqueness, suppose $\tilde{F}(s,t)$ is another bounded solution. Then

$$|F(s,t) - \tilde{F}(s,t)| \leqslant m \int_0^t |F(s,t-y) - \tilde{F}(s,t-y)| \, dG(y).$$

Letting the left side equal $V(t)$, we have

$$0 \leqslant V(t) \leqslant m \int_0^t V(t-y) \, dG(y). \tag{5}$$

It is easy to show that $m^n G^{*n}(t) \to 0$ for all $t>0$. (For $m \leqslant 1$ this is trivial; for $m>1$ see lemma 1 of section 4.) Hence iterating (5), we conclude that $V(t) \equiv 0$. □

3. Comparison with Galton-Watson Process; Embedded Generation Process; Extinction Probability

The process $\{Z(\delta n, \omega);\ n=0,1,2,\ldots\}$, $0<\delta<\infty$, will in general *not* be a branching process, this being so only in the Markov case of chapter III. However, there is another embedded branching process defined by

$\zeta_n(\omega)=$ *the number of particles in the nth generation of the* $Z(t,\omega)$ *process.*

Thus if we take the sample tree ω and transform it into a tree ω' having all its branches of unit length but otherwise identical to ω, then $\zeta_n(\omega)=Z_n(\omega')$, where Z_n is the usual Galton-Watson process. Let

$$f_n(s)=E(s^{\zeta_n}\,|\,\zeta_0=1).$$

Then f_n is the generating function of the number of particles in the nth generation of a Galton-Watson process, while $f_1(s)=f(s)$ is the same generating function as that governing particle production in the age-dependent process $Z(t)$. We will call $\zeta_n(\omega)$ the *embedded generation process.*

The following inequalities are basic in comparing the age-dependent and embedded generation processes. They are due to M. Goldstein (1971).

Lemma 1. *Let $F(s,t)$ be the generating function of the process determined by (f,G). Then the inequalities*

$$\begin{aligned} |q-f_n(s)|-|q-s|\,\theta^n G^{*n}(t) &\leqslant |q-F(s,t)| \qquad (1)\\ &\leqslant |q-f_n(s)|+|q-s|\sum_{k=0}^{n-1}\theta^k\left[G^{*k}(t)-G^{*(k+1)}(t)\right] \end{aligned}$$

are satisfied with $\theta=\gamma\equiv f'(q)$ when $0\leqslant s\leqslant q$, and with $\theta=m\equiv f'(1)$ when $q\leqslant s<1$.

Proof. (i) Suppose $0\leqslant s\leqslant q$; and let the iterates $F_n(s,t)$ be as defined in Section 2, with $F_0(s,t)\equiv s$. By induction it is clear from (2.1) that $F_n(s,t)$ is nondecreasing in n, and that $s\leqslant F_n\leqslant q$. Thus the proof of Theorem 2.1 tells us that

$$s\leqslant F_n(s,t)\uparrow F(s,t)\leqslant q, \qquad n\to\infty,\ 0\leqslant s\leqslant q. \qquad (2)$$

We claim first that

$$0\leqslant F(s,t)-F_n(s,t)\leqslant (q-s)\gamma^n G^{*n}(t). \qquad (3)$$

The proof of this fact is by induction. For $n=1$,

$$0 \leqslant F(s,t)-F_1(s,t)=\int_0^t \{f[F(s,t-y)]-f[F_0(s,t-y)]\}\,dG(y)$$

$$\leqslant \gamma \int_0^t \{F(s,t-y)-s\}\,dG(y) \leqslant \gamma(q-s)\,G(t),$$

where the first inequality follows from the mean value theorem, and the second from (2). Suppose (3) is true for $n=k$. Then applying the mean value theorem, and the induction hypothesis

$$0 \leqslant F(s,t)-F_{k+1}(s,t) \leqslant \gamma \int_0^t \{F(s,t-y)-F_k(s,t-y)\}\,dG(y)$$

$$\leqslant \gamma^{k+1}(q-s)\,G^{*(k+1)}(t).$$

Secondly we show that

$$0 \leqslant f_n(s)-F_n(s,t) \leqslant (q-s)\sum_{k=0}^{n-1} \gamma^k [G^{*k}(t)-G^{*(k+1)}(t)]. \tag{4}$$

Again proceeding by induction we note that

$$0 \leqslant f(s)-F_1(s,t)=[f(s)-s]\,[1-G(t)] \leqslant (q-s)\,[1-G(t)];$$

while if (4) is true for $n=k$, then

$$0 \leqslant f_{k+1}(s)-F_{k+1}(s,t)$$

$$=[f_{k+1}(s)-s]\,[1-G(t)]+\int_0^t \{f[f_k(s)]-f[F_k(s,t-y)]\}\,dG(y)$$

$$\leqslant (q-s)\,[1-G(t)]+\gamma\int_0^t \{f_k(s)-F_k(s,t-y)\}\,dG(y)$$

$$\leqslant (q-s)\,[1-G(t)]+\gamma(q-s)\int_0^t \sum_{j=0}^{k-1} \gamma^j [G^{*j}(t-y)-G^{*(j+1)}(t-y)]\,dG(y),$$

which implies (4).

Now combining (3), (4) and the fact that $F_n \leqslant F$, implies lemma 1 for $s \leqslant q$.

(ii) If $q \leqslant s < 1$ then (3) is to be replaced by

$$0 \leqslant F_n(s,t)-F(s,t) \leqslant (s-q)\,m^n\,G^{*n}(t), \tag{5}$$

and (4) by

$$0 \leqslant F_n(s,t)-f_n(s) \leqslant (s-q)\sum_{k=0}^{n-1} m^k [G^{*k}(t)-G^{*(k+1)}(t)]. \tag{6}$$

The proofs are exactly as before and the lemma follows. □

The extinction probability of $\{Z(t)\}$ can now be easily obtained.

Theorem 1. *If* $0 \leqslant s \leqslant q$ *then* $F(s,t)\uparrow q$ *as* $t\uparrow\infty$. *If* $q\leqslant s<1$ *then* $F(s,t)\downarrow q$ *as* $t\uparrow\infty$. *In particular*

$$\lim_{t\to\infty} P\{Z(t)=0\} = q. \tag{7}$$

Proof. $F_0(s,t)\equiv s$.

Induction on n in (2.1) shows at once that

$$F_n(s,t)\begin{cases} \geqslant s \text{ for } 0\leqslant s\leqslant q, \\ \leqslant s \text{ for } q\leqslant s\leqslant 1, \qquad n\geqslant 0. \end{cases} \tag{8}$$

If $0\leqslant s\leqslant q$, and $F_k(s,t)$ is nondecreasing in t for some $k\geqslant 0$, then by (2.1) and (8) we have for $t_2>t_1$

$$\begin{aligned} F_{k+1}(s,t_2)-F_{k+1}(s,t_1) &= \int_0^{t_1} \{f[F_k(s,t_2-y)]-f[F_k(s,t_1-y)]\}\,dG(y) \\ &\quad + \int_{t_1}^{t_2} \{f[F_k(s,t_2-y)]-s\}\,dG(y) \geqslant 0. \end{aligned}$$

Hence by induction $F_k(s,t)$ is nondecreasing in t for all k, and therefore so is $F(s,t)$. Similarly, if $q\leqslant s<1$, then $F_k(s,t)$, $k\geqslant 0$, and $F(s,t)$ are nonincreasing in t. Taking n large and then t large in (1), we get the theorem. □

Let us explore further the connection between the embedded generation process $\{\zeta_n(\omega)\}$ and $\{Z(t,\omega)\}$. For $j=1,2,\ldots,\zeta_k(\omega)$, define

$$\delta_{k,j}(t,\omega)=\begin{cases} 1 & \text{if the } j\text{th particle of the } k\text{th generation is alive at } t; \\ 0 & \text{otherwise.} \end{cases} \tag{9}$$

Let

$$\begin{aligned} Y_k(t,\omega) &\equiv \sum_{j=1}^{\zeta_k(\omega)} \delta_{k,j}(t,\omega) \qquad (10) \\ &= \text{the number of particles of the } k\text{th generation alive at time } t. \end{aligned}$$

Then we can write

$$Z(t,\omega) = \sum_{k=0}^{\infty} Y_k(t,\omega). \tag{11}$$

This obvious decomposition will be useful later (section 11 on the supercritical case). As an immediate illustration of how it can be applied, we obtain an expression for $\mu(t)=$ the mean number of particles at time t; namely

$$\mu(t) = \sum_{k=0}^{\infty} m^k [G^{*k}(t)-G^{*(k+1)}(t)]. \tag{12}$$

We need only note that

$$E(Y_k(t)|\zeta_k)=\zeta_k[G^{*k}(t)-G^{*(k+1)}(t)],$$

and that

$$E\zeta_k=m^k.$$

To see this observe that by (10) and symmetry among the particles

$$E(Y_k(t)|\zeta_k)=\zeta_k E\,\delta_{k,1}(t)=\zeta_k P(\delta_{k,1}(t)=1).$$

But a particle belonging to the kth generation is present at time t if and only if the sum of the lifetimes of its ancestors is $\leqslant t$, while the total of this ancestral lifetime and its own is $>t$. Thus $P(\delta_{k,1}(t)=1) = G^{*k}(t)-G^{*(k+1)}(t)$, proving (12).

We shall obtain (12) in section 5 as the solution of a renewal equation.

As another application one could use (11) to show that the ω-sets $\{\omega:\zeta_n(\omega)=0$ for some $n\}$, $\{\omega: Z(t,\omega)=0$ for some $t\}$ coincide with probability 1. This gives an immediate proof of the fact that the extinction probability for $Z(t)$ is the same as that for the embedded generation process.

4. Renewal Theory

The relevance of both linear and nonlinear renewal theory to the present work will become apparent throughout this chapter. We give here a brief resume of some results to which we shall want to refer later. Define the renewal function

$$U_\gamma(t)=\sum_{n=0}^{\infty}\gamma^n G^{*n}(t), \qquad t\geqslant 0, \tag{1}$$

where G is a distribution on $[0,\infty)$; G^{*n} is its n-fold convolution, $G^{*0}(t)=1$ for $t\geqslant 0$, and $0<\gamma<\infty$ is a constant.

Lemma 1. *If $\gamma G(0+)<1$, then $U_\gamma(t)<\infty$ for all $t<\infty$, and is bounded on finite t-intervals.*

Proof. Let $\{Y_i; i=1,2,\dots\}$ be independent random variables with distribution $G(t)$, and let $S_n=Y_1+\cdots+Y_n$, $n\geqslant 1$. Then

$$\begin{aligned}\gamma^n G^{*n}(t)&=\gamma^n P\{S_n\leqslant t\}=\gamma^n P\{e^{-\theta S_n}\geqslant e^{-\theta t}\} \quad \text{for } \theta\geqslant 0,\\ &\leqslant \gamma^n e^{\theta t}E(e^{-\theta S_n})=e^{\theta t}[\gamma E e^{-\theta Y_1}]^n.\end{aligned}$$

But $\gamma G(0+)<1$ and hence there exists a $\theta>0$ such that $r\equiv\gamma E e^{-\theta Y_1}<1$. Thus,

$$U_\gamma(t)\leqslant e^{\theta t}(1-r)^{-1}, \quad \text{proving the lemma.} \quad \square$$

It is easily checked that $U_\gamma(t)$ satisfies the integral equation

$$U_\gamma(t) = 1 + \gamma \int_0^t U_\gamma(t-y)\,dG(y), \qquad t \geqslant 0. \tag{2}$$

More generally, consider the equation

$$H(t) = \xi(t) + \gamma \int_0^t H(t-y)\,dG(y), \qquad t \geqslant 0, \tag{3}$$

where $\xi(t)$ is a given bounded measurable function on $[0, \infty)$, and $H(t)$ is the unknown function. Clearly

$$H(t) = (\xi * U_\gamma)(t) \equiv \int_0^t \xi(t-y)\,dU_\gamma(y) \tag{4}$$

satisfies (3). Note that (4) makes sense since $U_\gamma(\cdot)$ is a nondecreasing bounded function. If $H_1(t)$ and $H_2(t)$ are two solutions of (3) that are bounded on finite t-intervals then $h(t) = |H_1(t) - H_2(t)|$ satisfies

$$h(t) \leqslant \gamma \int_0^t h(t-y)\,dG(y). \tag{5}$$

Iterating (5), we obtain

$$h \leqslant h * \gamma G \leqslant h * \gamma^n G^{*n} \quad \text{for any } n. \tag{6}$$

By Lemma 1 $\gamma^n G^{*n}(t) \to 0$ as $n \to \infty$. Thus (6) and the boundedness of h imply that $h(t) = 0$. Summarizing the above, we have

Lemma 2. *$H \equiv \xi * U_\gamma$ is the unique solution of* (3) *which is bounded on finite intervals.*

When $\gamma = 1$, (3) is the classical renewal equation, and the following result is part of the standard textbook literature (see Feller, Vol.II (1966)).

Theorem 1. *Assume $\gamma = 1$ and let $\mu = \int_0^\infty t\,dG(t) \leqslant \infty$*[7].

(i) *If $\xi_0 = \lim_{t\to\infty} \xi(t)$ exists, then the solution of* (3) *satisfies*

$$\lim_{t\to\infty} \frac{H(t)}{t} = \frac{\xi_0}{\mu}. \tag{7}$$

(ii) *If ξ is directly Reimann integrable and G is nonlattice*[8], *then*

$$\lim_{t\to\infty} H(t) = \frac{1}{\mu} \int_0^\infty \xi(y)\,dy. \tag{8}$$

[7] If $\mu = \infty$ then the limits in (7) and (8) are zero.

[8] There is an analogous result for the lattice case.

A function ξ is directly Riemann integrable (R-integrable) if $\sum h\overline{m}_n(h)$ and $\sum h\underline{m}_n(h)$ converge absolutely for sufficiently small $h>0$, where $\overline{m}_n(h)=\sup_{nh\leqslant t\leqslant(n+1)h}\xi(t)$, and $\underline{m}_n(h)$ is the corresponding infimum; and if

$$h(\sum \overline{m}_n(h)-\sum \underline{m}_n(h))\to 0 \quad \text{as } h\to 0 .$$

Some sufficient conditions for direct R-integrability of ξ are: (i) $\xi\geqslant 0$, bounded, continuous, and

$$\sum \overline{m}_n(1)<\infty ;$$

(ii) $\xi\geqslant 0$, non-increasing, and R-integrable in the ordinary sense;
(iii) ξ is bounded by a directly R-integrable function;
(iv) ξ is constant on the intervals $(n,n+1)$ and absolutely integrable.

It will also be important to know the behavior of solutions of (3) when $\gamma\neq 1$. The asymptotic properties of the solution are frequently characterized by a key parameter (to be introduced below). In these cases an equation with $\gamma\neq 1$ can be reduced to one with $\gamma=1$.

Definition. *The Malthusian parameter for γ and G is the root, provided it exists, of the equation*

$$\gamma\int_0^\infty e^{-\alpha y}\,dG(y)=1 . \tag{9}$$

We denote it by $\alpha=\alpha(\gamma,G)$.

Due to the monotonicity of the left side of (9), such a root, when it exists, is always unique.

We note that when $\gamma\geqslant 1$, such a Malthusian parameter always exists (in this case $\alpha\geqslant 0$). If $\gamma<1$, then $\alpha(\gamma,G)$ may not exist (if it does, $\alpha<0$).

When the Malthusian parameter exists, then we can multiply (3) thru by $e^{-\alpha t}$, and letting

$$H_\alpha(t)=e^{-\alpha t}H(t);\qquad dG_\alpha(t)=\gamma e^{-\alpha t}\,dG(t);\qquad \xi_\alpha(t)=e^{-\alpha t}\xi(t), \tag{10}$$

get

$$H_\alpha(t)=\xi_\alpha(t)+\int_0^t H_\alpha(t-y)\,dG_\alpha(y). \tag{11}$$

This is again an equation of the form (3), but with $\gamma=1$, and we can then apply results like theorem 1 to obtain the properties of $H(t)$. (Note that G_α is a distribution function). For example, we get

Theorem 2. *If the Malthusian parameter $\alpha(\gamma, G)$ exists, if $e^{-\alpha t}\xi(t)$ is directly Riemann integrable, and if G is nonlattice, then the solution H of (3) satisfies*

$$H(t) \sim e^{\alpha t}\left\{\int_0^\infty e^{-\alpha y}\xi(y)\,dy\right\}\left\{\gamma\int_0^\infty y e^{-\alpha y}\,dG(y)\right\}^{-1}. \tag{12}$$

Proof. Make the reduction indicated in (10) and (11) and apply theorem 1 (ii). □

When $0<\gamma<1$, and $\xi(t)=1-G(t)$, this result was observed by Vinogradov (1964) under a second moment condition on G_α. He also proved a converse. Namely, if $H(t) \sim c\,e^{\alpha t}$ with $\alpha<0$, $0<\gamma<1$, then (9) has a solution α. (Of course if $\gamma>1$, then (9) always has a solution. In this case (12) goes back to Bellman and Harris (1952)).

If $\alpha(\gamma, G)$ does not exist, then the above method fails. Trivially $U_\gamma(t) \uparrow (1-\gamma)^{-1}$, and if ξ is directly R-integrable then

$$H(t) = (\xi * U_\gamma)(t) \to 0 \quad \text{as } t \to \infty .$$

However, we will need to know the rates of convergence of $U_\gamma(t)$ to $(1-\gamma)^{-1}$, and of $H(t)$ to 0.

There is a large class of distributions for which these rates can be given.

Definition. *The "sub-exponential" class $\mathscr{S}$ consists of all distribution functions G such that*

$$\lim_{t\to\infty} \frac{1-G^{*2}(t)}{1-G(t)} = 2 . \tag{13}$$

This class was introduced by V.P. Chistyakov (1964), and later (independently) in a more general setting by Chover, Ney, Wainger (1969). The latter paper is mainly concerned with local limit theorems for functions of measures; the former (and the present applications) with the tail behavior of a particular function of G (see (5.4) below).

Sufficient conditions for $G \in \mathscr{S}$ can be found in the above references. Examples of such G's are.

$$1-G(t) \sim t^{-k}, \qquad k>0 ;$$

$$1-G(t) \sim \exp\{-t^\beta\}, \qquad 0<\beta<1 ;$$

$$1-G(t) \sim \exp\left\{-\frac{t}{\log^2 t}\right\}.$$

These distributions all have tails which decay at a slower than exponential rate, and this seems to be the best qualitive description of $\mathscr{S}$. It is for this reason that we have called $\mathscr{S}$ the "sub-exponential" class.

Remark. In Chover, Ney, Wainger (1969) there is also described a larger class than $\mathscr{S}$, some of whose distributions have tails which do not decay at a slower than exponential rate, but which have associated renewal functions $U_\gamma(t)$ with behavior similar to theorem 3 below. We will not discuss this class here, however.

The following results are contained in the Chistyakov or C, N, W papers sited above. We are also thankful to H. Kesten for some helpful discussion (lemma 7 below). We start with some lemmas which help to describe the class $\mathscr{S}$.

Lemma 3 (Chistyakov). *If $G \in \mathscr{S}$, then*

$$\lim_{t\to\infty} \frac{1-G(t-A)}{1-G(t)} = 1 \quad \textit{for all } 0 < A < \infty . \tag{14}$$

Proof.

$$\frac{1-G^{*2}(t)}{1-G(t)} = 1 + \int_0^A \frac{1-G(t-y)}{1-G(t)}\, dG(y) + \int_A^t \frac{1-G(t-y)}{1-G(t)}\, dG(y)$$

$$\geqslant 1 + G(A) + \frac{1-G(t-A)}{1-G(t)} [G(t) - G(A)] .$$

Thus

$$1 \leqslant \frac{1-G(t-A)}{1-G(t)} \leqslant \left\{ \frac{1-G^{*2}(t)}{1-G(t)} - 1 - G(A) \right\} [G(t) - G(A)]^{-1} .$$

Now if $G \in \mathscr{S}$ then the right side $\to 1$ as $t \to \infty$, implying the lemma. □

Lemma 4. *If $G \in \mathscr{S}$, then*

$$\lim_{t\to\infty} \frac{1-G^{*n}(t)}{1-G(t)} = n . \tag{15}$$

Proof. The proof is by induction. The case $n=1$ is trivial, and $n=2$ is true by hypothesis. Suppose it true for arbitrary n. Then

$$\frac{1-G^{*(n+1)}(t)}{1-G(t)} = 1 + \frac{G(t) - G^{*(n+1)}(t)}{1-G(t)} = 1 + \int_0^t \frac{1-G^{*n}(t-y)}{1-G(t)}\, dG(y)$$

$$= 1 + \int_0^{t-A} \left\{ \frac{1-G^{*n}(t-y)}{1-G(t-y)} \, \frac{1-G(t-y)}{1-G(t)} \right\} dG(y) + \int_{t-A}^t \{-\}\, dG(y)$$

$$= 1 + I_1(t) + I_2(t) . \tag{16}$$

But $|((1-G^{*n}(t-y))/(1-G(t-y)))-n|$ can be made arbitrarly small for $0 \leqslant y \leqslant t-A$ and A sufficiently large, while

$$\int_0^{t-A} \frac{1-G(t-y)}{1-G(t)}\, dG(y) = \int_0^t - \int_{t-A}^t \frac{1-G(t-y)}{1-G(t)}\, dG(y)$$

$$= \frac{G(t)-G^{*2}(t)}{1-G(t)} - \int_{t-A}^t \frac{1-G(t-y)}{1-G(t)}\, dG(y)\,.$$

The first term $\to 1$ as $t \to \infty$ and the second term, being bounded by $(G(t)-G(t-A))/(1-G(t))$, goes to 0 by lemma 3. Hence

$$I_1(t) \to n \quad \text{as } t \to \infty\,.$$

Furthermore $I_2(t) \to 0$ since $(1-G^{*n}(t))/(1-G(t))$ is bounded and $\int_{t-A}^{t} ((1-G(t-y))/(1-G(t)))\, dG(y) \to 0$. This proves that the left side of (16) converges to $(n+1)$, completing the induction. □

A sufficient condition for $G \in \mathscr{S}$ is the following.

Lemma 5 (Chistyakov). *If*

$$\lim_{t \to \infty} \frac{1-G(tx)}{1-G(t)} \quad \textit{exists for } 0 < x \leqslant 1\,,$$

and is continuous at 1, *then* $G \in \mathscr{S}$.

Proof. Left as exercise.

The "sub-exponential" character of $\mathscr{S}$ is illustrated by

Lemma 6. *If* (14) *holds, then*

$$\lim e^{st}[1-G(t)] = \infty \quad \textit{for all } s > 0\,. \tag{17}$$

Proof. Left as exercise. (Thus (17) holds for all $G \in \mathscr{S}$.)

The main tool needed for the proof of theorem 3 below is

Lemma 7. *If* $G \in \mathscr{S}$, *then given any* $\varepsilon > 0$ *there is a* $D < \infty$ *such that*

$$\frac{1-G^{*n}(t)}{1-G(t)} \leqslant D(1+\varepsilon)^n \quad \textit{for all } n \textit{ and } t\,. \tag{18}$$

Proof. Let

$$\alpha_n = \sup_{t \geqslant 0} \frac{1-G^{*n}(t)}{1-G(t)}\,, \tag{19}$$

and note that $1-G^{*(n+1)}=1-G+G*(1-G^{*n})$. Then for any $T<\infty$,

$$\alpha_{n+1} \leqslant 1 + \sup_{0\leqslant t\leqslant T} \int_0^t \frac{1-G^{*n}(t-y)}{1-G(t)} dG(y)$$

$$+ \sup_{t\geqslant T} \int_0^t \frac{1-G^{*n}(t-y)}{1-G(t-y)} \cdot \frac{1-G(t-y)}{1-G(t)} dG(y)$$

$$\leqslant 1 + A_T + \alpha_n \sup_{t\geqslant T} \frac{G(t)-G^{*2}(t)}{1-G(t)}, \tag{20}$$

where $A_T=[1-G(T)]^{-1}<\infty$. Now since $G\in\mathscr{S}$ we can, given any $\varepsilon>0$, choose T such that

$$\alpha_{n+1} \leqslant 1+A_T+\alpha_n(1+\varepsilon).$$

Hence

$$\alpha_n \leqslant (1+A_T)\varepsilon^{-1}(1+\varepsilon)^n. \tag{21}$$

implying (18). □

It is now easy to describe the tail behavior of $U_\gamma(t)$.

Theorem 3. *If* $G\in\mathscr{S}$, *and* $0<\gamma<1$, *then*

$$\lim_{t\to\infty} \frac{(1-\gamma)^{-1}-U_\gamma(t)}{1-G(t)} = \frac{\gamma}{(1-\gamma)^2}. \tag{22}$$

Proof.

$$\frac{(1-\gamma)^{-1}-U_\gamma(t)}{1-G(t)} = \sum_{n=0}^{\infty} \gamma^n \frac{1-G^{*n}(t)}{1-G(t)}. \tag{23}$$

Choose ε so that $\gamma(1+\varepsilon)<1$. By lemma 7 and the dominated convergence theorem, we can take the limit as $t\to\infty$ inside the sum in (23). Lemma 4 then implies (22). □

5. Moments

In order to derive an equation for

$$\mu(t) = \left.\frac{\partial F(s,t)}{\partial s}\right|_{s=1} \equiv EZ(t),$$

we need merely justify differentiating under the integral in the basic equation (1.1). For $0\leqslant s<1$ the differentiation is clearly permissible and yields

$$\frac{\partial F(s,t)}{\partial s} = 1 - G(t) + \int_0^t f'[F(s,t-y)] \frac{\partial F(s,t-y)}{\partial s} dG(y)$$
$$\leqslant 1 - G(t) + m \int_0^t \frac{\partial F(s,t-y)}{\partial s} dG(y), \tag{1}$$

where $m = f'(1)$.

Iterating the inequality we obtain

$$\frac{\partial F(s,t)}{\partial s} \leqslant [1 - G(t)] * U_m(t), \tag{2}$$

where U_m is given in (4.1). Letting $s \uparrow 1$ through real values, $\partial F(s,t)/\partial s$ converges to $\mu(t)$, which by lemma 4.1 and (2) must be bounded on finite t-intervals. Hence we may take the limit $s \uparrow 1$ through the integral on the right side of (1) and get

Theorem 1. *$EZ(t) \equiv \mu(t)$ is the unique solution of*

$$\mu(t) = [1 - G(t)] + m \int_0^t \mu(t-y) dG(y) \tag{3}$$

which is bounded on finite t-intervals.

(The uniqueness follows from lemma 4.2.)

From lemma 4.2 it also follows that

$$\mu(t) = [1 - G(t)] * U_m(t) = \sum_{k=0}^{\infty} m^k [G^{*k}(t) - G^{*(k+1)}(t)] \tag{4}$$

as we have seen in (3.12). Similar calculations can be used to derive higher moments, such as $E[Z(t)]^2$, $E[Z(t_1)Z(t_2)]$, $E[Z(t)]^3$, etc. The procedure is to write down an equation similar to (1.1), e.g. for $F(s_1, s_2; t_1, t_2) = E\{s_1^{Z(t_1)} s_2^{Z(t_2)}\}$, prove an existence and uniqueness theorem analogous to theorem 2.1, and then pass to an equation for the moments as above. This leads e.g. to

Theorem 2. *$EZ(t_1)Z(t_2) \equiv \mu_2(t_1, t_2)$ is the unique solution (bounded on finite intervals) of*

$$\mu_2(t_1, t_2) = r(t_1, t_2) + m \int_0^{t_1} \mu_2(t_1 - y, t_2 - y) dG(y), \tag{5}$$

where

$$r(t_1, t_2) = 1 - G(t_2) + m_2 \int_0^{t_1} \mu(t_1 - y) \mu(t_2 - y) dG(y) + m \int_{t_1}^{t_2} \mu(t_2 - y) dG(y),$$

and

$$m_2 = E(Z_1)(Z_1 - 1).$$

If we fix $t_2-t_1=\Delta$, set $t_1=t$ and write $\mu_2(t,t+\Delta)=\mu_2(t|\Delta)$, $r(t,t+\Delta)=r(t|\Delta)$, then we see that $\mu_2(t|\Delta)$ satisfies an ordinary renewal equation in the variable t, and hence

$$\mu_2(t|\Delta)=r(t|\Delta)*U_m(t). \tag{5'}$$

Using the renewal theorems of section 4, we can easily describe the asymptotic behavior of the moments. For example theorems 1 and 4.2, together with the observation that $e^{-\alpha t}[1-G(t)]$ is directly R-integrable (see complements) yield

Theorem 3A[9]. *If* $m=1$ *then* $\mu(t)\equiv 1$. *If* $m>1$ *and G is non-lattice*[10], *then*

$$\mu(t)\sim c' e^{\alpha t}, \qquad t\to\infty, \tag{6}$$

where $\alpha>0$ *is the Malthusian parameter for* (m,G) *and*

$$c'=\frac{\int_0^\infty e^{-\alpha y}[1-G(y)]\,dy}{m\int_0^\infty y e^{-\alpha y}\,dG(y)}=\frac{m-1}{\alpha m^2\int_0^\infty y e^{-\alpha y}\,dG(y)}.$$

If $m<1$, then we are in the setting of theorem 4.2 or 4.3 and we obtain

Theorem 3B. (i) *If* $m<1$, *if the Malthusian parameter* $\alpha(m,G)$ *exists, and if* $\int_0^\infty y e^{-\alpha y}\,dG(y)<\infty$, *then* (6) *holds (with* $\alpha<0$).

(ii)[11] *If* $m<1$, *and G is in the sub-exponential class* $\mathscr{S}$, *then*

$$\mu(t)\sim\frac{1-G(t)}{1-m}.$$

(We leave these calculations to the complements.)

Similarly we can describe the behavior of the higher moments. The equation for the nth moment will involve the moments of order $1,\ldots,n-1$ in a nonlinear manner, but will be linear in the nth moment. For example, if $m>1$ we get from (5)

$$\begin{aligned} e^{-2\alpha t}\mu_2(t,t+\Delta)&=e^{-2\alpha t}r(t,t+\Delta)\\ &\quad+m\int_0^t e^{-2\alpha(t-y)}\mu_2(t-y,t+\Delta-y)\,e^{-2\alpha y}\,dG(y). \end{aligned} \tag{6'}$$

---

[9] Bellman and Harris (1952), Harris (1963).

[10] There is again an obvious lattice analog.

[11] Chistyakov (1964) for m sufficiently small, and Chover, Ney, Wainger (1969), (1971).

But

$$\lim_{t\to\infty} e^{-2\alpha t} r(t,t+\Delta) = m_2 \left(\frac{\int_0^\infty e^{-\alpha y}[1-G(y)]\,dy}{m\int_0^\infty y e^{-\alpha y} dG(y)} \right)^2 e^{\alpha\Delta} \int_0^\infty e^{-2\alpha y} dG(y)$$

$$= m_2 c'^2 e^{\alpha\Delta} \int_0^\infty e^{-2\alpha y} dG(y) = c'' \quad \text{(say)},$$

and hence (see also complement 24)

$$\lim_{t\to\infty} e^{-2\alpha t} \mu_2(t,t+\Delta) = \frac{c''}{1-m\int_0^\infty e^{-2\alpha y} dG(y)}. \tag{7}$$

The above method of differentiating the basic equation (1.1) is applicable only to integral moments. Although we can *not* obtain an expression either explicitly or in terms of an integral equation like (3) for $E(Z(t))^\alpha$, when $\alpha > 1$, is not an integer, we can obtain necessary and sufficient conditions for its finiteness. In fact it is possible to do this not just for $(Z(t))^\alpha$, but for a wider class of convex functions $\phi(Z(t))$.

Theorem 4 (Athreya (1968)). *Let $\phi(x)$ be a measurable function from $R^+ = [0,\infty)$ to R^+ having the property that there exist constants $c \geqslant 0$ and $K > 0$ such that*

(i) *$\phi(x)$ is convex on $[c,\infty)$,*

(ii) *$\phi(xy) \leqslant K\phi(x)\phi(y)$ for all x, y in $[c,\infty)$,*

(iii) *$\phi(x)$ is bounded in $[0,c)$.*

Let $t_0 = \inf\{x: G(x+) > 0\}$. Then

$$E\{\phi[Z(t)]\} < \infty \quad \text{for any } t \geqslant t_0 \tag{8}$$

if and only if

$$\sum p_j \phi(j) < \infty. \tag{9}$$

As our most important application of this theorem we note that $\phi(x) = x^\alpha |\log x|^\beta$, $\alpha \geqslant 1$, $\beta \geqslant 0$, satisfies the hypotheses (i), (ii), (iii), since $\phi''(x) \geqslant 0$ and $(\log xy)^\beta (\log x \log y)^{-\beta} \to 0$ as $x, y \to \infty$. Hence we have

Corollary 1. *For any constants $\alpha \geqslant 1$, $\beta \geqslant 0$*

$$E\{Z^\alpha(t)|\log Z(t)|^\beta\} < \infty \quad \text{for any } t \geqslant t_0$$

if and only if

$$\sum p_j j^\alpha |\log j|^\beta < \infty.$$

Remark. Note that theorem III. 6.2, whose proof was postponed, is a special case of theorem 4.

The proof of Theorem 4 will take up the rest of this section. It will be broken into several lemmas.

Lemma 1. *It is no loss of generality to take* $c=0$ *in the hypothesis of theorem 4, and to assume that* $\phi(x)$ *is nondecreasing and* >1.

Proof. Let ϕ be a function sytisfying the original hypothesis. It is sufficient to show that there exists a function $\tilde{\phi}$ satisfying the *additional* hypothesis of the present lemma, and such that, for any probability measure μ on R^+

$$\int_0^\infty \phi(x)\,\mu(dx)<\infty \quad \text{if and only if} \quad \int_0^\infty \tilde{\phi}(x)\,\mu(dx)<\infty. \tag{10}$$

We need only consider unbounded ϕ. Then since ϕ is convex on $[c,\infty)$, bounded on $[0,c)$, and unbounded above, there exists a $c'\geqslant c$ such that

a) $$\phi(c')\geqslant \sup_{x\leqslant c'}\phi(x)\geqslant 1,$$

and

b) $$\phi \text{ is increasing on } [c',\infty).$$

Define

$$\tilde{\phi}(x)=\begin{cases}\phi(c') & \text{for } x\leqslant c',\\ \phi(x) & \text{for } x>c'.\end{cases} \tag{11}$$

Direct verification shows that this is the desired function, and proves lemma 1. □

As a tool in the proof of the theorem, we need to extend to age-dependent processes the notion of split times introduced in chapter III for the Markov case. These are defined to be the sequence of times of death of the particles. By Theorem 1.1 we know that the number of such split times in any finite interval is finite with probability one, and hence these times can be ordered. (If it is possible for more than one particle to die at a given time with positive probability, we give any fixed order to such particles.) We again let $\tau_n(\omega)$ be the time of the nth death in the family tree ω. Let $N(\omega)=$ the total number of deaths in the family tree ω, and set $\tau_n(\omega)=\infty$ for $n>N(\omega)$. Then $\tau_n(\omega)\to\infty$ a.s. as $n\to\infty$. Define

$$X_n(t,\omega)=\begin{cases}Z(t,\omega) & \text{if } \tau_n>t,\\ 0 & \text{if } \tau_n\leqslant t,\end{cases}$$

and

$$\xi_j(\omega)=1+Z(\tau_j+,\omega)-Z(\tau_j-,\omega),\qquad j\geqslant 1,$$
$$=\text{number of particles produced in the } j\text{th split.}$$

Clearly

$$X_n(t) = 1 + \sum_{j=1}^{n} \xi_j, \tag{12}$$

and

$$X_n(t) \uparrow Z(t) \quad \text{monotonely, as } n \to \infty.$$

Hence by the monotone convergence theorem

$$\lim_{n\to\infty} E\,\phi(X_n(t)) = E\,\phi(Z(t)). \tag{13}$$

Our plan is to show that (9) implies

$$\sup_n E\,\phi(X_n(t)) < \infty \quad \text{for any } t > t_0; \tag{14}$$

and conversely that $E\,\phi(Z(t)) < \infty$ implies $E\,\phi(\xi_1) < \infty$, and hence (9). We start with the first implication.

Lemma 2. *If* $\sum p_j \phi(j) < \infty$, *then* $E\,\phi(X_n(t)) < \infty$ *for every n and t.*

Proof. From (12) and the properties of ϕ in (i), (ii), (iii) and in lemma 1,

$$\begin{aligned}\phi(X_n(t)) &\leq \frac{1}{2}\,\phi(2) + \frac{1}{2}\,\phi\left(2\sum_{j=1}^{n} \xi_j\right) \leq \frac{1}{2}\,\phi(2) + \left(\frac{K}{2}\right)\phi(2n)\,\phi\left(\sum_{j=1}^{n} \frac{\xi_j}{n}\right)\\ &\leq \frac{1}{2}\,\phi(2) + \left(\frac{K}{2n}\right)\phi(2n) \sum_{j=1}^{n} \phi(\xi_j).\end{aligned}$$

Taking expectations implies the lemma.

Lemma 3. *Let* $m_n(t) = E\,\phi(X_n(t))$. *If* $\sum p_j \phi(j) < \infty$, *then the sequence* $\{m_n(t); n = 1, 2, \ldots\}$ *satisfies*

$$m_{n+1}(t) \leq c_1(1 - G(t)) + c_2 \int_0^t m_n(t-u)\,dG(u),$$

where $0 \leq c_1, c_2 < \infty$ *are constants.*

Proof. For $n \geq 1$

$$\begin{aligned}E\,\phi(X_{n+1}(t)) &= E\{\phi(X_{n+1}(t)); \tau_1 > t\} + E\{\phi(X_{n+1}(t)); \tau_1 \leq t\}\\ &= \phi(1)\,[1 - G(t)] + E\{\phi(X_{n+1}(t)); \tau_1 \leq t\}.\end{aligned} \tag{15}$$

Suppose that $\tau_1 \leq t$. Then the $(n+1)$st split can occur after time t only if each of the ξ_1 particles born at time τ_1 has a line of descent in which the nth split occurs after an additional time of at least $t - \tau_1$. Thus

$$X_{n+1}(t) \leq \sum_{j=1}^{\xi_1} \tilde{X}_n^{(j)}(t - \tau_1),$$

where $\tilde{X}_n^{(j)}(u)$, $j=1,\ldots,\xi_1$, are independent copies of $X_n(u)$, and the sum is taken to be zero when $\xi_1=0$. Consequently

$$E\{\phi(X_{n+1}(t);\tau_1\leqslant t\}\leqslant\int_0^t\left\{\phi(0)p_0+\sum_{k=1}^{\infty}p_kE\phi\left[\sum_{j=1}^{k}\tilde{X}_n^{(j)}(t-u)\right]\right\}dG(u). \tag{16}$$

But by hypothesis (ii) of theorem 4 and the convexity of ϕ

$$\phi\left\{\sum_{j=1}^{k}\tilde{X}_n^{(j)}(t-u)\right\}\leqslant\frac{K}{k}\phi(k)\sum_{j=1}^{k}\phi(\tilde{X}_n^{(j)}(t-u)),$$

and hence the left side of (16) is

$$\leqslant\int_0^t\left\{\phi(0)p_0+K\left(\sum_{k=1}^{\infty}p_k\phi(k)\right)m_n(t-u)\right\}dG(u).$$

Since $\phi(0)p_0+\sum_{k=1}^{\infty}p_k\phi(k)=E\phi(\xi_1)<\infty$ by assumption, and $m_n(t-u)\geqslant\phi(0)\geqslant 1$, we see that there exists a constant $0\leqslant c_2<\infty$ such that

$$E\{\phi(X_{n+1}(t);\tau_1\leqslant t\}\leqslant c_2\int_0^t m_n(t-u)\,dG(u).$$

Inserting this inequality in (15) yields lemma 3. □

Now for any finite, positive c_1, c_2 we know from lemma 4.2 that the equation

$$m(t)=c_1[1-G(t)]+c_2\int_0^t m(t-u)\,dG(u) \tag{17}$$

has a unique nonnegative solution $\overline{m}(t)$, which is bounded on compact sets. But note that $m_1(t)=\phi(1)[1-G(t)]+\phi(0)G(t)\leqslant\overline{m}(t)$ for sufficiently large c_1, c_2. Choosing such (c_1, c_2), and iterating (17), we see that $m_n(t)\leqslant\overline{m}(t)$ for all n and this implies (14).

Turning to the converse part of theorem 4, we again start with two lemmas.

Lemma 4. *Let $\{\delta_i; i=1,2,\ldots\}$ be a sequence of independent Bernoulli variables with*

$$P\{\delta_i=1\}=p=1-P\{\delta_i=0\},\qquad 0<p<1;\quad \textit{and let } N\geqslant 0$$

be an integer-valued random variable independent of the δ_i's; and set $R_n=\sum_{i=1}^{n}\delta_i$. Let $\phi(x)$ be any nondecreasing function such that

$$\lim_{x\to\infty}\phi(x)=\infty\quad \textit{and}\ \liminf_n\frac{E\phi(R_n)}{\phi(n)}=c>0.$$

Then

$$E\phi(R_N)<\infty \quad \textit{if and only if } E\phi(N)<\infty .$$

Proof. Since $R_N \leqslant N$, the "if" part is obvious. To prove the converse, note that

$$E\phi(R_N)=\sum_{n=0}^{\infty} P(N=n)E\phi(R_n).$$

By hypothesis there exists an n_0 such that

$$E\phi(R_n) \geqslant \frac{c}{2}\phi(n) \quad \text{for } n \geqslant n_0 .$$

Hence

$$E\phi(R_N) \geqslant \frac{c}{2}\sum_{n=n_0}^{\infty} P(N=n)\phi(n),$$

and the conclusion follows. □

Lemma 5. *Suppose that* ϕ *satisfies the hypotheses of theorem 4, the additional hypotheses allowed by lemma 1, and also that* $\phi(x)\to\infty$ *as* $x\to\infty$. *Let* R_n *be as defined in lemma* 4. *Then* $\liminf_n (E\phi(R_n)/\phi(n))=c>0$.

Proof. Since $\phi(n)\leqslant K\phi(n/R_n)\phi(R_n)$,

$$\frac{E\phi(R_n)}{\phi(n)} \geqslant \frac{1}{K}E\left[\phi\left(\frac{n}{R_n}\right)\right]^{-1}.$$

Since $(\phi)^{-1}$ is a bounded, continuous nonnegative function, we conclude by the strong law of large numbers and the bounded convergence theorem, that

$$\lim\frac{E\phi(R_n)}{\phi(n)} \geqslant \frac{1}{K}\frac{1}{\phi\left(\frac{1}{p}\right)}>0,$$

implying the lemma. □

Finally, we complete the proof of theorem 4 by showing that (8) implies (9). Suppose that $E\phi(Z(t))<\infty$ for some $t>t_0$. Fix this t.

On the set $\{\tau_1 \leqslant t\}$, note that $Z(t)=\sum_{j=1}^{\xi_1} \tilde{Z}^{(j)}(t-\tau_1)$, where $\{\tilde{Z}^{(j)}(s), s \geqslant 0\}$ are independent copies of $\{Z(t); t \geqslant 0\}$; Hence

$$\begin{aligned}\infty > E\,\varphi(Z(t)) &= E[\phi(Z(t)); \tau_1 > t] \\ &\quad + E[\phi(Z(t)); \tau_1 \leqslant t, \xi_1 = 0] + E[\phi(Z(t)); \tau_1 \leqslant t, \xi_1 \geqslant 1] \\ &= \phi(1)[1-G(t)] + \phi(0)\,p_0\,G(t) \\ &\quad + \int_0^t E\left\{\phi\left[\sum_{j=1}^{\xi_1} \tilde{Z}^{(j)}(t-u)\right]; \xi_1 \geqslant 1\right\} dG(u).\end{aligned}$$

Thus there must be a $u_0 \in [0, t-t_0]$ such that

$$E\left\{\phi\left[\sum_{j=1}^{\xi_1} \tilde{Z}^{(j)}(t-u_0)\right]; \xi_1 \geqslant 1\right\} < \infty. \tag{18}$$

Now let

$$\delta_j = \begin{cases} 1 & \text{if } \tilde{Z}^{(j)}(t-u_0) > 0, \\ 0 & \text{otherwise}. \end{cases}$$

Then $\sum_{j=1}^{\xi_1} \delta_j \leqslant \sum_{j=1}^{\xi_1} \tilde{Z}^{(j)}(t-u_0)$, and hence by (18)

$$E\,\phi\left(\sum_{j=1}^{\xi_1} \delta_j\right) < \infty. \tag{19}$$

Furthermore $0 < P\{\delta_j = 1\} < 1$. Hence applying lemmas 5 and 4 (in that order) to (19) we conclude that $E\,\phi(\xi_1) < \infty$, which completes the proof of theorem 4.

6. Asymptotic Behavior of $F(s, t)$ in the Critical Case

The importance of knowing the asymptotic behavior of generating functions in the study of branching processes has been apparent through this book. In this section we give the relevant result for the critical process, and in sections 7 and 8 we discuss the noncritical cases.

Theorem 1. (Goldstein (1971)). *If $m=1$, $\sigma^2 = f''(1) < \infty$, $\mu = \int t\, dG(t) < \infty$, and $t^2[1-G(t)] \to 0$ as $t \to \infty$, then*

$$\lim_{t\to\infty} \left[\frac{\sigma^2}{2\mu}(1-s)t + 1\right]\left[\frac{1-F(s,t)}{1-s}\right] = 1$$

uniformly for $0 \leqslant s < 1$.

Stronger moment assumptions can be converted into sharper remainder estimates. For example, if $s=0$ then $1-F(0,t)=P\{Z(t)>0\}$ and one has

Theorem 2. *If* $m=1$, $\sum n^{3+\varepsilon} p_n<\infty$ *for some* $\varepsilon>0$, *and* $\int t^{4+\delta} dG(t)<\infty$ *for some* $\delta>0$, *then*

$$\frac{1}{P\{Z(t)>0\}}=\frac{a}{\mu}t+\left(\frac{a^2-b}{a}-\frac{a\hat{\tau}}{\mu^2}\right)\log t+c+O(t^{-d}),$$

where $\hat{\tau}=$the variance of G, $a=f''(1)/2$, $b=f'''(1)/6$, *and* $c,d>0$ *are constants.*

Of course, theorem 1 implies that if $f''(1)<\infty$ and $\int t\,dG(t)<\infty$, then

$$P\{Z(t)>0\}\sim\frac{\mu}{a}\frac{1}{t}.$$

This fact was previously proved by Chover, Ney (1968) using a different technique; and by Sevastyanov (1964) under third moment assumptions on f and G.

The proof of theorem 2 can be found in Chover, Ney (1968). Here we only give the proof of the first theorem, which follows from the comparison lemma of section 3.

Proof of theorem 1. The proof is broken into several lemmas. We first restate lemma 3.1 in the particular form it takes when $m=1$.

Lemma 1. *If* $m=1$, *then*

$$1-f_n(s)-(1-s)G^{*n}(t)\leqslant 1-F(s,t)\leqslant 1-f_n(s)+(1-s)[1-G^{*n}(t)]. \tag{1}$$

The next lemma is an analog of theorem 1 for Galton-Watson processes.

Lemma 2. *If* $m=1$ *and* $f''(1)=\sigma^2<\infty$ *then*

$$A_n(s)\equiv\left[\frac{\sigma^2}{2}(1-s)n+1\right]\left[\frac{1-f_n(s)}{1-s}\right]\to 1 \quad \textit{as } n\to\infty,$$

uniformly for $0\leqslant s<1$.

Proof. Let

$$B_n(s)=\frac{2}{\sigma^2 n}\left[\frac{1}{1-f_n(s)}-\frac{1}{1-s}\right].$$

Then by the basic lemma of section I.9

$$\lim_{n\to\infty} B_n(s)=1$$

uniformly for $0 \leqslant s < 1$. But $B_n(s) \geqslant 0$ and hence

$$|A_n(s)-1| = \left| \frac{\frac{\sigma^2}{2}(1-s)n}{\frac{\sigma^2}{2}(1-s)n + \frac{1}{B_n(s)}} \right| \left| \frac{1-B_n(s)}{B_n(s)} \right| \leqslant \left| \frac{1-B_n(s)}{B_n(s)} \right| \to 0$$

as $n \to \infty$ uniformly for $0 \leqslant s < 1$. □

Lemma 3. *If* $\mu = \int t\,dG(t) < \infty$ *and* $t^2[1-G(t)] \to 0$ *as* $t \to \infty$, *and if* $\varepsilon > 0$, *then*

(i) *If* $n = [(t/\mu)(1+\varepsilon)]$, *then* $t\,G^{*n}(t) \to 0$;
(ii) *If* $n = [(t/\mu)(1-\varepsilon)]$, *then* $t[1-G^{*n}(t)] \to 0$;

($[x]$ is the largest integer $\leqslant x$).

Proof. By theorem 4 of L. Baum, M. Katz (1965)

$$\lim_{n \to \infty} n\,G^{*n}(n\mu - n\delta) = 0$$

for every $\delta > 0$. Thus it is sufficient to show that for appropriate $\delta > 0$ (and sufficiently large t) we have $t \leqslant n\mu - n\delta$, or $0 < \delta \leqslant \mu - t/n$. But for the choice of n in (i), $n \geqslant (t/\mu)(1+\varepsilon) - 1$, or equivalently

$$\mu - \frac{t}{n} \geqslant \mu \left[1 - \left\{ 1 + \varepsilon - \frac{\mu}{t} \right\}^{-1} \right],$$

and hence it is sufficient to have δ satisfy

$$0 < \delta < \mu \left[1 - \left(1 + \varepsilon - \frac{\mu}{t} \right)^{-1} \right].$$

But the last term converges to $\mu\varepsilon/(1+\varepsilon) > 0$ as $t \to \infty$, and hence δ may be chosen as required for large t. This proves (i).

To prove (ii) we again use a result from Baum and Katz, to the effect that for every $\delta > 0$

$$\lim_{n \to \infty} n[1 - G^{*n}(n\mu + n\delta)] = 0.$$

Hence we only need to show that for $n = [(t/\mu)(1-\varepsilon)]$ and some δ,

$$t \geqslant n\mu + n\delta.$$

But $n \leqslant (t/\mu)(1-\varepsilon)$ can be rewritten as

$$\frac{\mu\varepsilon}{1-\varepsilon} \leqslant \frac{t}{n} - \mu,$$

and it is thus sufficient to choose

$$\delta < \frac{\mu\varepsilon}{1-\varepsilon}. \qquad \square$$

To complete the proof of theorem 2 let $n = [(t/\mu)(1+\varepsilon)]$, $0 < \varepsilon$. Then

$$t \geqslant \frac{n\mu}{1+\varepsilon} = \mu n \left(1 - \frac{\varepsilon}{1+\varepsilon}\right).$$

Using this fact, the left side of (1), and $f_n(s) \geqslant f_n(0)$, we get

$$\begin{aligned} &\left[\frac{\sigma^2}{2}(1-s)n+1\right]\left[\frac{1-f_n(s)}{1-s}\right] - \frac{\varepsilon}{1+\varepsilon}\,\frac{\sigma^2}{2} n[1-f_n(0)] \\ &\leqslant \left[\frac{\sigma^2}{2\mu}(1-s)t+1\right]\left[\frac{1-f_n(s)}{1-s}\right] \\ &\leqslant \left[\frac{\sigma^2}{2\mu}(1-s)t+1\right]\left[\frac{1-F(s,t)}{1-s}\right] + \left(\frac{\sigma^2}{2\mu}(1-s)t+1\right) G^{*n}(t). \end{aligned} \tag{2}$$

But by lemma 2 and theorem I.9.1 the left hand side of (2) goes to $1-(\varepsilon/(1+\varepsilon))$ as $t \to \infty$, and by lemma 3 with $n = [(t/\mu)(1+\varepsilon)]$,

$$\left(\frac{\sigma^2}{2\mu}(1-s)t+1\right) G^{*n}(t) \to 0 \quad \text{as } t \to \infty.$$

Hence since ε is arbitrary

$$\liminf_{t\to\infty} \left[\frac{\sigma^2}{2\mu}(1-s)t+1\right]\left[\frac{1-F(s,t)}{1-s}\right] \geqslant 1 \tag{3}$$

uniformly for $0 \leqslant s < 1$.

In the other direction, taking $n = [(t/\mu)(1-\varepsilon)]$,

$$\begin{aligned} &\left[\frac{\sigma^2}{2\mu}(1-s)t+1\right]\left[\frac{1-F(s,t)}{1-s}\right] \\ &\leqslant \left[\frac{\sigma^2}{2\mu}(1-s)t+1\right]\left[\frac{1-f_n(s)}{1-s}\right] + \left[\frac{\sigma^2}{2\mu}(1-s)t+1\right][1-G^{*n}(t)] \\ &\leqslant \left[\frac{\sigma^2}{2}(1-s)n+1\right]\left[\frac{1-f_n(s)}{1-s}\right] + \left[\frac{\sigma^2}{2}\left(\frac{n\varepsilon+1}{1-\varepsilon}\right)+1\right][1-f_n(0)] \\ &\quad + \left[\frac{\sigma^2}{2\mu}(1-s)t+1\right][1-G^{*n}(t)], \end{aligned}$$

where we have used the fact that

$$t \leqslant \frac{\mu(n+1)}{1-\varepsilon} = \mu n + \frac{\mu}{1-\varepsilon}(n\varepsilon+1)$$

since $n \geqslant (t/\mu)(1-\varepsilon)-1$. But by lemma 3

$$\lim_{t\to\infty}\left(\frac{\sigma^2}{2\mu}(1-s)t+1\right)[1-G^{*n}(t)]=0,$$

and hence, applying lemma 2 and theorem I.9.1, as before, we see that

$$\limsup_{t\to\infty}\left[\frac{\sigma^2}{2\mu}(1-s)t+1\right]\left[\frac{1-F(s,t)}{1-s}\right]\leqslant 1$$

uniformly for $0\leqslant s<1$.

The theorem follows from (3) and (4). □

7. Asymptotic Behavior of $F(s,t)$ when $m\neq 1$: The Malthusian Case

Traditionally, the term "Malthusian parameter" has been applied only to the solution $\alpha(m,G)$, of $m\hat{G}(\alpha)=1$, where $m=f'(1)>1$, and $\hat{G}(\alpha)=\int e^{-\alpha t}dG(t)$. In that case we have seen in theorem 5.3A that $EZ(t)\sim c'e^{\alpha t}$, with $\alpha>0$; and we will also show in section (11) that $Z(t)/c'e^{\alpha t}$ converges to a random variable. (When $m<1$ we saw that $\mu(t)\sim c'e^{\alpha t}$ continues to hold, this time with $\alpha<0$, provided $\alpha=\alpha(m,G)$ exists.)

Thus the customary meaning of Malthusian parameter (see T. Harris (1963), R. A. Fisher (1930)) is that it is the constant in the exponential function describing the *growth rate of the population.* In section 4, however we defined such a parameter more generally for any (γ,G), and saw that it described the asymptotic behavior of the solution of the linear integral equation (4.3).

In this section we extend such results to non-linear equations (1.1). We caution the reader to note that *the term Malthusian parameter is here used in the broad sense* as the root of $\gamma\hat{G}(\alpha)=1$ for some $\gamma>0$ and *describes the convergence rate of the solution of an equation, not necessarily the population growth.*

In fact the main result below is a direct generalization of the geometric convergence rate of the generating functions of a Galton-Watson process (corollary I.11.1).

Theorem 1. *If* $m\neq 1$, $0<\gamma=f'(q)$, G *is non-lattice*[12], *the Malthusian parameter* $\alpha=\alpha(\gamma,G)$ *exists, and* $\mu_\alpha=\int t\,dG_\alpha(t)<\infty$, *then*[13]

$$\lim_{t\to\infty} e^{-\alpha t}(q-F(s,t))\equiv Q(s) \tag{1}$$

[12] As usual, we leave the lattice analogue of this theorem to the reader.

[13] To completely maintain the correspondence with section I.11 we should really take $Q(s)\equiv\lim e^{-\alpha t}(F(s,t)-q)$; but the other sign is a little more convenient here.

exists for $0\leqslant s<1$. *Furthermore*

$$Q(s)\equiv 0 \quad \textit{if and only if} \quad m<1 \quad \textit{and} \ \sum p_j j\log j=\infty. \tag{2}$$

If $m>1$ *or* $\sum p_j j\log j<\infty$ *then* $Q(s)\neq 0$ *for* $s\neq q$. (Of course $Q(q)=0$.)

As in the Galton-Watson case (section I.11) one could give a proof based on behavior of the derivative $(\partial/\partial s)F(s,t)$, but we will consider $F(s,t)-q$ directly, as the present method will then generalize to cases where the derivatives do not exist. In the subcritical case, most of the following proof is essentially due to T. Ryan (1968). (In this case it was previously proved by O. P. Vinogradov (1964), that $1-F(0,t)\sim c\,e^{\alpha t}$, but he does not seem to have shown that $c\neq 0$.)

Proof of theorem 1. From the basic equation (1.1)

$$q-F(s,t)=(q-s)(1-G(t))+\int_0^t (q-f(F(s,t-u))dG(u). \tag{3}$$

Fix $0\leqslant s\leqslant 1$ throughout. Set

$$\begin{aligned} H(t)&=q-F(s,t),\\ \xi_1(t)&=(q-s)(1-G(t)),\\ \xi_2(t)&=\int_0^t (q-f(F(s,t-u))-\gamma H(t-u))dG(u),\\ \xi(t)&=\xi_1(t)+\xi_2(t). \end{aligned}$$

We may now rewrite (3) as

$$H(t)=\xi(t)+\gamma\int_0^t H(t-u)\,dG(u). \tag{4}$$

If $e^{-\alpha t}\xi(t)$ is shown to be directly Riemann integrable then, by Theorem 4.2, (1) will follow.

The direct Riemann integrability of $e^{-\alpha t}\xi_1(t)$ is left to the complements. Turning to $e^{-\alpha t}\xi_2(t)$, we first shall establish its ordinary integrability.

Lemma 1. (i) *For any* $\alpha'>\alpha=\alpha(\gamma,G)$,

$$\sup_{t\geqslant 0} e^{-\alpha' t}|H(t)|=K_{\alpha'}<\infty.$$

(ii) *If* $m<1$ *then*

$$\sup_{\substack{t\geqslant 0\\ 0\leqslant s<1}} e^{-\alpha t}|H(t)|<\infty.$$

Proof. Since $\alpha=\alpha(\gamma,G)$ exists, $\alpha(\gamma',G)$ exists for any $\gamma'\geqslant\gamma$ due to the continuity of $\hat{G}(\alpha)$. Given an $\alpha'>\alpha$, there exists a γ' such that $\alpha'=\alpha(\gamma',G)$. By Theorem 3.1 $F(s,t)\to q$ and hence there exists a t_0 such that for $t\geqslant t_0$

$$|f(F(s,t))-q|<\gamma'|F(s,t)-q|. \tag{5}$$

From (3) and (5)

$$|H(t)|\leqslant \xi_1(t)+\gamma'\int_0^{t-t_0}|H(t-u)|\,dG(u)+(G(t)-G(t-t_0))$$

$$\leqslant 1-G(t-t_0)+\gamma'\int_0^t|H(t-u)|\,dG(u).$$

Since t_0 is fixed, (i) follows by a comparison argument from Theorem 4.2.

For (ii) use the simpler bound on (3)

$$|H(t)|\leqslant 1-G(t)+m\int_0^t|H(t-u)|\,dG(u). \qquad \square$$

Lemma 2. *Under the hypothesis of Theorem 1*

$$\int_0^\infty e^{-\alpha t}|\xi_2(t)|\,dt<\infty. \tag{6}$$

Proof. Suppose $m>1$. Since $q-F(s,t)$ has the same sign for all t, so does $q-f(F(s,t))-\gamma(q-F(s,t))$; and hence it suffices to show that

$$I\equiv\left|\int_0^\infty e^{-\alpha t}\,\xi_2(t)\,dt\right|<\infty.$$

On interchanging orders of integration (which is justified since the integrand has the same sign for all t and u) we see that

$$I=\left(\int_0^\infty e^{-\alpha u}dG(u)\right)\left|\int_0^\infty e^{-\alpha t}[q-f(F(s,t))-\gamma(q-F(s,t))]\,dt\right|.$$

But for $x\leqslant 1-\eta$, $\eta>0$

$$|q-f(x)-\gamma(q-x)|\leqslant c|x-q|^2, \tag{7}$$

for some $0<c<\infty$, and hence by lemma 1 (i)

$$|I|\leqslant\gamma^{-1}cK_{\alpha'}^2\int_0^\infty e^{-\alpha t}e^{2\alpha' t}\,dt \quad \text{for any } \alpha'>\alpha.$$

We may now choose α' so that $2\alpha'<\alpha$ (both α and α' are negative), and hence $|I|<\infty$.

Suppose $m<1$. Multiplying (4) by $e^{-\alpha t}$ we get

$$H_\alpha(t)=\xi_{1\alpha}(t)+\xi_{2\alpha}(t)+\int_0^t H_\alpha(t-u)\,dG_\alpha(u), \tag{8}$$

with the notation $e^{-\alpha t}a(t)=a_\alpha(t)$ and $dG_\alpha(u)=\gamma e^{-\alpha u}dG(u)$ as before. By lemma 1 (ii) $H_\alpha(t)$ is bounded. Also $\xi_{1\alpha}(t)$ is bounded, and thus $\xi_{2\alpha}(t)$ is bounded. Taking Laplace transforms, we see from (7) that

$$\hat{H}_\alpha(\theta)=\hat{\xi}_{1\alpha}(\theta)+\hat{\xi}_{2\alpha}(\theta)+\hat{H}_\alpha(\theta)\hat{G}_\alpha(\theta)$$

or

$$\hat{H}_\alpha(\theta)\,(1-\hat{G}_\alpha(\theta))+(-\hat{\xi}_{2\alpha}(\theta))=\hat{\xi}_{1\alpha}(\theta). \tag{9}$$

Since $q-F(s,t)$ and $q-f(F(s,t))-\gamma(q-F(s,t))$ have opposite signs for all t it follows that $H(t)$, $-\xi_2(t)$ and $\xi_1(t)$ all have the same sign. But we know that

$$\int|\xi_{1\alpha}(t)|\,dt<\infty.$$

Hence, letting $\theta\downarrow 0$ in (9) we see that we must have

$$I=\lim_{\theta\downarrow 0}|\hat{\xi}_{2\alpha}(\theta)|<\infty. \quad \square$$

Lemma 3. *$e^{-\alpha t}\xi_2(t)$ is directly Riemann integrable.*

Proof. If $m>1$ then since $q-f(F(s,t))-\gamma(q-F(s,t))$ has the same sign for all t, we can bound $e^{-\alpha t}\xi_2(t)$ by $c e^{-\alpha t}\int_0^t e^{2\alpha'(t-u)}dG(u)$, which can be shown to be directly Riemann integrable for all $\alpha<\alpha'<\alpha/2$ by using condition (i) after theorem 4.1.

Suppose $m<1$. Then $q=1$, and for $n\leqslant t<n+1$ we have

$$e^{-\alpha t}|\xi_2(t)|\leqslant e^{-\alpha}e^{-\alpha n}|\xi_2(n)|+e^{-\alpha}e^{-\alpha n}(1-G(n)).$$

But we have already seen that $e^{-\alpha t}|\xi_2(t)|$ and $e^{-\alpha t}(1-G(t))$ are integrable, and hence

$$\sum_n e^{-\alpha n}|\xi_2(n)|+\sum_n e^{-\alpha n}(1-G(n))<\infty.$$

Now use condition (i) listed after theorem 4.1. $\square$

This establishes (1), and also identifies $Q(s)$ to be

$$Q(s)=\frac{1}{\mu_\alpha}\int_0^\infty e^{-\alpha t}\xi(t)\,dt,$$

where $\mu_\alpha = \int_0^\infty t\,dG_\alpha(t)$. Thus

$$\begin{aligned}\mu_\alpha Q(s) &= (q-s)\int_0^\infty e^{-\alpha t}[1-G(t)]\,dt \\ &\quad + \left(\int_0^\infty e^{-\alpha u}\,dG(u)\right)\left(\int_0^\infty e^{-\alpha t}\{q-f[F(s,t)]-\gamma(q-F(s,t))\}\,dt\right) \qquad (10)\\ &= (q-s)\frac{\gamma-1}{\alpha\gamma} + \frac{1}{\gamma}\int_0^\infty e^{-\alpha t}\{q-f[F(s,t)]-\gamma(q-F(s,t))\}\,dt\,.\end{aligned}$$

It thus remains only to determine the conditions under which $Q(s)=0$. If $m>1$ and $s>q$, then the right hand side of (10) is negative and hence $Q(s)<0$. Furthermore note that $Q(s)$ is monotone decreasing, since it is defined as the limit of a sequence of monotone functions. Thus, if $m>1$ and $s_0<q$ and if $Q(s_0)=0$, then the monotonicity of Q and the fact that $Q(q)=0$ imply that $Q(s)=0$ for s in (s_0,q). The analyticity of $Q(s)$ will then force $Q(s)\equiv 0$ for s in $[0,1)$; contradicting the fact that $Q(s)<0$ for $s>q$ which we have just seen. Thus

$$\text{if } m>1 \quad \text{then } Q(s)\neq 0 \quad \text{for } s\neq q\,. \tag{11}$$

Now suppose that $m<1$. Here $q=1$, $\gamma=m$ and $Q(s)\geqslant 0$ in $[0,1)$. Suppose $Q(s)>0$ for some s in $[0,1)$. We have seen that

$$0<I=\int_0^\infty e^{-\alpha t}[m(1-F(s,t))-(1-f(F(s,t))]\,dt<\infty\,.$$

Setting $A(x)=m-(1-f(1-x))/x$ as in section I.10,

$$I=\int_0^\infty e^{-\alpha t}(1-F(s,t))A(1-F(s,t))\,dt\,. \tag{12}$$

By hypothesis $(1-F(s,t))e^{-\alpha t}\to c=Q(s)>0$, and hence we can pick a t_0 such that

$$\frac{c}{2}e^{\alpha t}\leqslant 1-F(s,t) \quad \text{for } t\geqslant t_0\,.$$

Since $m(1-x)-(1-f(x))$ is decreasing and nonnegative

$$\begin{aligned} I &\geqslant \frac{c}{2}\int_{t_0}^\infty e^{-\alpha t}e^{\alpha t}A\left(\frac{c}{2}e^{\alpha t}\right)dt \\ &= -\frac{c}{2\alpha}\int_0^{\left(\frac{c}{2}\right)e^{\alpha t_0}}\frac{A(u)}{u}\,du>0 \quad \text{(remember } \alpha<0)\,.\end{aligned}$$

Thus $c>0$ implies $\int_0^\eta (A(u)/u)\,du<\infty$ for some $\eta>0$. This, by corollary I.10.2, implies $\sum p_j j \log j<\infty$.

Conversely suppose $m<1$ and $\sum p_j j \log j<\infty$. We claim that

$$\lim_{s\uparrow 1} \frac{Q(s)}{1-s} = \mu_\alpha^{-1} \int_0^\infty e^{-\alpha t}[1-G(t)]\,dt\,. \tag{13}$$

This implies that $Q(s)>0$ for s sufficiently close to 1, and hence by monotonicity that $Q(s)>0$ for $0 \leqslant s<1$. This, with (11), completes the proof of the theorem.

Referring back to (10) with $\gamma=m$ and $q=1$, we see that to prove (13) it is sufficient to show that

$$\lim_{s\uparrow 1} \int_0^\infty e^{-\alpha t}\left\{\frac{m(1-F(s,t))-(1-f[F(s,t)])}{(1-s)}\right\} dt=0\,; \tag{14}$$

or, by the definition of $A(\cdot)$ that

$$\lim_{s\uparrow 1} \int_0^\infty e^{-\alpha t} \frac{[1-F(s,t)]\,A[1-F(s,t)]}{1-s}\, dt=0\,. \tag{15}$$

But $(1-s)^{-1}\,[1-F(s,t)] \leqslant \mu(t) \leqslant K e^{\alpha t}$, where K is independent of s in $[0,1)$. Since $A(\cdot)$ is nondecreasing, we thus have

$$e^{-\alpha t} \frac{1-F(s,t)}{1-s} A[1-F(s,t)] \leqslant e^{-\alpha t} K e^{\alpha t} A(K e^{\alpha t})\,.$$

But applying corollary I.10.2 again, the hypothesis $\sum p_j j \log j<\infty$ implies

$$\int_0^\infty A(K e^{\alpha t})\,dt = -\frac{1}{\alpha}\int_0^K \frac{A(u)}{u}\,du<\infty\,.$$

By the dominated convergence theorem we may hence take the limit under the integral in (15). Since the integrand $\downarrow 0$ as $s \uparrow 1$, we are finished. □

Remark 1. We have in fact also shown that if $m<1$ and $\sum p_j j \log j<\infty$, then

$$Q'(1-)=c'=\mu_\alpha^{-1} \int_0^\infty e^{-\alpha t}[1-G(t)]\,dt\,.$$

Similarly one can show that if $m>1$, then $Q'(q)=c'>0$.

8. Asymptotic Behavior of $F(s, t)$ when $m \neq 1$: Sub-Exponential Case

When G is in the sub-exponential class defined in section 4, then $q-F(s,t)$, like the mean $\mu(t)$, behaves like the tail of G. The following theorem was proved under a second moment assumption on f, and for $\gamma < C^{-1}$, $C = \sup\{(G(t)-G^{*2}(t))/(1-G(t))\}$, by Chistyakov (1964), and in a more general setting by Chover, Ney, Wainger (1969), (1972).

Theorem 1. *If* $m \neq 1$, $\gamma = f'(q)$, *and* $G \in \mathscr{S}$, *then*

$$\lim_{t\to\infty} \frac{q-F(s,t)}{1-G(t)} = \frac{q-s}{1-\gamma}, \qquad 0 \leqslant s < 1. \tag{1}$$

Proof. Suppose first that $s<q$, and write

$$q-F(s,t) = (q-s)[1-G(t)] + \int_0^t \{q-f(q-[q-F(s,t-y)])\}\, dG(y).$$

Pick any $\varepsilon>0$. Since $q-F(s,t)\to 0$ as $t\to\infty$ (see theorem 3.1), and $q-f(s) = (q-s) f'(q) + o(q-s)$, we can find a $t_0<\infty$ such that

$$q-f(q-[q-F(s,t-y)]) \leqslant (\gamma+\varepsilon)[q-F(s,t-y)]$$

for $0 \leqslant y \leqslant t-t_0$, $t \geqslant t_0$. Thus

$$q-F(s,t) \leqslant (q-s)[1-G(t)] + \int_{t-t_0}^{t} \{q-f(q-[q-F(s,t-y)])\}\, dG(y)$$
$$+(\gamma+\varepsilon)\int_0^{t-t_0} [q-F(s,t-y)]\, dG(y);$$

and hence

$$q-F(s,t) \leqslant (q-s)[1-G(t)] + r(t) + (\gamma+\varepsilon)\int_0^t [q-F(s,t-y)]\, dG(y),$$

where

$$r(t) = \int_{t-t_0}^{t} |q-f(q-[q-F(s,t-y)])|\, dG(y) - (\gamma+\varepsilon)\int_{t-t_0}^{t} |q-F(s,t-y)|\, dG(y).$$

Observe that

$$r(t) = O[G(t)-G(t-t_0)].$$

If $s>q$, then a similar inequality holds with $q-F$ replaced by $F-q$ and $q-s$ by $s-q$. Also there is a similar lower bound for $q-F$ with $(\gamma+\varepsilon)$ replaced by $(\gamma-\varepsilon)$. Thus, letting $x(t) = |q-F(s,t)|$ and $R(t) = |q-s|[1-G(t)] + r(t)$, we see that

$$R(t) + (\gamma-\varepsilon)\, x(t) * G(t) \leqslant x(t) \leqslant R(t) + (\gamma+\varepsilon)\, x(t) * G(t).$$

A standard iteration argument now yields

$$R(t) * U_{\gamma-\varepsilon}(t) \leqslant x(t) \leqslant R(t) * U_{\gamma+\varepsilon}(t), \tag{2}$$

where $U_\gamma(t)$ is defined in (4.1). We have already observed in section 5 that

$$[1-G(t)] * U_\gamma(t) \sim \frac{1-G(t)}{1-\gamma},$$

and hence

$$[1-G(t)] * |q-s|\, U_{\gamma\pm\varepsilon}(t) \sim \frac{|q-s|}{1-(\gamma\pm\varepsilon)}[1-G(t)].$$

Since ε is arbitrary, we see by (2) and the definition of $R(t)$, that it is sufficient to prove that

$$[G(t)-G(t-t_0)] * U_\gamma(t) = o[1-G(t)]. \tag{3}$$

But $G(t)-G(t-t_0)=o(1-G(t))$ since $G\in\mathscr{S}$, and this with theorem 4.3 implies that

$$U_\gamma(t)-U_\gamma(t-t_0) = o[1-G(t)]. \tag{4}$$

From (4), the definition U_γ, and a little manipulation of convolutions, one obtains (3). Details are left as an exercise. □

9. The Exponential Limit Law in the Critical Case

When $m=1$ we have an analog of the exponential limit law of Section I.9.

Theorem (Goldstein (1971)). *If* $m=1$, $f''(1)=\sigma^2<\infty$, $\int_0^\infty t\,dG(t)=\mu<\infty$, *and* $t^2[1-G(t)]\to 0$ *as* $t\to\infty$, *then*

$$\lim_{t\to\infty} P\left\{\frac{Z(t)}{t} \leqslant x \,|\, Z(t)>0\right\} = 1-e^{-\left(\frac{2\mu}{\sigma^2}x\right)}, \qquad x\geqslant 0.$$

Proof. The Laplace transform of the distribution of $Z(t)/t$, conditioned on non-extinction is

$$E\left[e^{-u\frac{Z(t)}{t}}\,\middle|\, Z(t)>0\right] = 1-\frac{1-F(e^{-\frac{u}{t}},t)}{1-F(0,t)}$$

$$= 1-\frac{\left[\frac{\sigma^2}{2\mu}(1-e^{-\frac{u}{t}})t+1\right]\left[\frac{1-F(e^{-\frac{u}{t}},t)}{1-e^{-\frac{u}{t}}}\right]}{\left[\frac{\sigma^2}{2\mu}t+\frac{1}{1-e^{-\frac{u}{t}}}\right][1-F(0,t)]}.$$

By Theorem 6.1 the numerator goes to 1 as $t\to\infty$, while the denominator

$$=\left[\frac{\sigma^2}{2\mu}+\frac{1}{u+o(1)}\right]t[1-F(0,t)]\to 1+\frac{2\mu}{u\sigma^2}.$$

Thus

$$\lim_{t\to\infty} E\left[e^{-u\frac{Z(t)}{t}}\,|\,Z(t)>0\right]=\frac{1}{1+\dfrac{u\sigma^2}{2\mu}},$$

and our theorem follows from the continuity theorem for Laplace transforms (see Feller, Vol. II, p. 408). □

10. The Limit Law for the Subcritical Age-Dependent Process

From the results of sections 7 and 8 we can easily derive the analogs for age-dependent processes of the Yaglom theorem for the Galton-Watson process.

Theorem 1 (Ryan, 1968). *Let* $\{Z(t); t\geqslant 0\}$ *be an age-dependent branching process with* $m<1$ *and* $\sum_j (j\log j)p_j<\infty$. *Assume that the lifetime distribution G is such that the Malthusian parameter* $\alpha(\gamma,G)$ *exists and* $\int_0^\infty te^{-\alpha t}dG(t)<\infty$. *Then,*

$$\lim_{t\to\infty} P\{Z(t)=k\,|\,Z(t)>0\}=b_k$$

exists for all $k\geqslant 1$, $\sum_{k=1}^\infty b_k=1$ *and* $\sum_{k=1}^\infty kb_k<\infty$.

Proof. Note that

$$\mathscr{B}(s,t)\equiv\sum_1^\infty s^k P\{Z(t)=k\,|\,Z(t)\neq 0\}=1-\frac{1-F(s,t)}{1-F(0,t)}.$$

By Theorem 7.1

$$1-\mathscr{B}(s,t)=\frac{e^{-\alpha t}(1-F(s,t))}{e^{-\alpha t}(1-F(0,t))}\to\frac{Q(s)}{Q(0)}.$$

Thus $\mathscr{B}(s,t)\to\mathscr{B}(s)\equiv 1-Q(s)/Q(0)$. Clearly $\mathscr{B}(s)$ is a power series with nonnegative coefficients. Furthermore we see from (7.13) that

$$\lim_{s\uparrow 1} Q(s)=0 \quad\text{and hence}\quad \lim_{s\uparrow 1}\mathscr{B}(s)=1.$$

Define b_k by $\mathscr{B}(s) \equiv \sum_1^\infty b_k s^k$. To complete the proof we need to show $\mathscr{B}'(1-) < \infty$. But $\mathscr{B}'(s) = -Q'(s)/Q(0)$, and we have already observed in remark 7.1 that

$$Q'(1-) = \frac{\mu_\alpha^{-1}(1-m)}{m(-\alpha)} < \infty.$$

Thus

$$\lim_{s \uparrow 1} \mathscr{B}'(s) = \frac{1-m}{m(-\alpha)} \cdot \frac{\mu_\alpha^{-1}}{Q(0)} < \infty. \qquad \square$$

Remark. The hypothesis of Theorem 1 includes $\sum j \log j p_j < \infty$. The reader may recall that this was not needed in the discrete time case, except to make $\sum k b_k < \infty$. The problem is open in the age-dependent case. A similar situation exists in the supercritical case when $\sum j \log j p_j = \infty$. (See section 11.)

When the lifetime distribution has sub-exponential tails then we have the interesting fact, contrasted to theorem 1, that the limit distribution is degenerate at 1. (Under the extra restrictions indicated in the first paragraph of section 8, this was proved by Chistyakov (1964).)

Theorem 2. *If $Z(t)$ is an age-dependent process with $m<1$ and G is in $\mathscr{S}$, then*

$$\lim_{t \to \infty} P\{Z(t)=1 \mid Z(t)>0\} = 1.$$

Proof. By theorem 8.1, with $q=1$, $f'(q)=m$ one has

$$\lim_{t \to \infty} \sum P\{Z(t)=k \mid Z(t)>0\} s^k = s. \qquad \square$$

Remark. Of course in the setting of theorem 1 we have

$$P\{Z(t)>0\} \sim c e^{\alpha t},$$

while if $G(t) \in \mathscr{S}$

$$P\{Z(t)>0\} \sim (1-m)^{-1}[1-G(t)].$$

11. Limit Theorems for the Supercritical Case

When $m>1$, and $f''(1)<\infty$, then an argument using the first and second moment of $Z(t)$ can be used to show that $e^{-\alpha t} Z(t)$, or equivalently $Z(t)/\mu(t)$, converges in mean square to a non-degenerate random variable (α is the Malthusian parameter for (m, G)). Although we will shortly give stronger results, the mean convergence is worth noting because of its simplicity, and the ease with which it extends to more complicated models. It was first observed by Bellman and Harris (1952), and is also discussed in chapter VI of Harris (1963).

Theorem 1. *If* $m>1$, $f''(1)<\infty$ *and* G *is non-lattice then* $W(t)=Z(t)/c' e^{\alpha t}$ *(where* c' *is as in (5.6)) converges in mean square to a non-degenerate random variable* W.

Proof. By (5.7)

$$E[W(t+\Delta)-W(t)]^2 \to 0 \quad \text{as } t\to\infty \tag{1}$$

uniformly for $\Delta\geqslant 0$, and this implies the mean square convergence of $W(t)$. The non-degeneracy of the limit variable can be seen from the fact that it has a positive variance. The latter is computed explicitly on p. 146 of Harris (1963). □

Harris (1963) and Jagers (1969) have shown that

$$\int_0^\infty E[W(t)-W]^2\,dt<\infty, \tag{2}$$

and thence prove a.s. convergence. The following extension of some of the results of section I.10 is somewhat sharper, and completes our study of limit laws for age-dependent processes.

Theorem 2. *Assume that* $m>1$.

(i) *If* $\sum p_j j\log j=\infty$ *then* $W(t)\equiv Z(t)/c' e^{\alpha t}\to 0$ *in probability.*

(ii) *If* $\sum p_j j\log j<\infty$ *then* $W(t)$ *converges in distribution to a non-negative random variable* W *having the following properties:*

a) $EW=1$.

b) $\varphi(u)=Ee^{-uW}$, $u\geqslant 0$, *is the unique solution of the equation*

$$\varphi(u)=\int_0^\infty f[\varphi(ue^{-\alpha y})]\,dG(y) \tag{3}$$

in the class

$$\mathsf{C}=\left\{\varphi\colon \varphi(u)=\int_0^\infty e^{-ut}\,dF(t),\ F(0+)<1,\ \int_0^\infty t\,dF(t)=1\right\}.$$

c) $P(W=0)=q\equiv P\{Z(t)=0$ *for some* $t\}$.

d) *The distribution of* W *is absolutely continuous on* $(0,\infty)$.

Proof. The proof parallels that of the corresponding result for the Galton-Watson process in chapter I, and was first given by Athreya (1969). (For earlier work along these lines see Levinson (1960).) The key idea is to identify the right side of the integral equation (3) as the expectation of a random variable, and then to use properties of i.i.d. random variables. Some parts of the proof are quite lengthy, and we will therefore omit the argument pertaining to the existence of a solution of (3) in C, and the absolute continuity. The reader is referred to the above reference for these proofs. The rest of the argument is broken into several lemmas.

Lemma 1 (Uniqueness). *There is at most one solution of* (3) *in* C.

Proof. Suppose φ_1 and φ_2 are in C and are solutions of (3). Let $\psi(u)=u^{-1}|\varphi_1(u)-\varphi_2(u)|$ for $u>0$. Using (3) and the mean value theorem we get

$$\psi(u)\leqslant\int_0^\infty \psi(u e^{-\alpha x})\,dG_\alpha(x), \tag{4}$$

where G_α is the function defined in section 4, i. e.

$$G_\alpha(x)=m\int_0^x e^{-\alpha y}\,dG(y).$$

We can rewrite (4) as $\psi(u)\leqslant E\psi(u e^{-\alpha X})$ where X is a random variable with distribution function G_α. Iteration yields $\psi(x)\leqslant E\psi(u e^{-\alpha S_n})$ where S_n is the nth partial sum of a sequence of independent random variables $\{X_i\colon i=1,2,\ldots\}$ with G_α as their common distribution function. By the strong law of large numbers $S_n\to\infty$ with probability one. Since $\varphi_1,\varphi_2\in\mathsf{C}$, ψ is bounded and $\lim_{x\downarrow 0}\psi(x)=0$. Hence the bounded convergence theorem yields $\psi(u)\leqslant\lim_n E\psi(u e^{-\alpha S_n})=0$. □

Now let $\varphi(t)$ denote the unique solution of (3) in C. Also define the functions

$$F_1(u,t)=E e^{-uW(t)},$$

$$H(u,t)=\frac{\mu(t)}{c'\,e^{\alpha t}}-\frac{1-F_1(u,t)}{u}.$$

Lemma 2. *If* $\sum p_j j\log j<\infty$, *then*

$$\lim_{u\downarrow 0}\sup_{t\geqslant 0}|H(u,t)|=0. \tag{5}$$

Proof. Recall the function $A(x)$ introduced in Section I.10, namely $A(x)=m-x^{-1}[1-f(1-x)]$, $0<x\leqslant 1$, $A(0)=0$. From the integral equation for $F(s,t)$, the definition of $A(x)$, and the facts that $H(u,t)=u^{-1}E[uW(t)-1+e^{-uW(t)}]$ and $1-e^{-x}\leqslant x$ for $x>0$ we get the inequality

$$\begin{aligned}0\leqslant H(u,t)\leqslant u(2c'^2)^{-1}+\int_0^t u^{-1}\{&u e^{-\alpha y} m H(u e^{-\alpha y},t-y)\\ &+m[1-F_1(u e^{-\alpha y},t-y)]\,A[1-F_1(u e^{-\alpha y},t-y)]\}\,dG(y).\end{aligned} \tag{6}$$

For $T>0, u>0$, let

$$H_T(u)=\sup_{0\leqslant t\leqslant T} H(u,t). \tag{7}$$

We shall show that there exist constants c_1 and v in $(0,\infty)$ such that

$$H_T(u)\leqslant u c'^{-2}+2c_1 A(c_1 u)+H_T(u e^{-\alpha v}). \tag{8}$$

Since $H_T(u)\to 0$ as $u\to 0$, iteration of (8) yields

$$H_T(u)\leqslant H(u)\equiv u c'^{-2}(1-e^{-\alpha v})^{-1}+2c_1\sum_{r=0}^{\infty} A(c_1 u e^{-r\alpha v}). \tag{9}$$

By corollary I.10.2, the last sum converges since $\sum p_j j\log j<\infty$. The right side of (9) being independent of T we have that

$$\lim_{u\downarrow 0}\sup_{t\geqslant 0}|H(u,t)|\leqslant \lim_{u\downarrow 0} H(u)=0.$$

It thus remains only to establish (8). Non-negativity of $H(u,t)$ implies

$$0\leqslant (u e^{-\alpha y})^{-1}\left[1-F_1(u e^{-\alpha y},t-y)\right]\leqslant (c' e^{\alpha(t-y)})^{-1}\mu(t-y)\leqslant c_1, \tag{10}$$

where c_1 is a constant independent u, y, t, whose existence is assured by Theorem 5.3A. For X as in the proof of lemma 1, there exists a v such that $P(X\leqslant v)=2^{-1}$. Now, for each u and T, $H(u,t)$ being continuous in t, there exists a $t_0=t_0(u,T)$ such that $H(u,t_0)=H_T(u)$. Using (6) with $t=t_0$ and then (10), we obtain

$$H_T(u)=H(u,t_0)\leqslant u(2c'^2)^{-1}+c_1 A(c_1 u)+\int_0^{t_0} H(u e^{-\alpha y},t_0-y)\,dG_\alpha(y). \tag{11}$$

Observe that, for any t, $H(u,t)$ is nondecreasing in u and so,

$$H(u e^{-\alpha y},t_0-y)\leqslant H_T(u) \quad \text{for } y \text{ in } [0,v],$$

and

$$H(u e^{-\alpha y},t_0-y)\leqslant H_T(u e^{-\alpha v}) \quad \text{for } y \text{ in } [v,t_0].$$

Thus (11) yields

$$H_T(u)\leqslant u(2c'^2)^{-1}+c_1 A(c_1 u)+H_T(u)\,G_\alpha(v)+H_T(u e^{-\alpha v})\,(1-G_\alpha(v)),$$

which implies (8) since $G_\alpha(v)=1-G_\alpha(v)=2^{-1}$. This proves lemma 2. □

Introduce the functions

$$K(u,t)=u^{-1}|F_1(u,t)-\varphi(u)|, \qquad K_T(u)=\sup_{t\geqslant T} K(u,t).$$

and

$$K(u)=\lim_{t\geqslant 0}\sup K(u,t).$$

Lemma 3. *If* $\sum p_j j\log j<\infty$, *then*

(i) $\lim_{u\downarrow 0} K(u)=0$

and

(ii) $K(u)\leqslant E\,K(u e^{-\alpha X})$,

where X *is a random variable with distribution* G_α.

Proof. Clearly

$$|K(u,t)|\leqslant |u^{-1}(F_1(u,t)-1)-\mu(t)\,(c' e^{\alpha t})^{-1}| + |\mu(t)\,(c' e^{\alpha t})^{-1}-1|+|1+u^{-1}(1-\varphi(u))|.$$

Now use lemma 2, theorem 5.3A and the fact that $\varphi \in \mathbf{C}$ to conclude that the three terms on the right go to zero as $t \to \infty$ and $u \to 0$ in that order. This proves part (i).

To prove (ii) use the fundamental integral equation (1.1) satisfied by $F(s,t)$, the equation (3) satisfied by $\varphi(u)$, and the definition of $A(x)$, to get the inequality

$$\begin{aligned} |K(u,t)| &\leqslant \int_0^t (u e^{-\alpha y})^{-1} |f(F_1(u e^{-\alpha y}, t-y)) - f(\varphi(u e^{-\alpha y}))| e^{-\alpha y} dG(y) \\ &\quad + \int_t^\infty (u e^{-\alpha y})^{-1} |e^{-u(c' e^{\alpha t})^{-1}} - f(\varphi(u e^{-\alpha y}))| e^{-\alpha y} dG(y) \\ &= I_1 + I_2 \quad \text{(say)}. \end{aligned}$$

Dominate I_2 by

$$\begin{aligned} I_2 &\leqslant \int_t^\infty (u e^{-\alpha y})^{-1} |e^{-u(c' e^{\alpha t})^{-1}} - 1| e^{-\alpha y} dG(y) \\ &\quad + \int_t^\infty (u e^{-\alpha y})^{-1} |1 - f(\varphi(u e^{-\alpha y}))| e^{-\alpha y} dG(y) \\ &\leqslant (c' e^{\alpha t})^{-1} + m \int_t^\infty (u e^{-\alpha y})^{-1} (1 - \varphi(u e^{-\alpha y})) e^{-\alpha y} dG(y) \end{aligned}$$

(since $(1-e^{-x}) \leqslant x$ for $x>0$ and the mean value theorem applies to f)

$$\leqslant (c' e^{\alpha t})^{-1} + (1 - G_\alpha(t)) \quad (\text{since } u^{-1}(1-\varphi(u)) \leqslant 1 \text{ for all } u>0).$$

For $t > T$ write

$$I_1 = \int_0^{t-T} + \int_{t-T}^t = I_{11} + I_{12} \quad \text{(say)}.$$

Since $K(0+)=0$ an application of mean value theorem to f gives the bound (for $t > 2T$)

$$|I_{12}| \leqslant C_2(1 - G_\alpha(T)), \quad \text{where } C_2 \text{ is some constant.}$$

Again by the mean value theorem applied to f, we have for any $t > T$

$$|I_{11}| \leqslant \int_0^{t-T} K(u e^{-\alpha y}, t-y) \, dG_\alpha(y) \leqslant \int_0^{t-T} K_T(u e^{-\alpha y}) \, dG_\alpha(y) \leqslant E K_T(u e^{-\alpha X}).$$

Thus, for $t > 2T$, we get

$$|K(u,t)| \leqslant (c' e^{\alpha t})^{-1} + (1 - G_\alpha(t)) + C_2(1 - G_\alpha(T)) + E K_T(u e^{-\alpha X})$$

or

$$K_{2T}(u) \leqslant (c' e^{\alpha 2T})^{-1} + (C_2 + 1)(1 - G_\alpha(T)) + E K_T(u e^{-\alpha X}).$$

On letting $T \to \infty$ this yields (ii) by the bounded convergence theorem. □

Iterating the inequality in lemma 3 (ii) we see that for all $n>1$

$$K(u) \leqslant E\,K(u\,e^{-\alpha S_n}), \tag{12}$$

where S_n is a sum of n independent random variables with distribution G_α. But the right side of (12) goes to $K(0+)$ as $n\to\infty$, and $K(0+)=0$. Hence

$$K(u)=0, \qquad u\geqslant 0, \tag{13}$$

and thus

$$\lim_{t\to\infty} E\,e^{-uW(t)} = \varphi(u), \qquad u\geqslant 0. \tag{14}$$

Letting W be a random variable with Laplace transform φ, this proves the convergence assertion in part (ii) of the theorem. That $EW=1$ follows from the fact that φ is in C.

Finally $P\{W=0\}=q$ follows at once on letting $u\to\infty$ in (3), by noting that $P\{W=0\}<1$ must be a root of $t=f(t)$. (That $P\{W=0\}\neq 1$ follows from the fact that $EW=1$.)

This completes the proof of part (ii) except for the existence and absolute continuity parts previously mentioned.

Turning to part (i) we will exploit the embedded Galton-Watson process $\{\zeta_n;\, n=0,1,2,\ldots\}$ defined in Section 3.

We know from Chapter I that when $\sum_{j=2}^{\infty} p_j j\log j=\infty$, $\lim_n \zeta_n(\omega)m^{-n}=0$ a.s. Let $(\Omega, \mathbb{F}, P)$ be our basic probability space. Let $\mathscr{G}$ be the sub σ-algebra of $\mathbb{F}$ generated by $\zeta_0(\omega), \zeta_1(\omega),\ldots, \zeta_n(\omega),\ldots$. Let η_1, η_2 and ε be three arbitrary numbers in $(0,1)$. Since $\lim_n \zeta_n(\omega)m^{-n}=0$ a.s., there exists, by Egoroff's theorem, a set A in $\mathscr{G}$, and an integer N such that $P(\mathsf{A})>1-\eta_1$, and such that $\zeta_n(\omega)m^{-n}<\eta_2$ for all ω in A and $n\geqslant N$. Then

$$P(\omega\colon Z(t,\omega)>\varepsilon\mu(t)) \leqslant \eta_1 + P(\omega\colon Z(t,\omega)>\varepsilon\mu(t), \omega\in\mathsf{A}). \tag{15}$$

But

$$\begin{aligned} P(\omega\colon Z(t,\omega)>\varepsilon\mu(t), \omega\in\mathsf{A}) &\leqslant P\Big(\omega\colon \sum_{k\leqslant N} Y_k(t,\omega)>\varepsilon 2^{-1}\mu(t), \omega\in\mathsf{A}\Big) \\ &\quad + P\Big(\omega\colon \sum_{k>N} Y_k(t,\omega)>\varepsilon 2^{-1}\mu(t), \omega\in\mathsf{A}\Big) \\ &= B_1(t)+B_2(t) \quad \text{(say)}, \end{aligned}$$

where $Y_k(t,\omega)$ is as in Section 3.

Now

$$B_1(t)\leqslant 2\varepsilon^{-1}(\mu(t))^{-1} E\Big(\sum_{k\leqslant N} Y_k(t,\omega)\Big) \leqslant 2(\varepsilon\mu(t))^{-1}\Big(\sum_{k=0}^{N} m^k\Big).$$

Since $\mu(t)\to\infty$ as $t\to\infty$ and ε and N are fixed we see that $\lim_{t\to\infty} B_1(t)=0$.

Next, let $\mathsf{D}=\left\{\omega\colon \sum_{k>N} Y_k(t,\omega)>\varepsilon 2^{-1}\mu(t)\right\}$. Then, since A is in $\mathscr{G}$, $B_2(t) \leqslant E[\chi_{\mathsf{A}}(\omega)E(\chi_{\mathsf{D}}(\omega)|\mathscr{G})]$ where χ_{A} and χ_{D} are, respectively, the indicator functions of the two sets A and D. Thus

$$\begin{aligned} B_2(t) &\leqslant 2(\varepsilon\mu(t))^{-1} E\left(\chi_{\mathsf{A}}(\omega)\sum_{k>N}\zeta_k(\omega)\,p_k(t)\right) \quad \text{where } p_k(t)=G^{*k}(t)-G^{*(k+1)}(t) \\ &\leqslant 2(\varepsilon\mu(t))^{-1}\eta_2\left(\sum_{k>N} m^k p_k(t)\right) \\ &\leqslant 2\varepsilon^{-1}\eta_2 \quad \left(\text{since } \sum_{k>N} m^k p_k(t)<\mu(t)\right), \end{aligned}$$

leading to the estimate

$$\limsup_{t\to\infty} P(\omega\colon Z(t,\omega)>\varepsilon\mu(t)) \leqslant \eta_1+2\varepsilon^{-1}\eta_2.$$

The proof is complete since η_1 and η_2 are arbitrary. □

There is also the following generalization of Theorem I.10.2.

Theorem 3. *Let W be as in Theorem 2. Then for any $p>0$, $E W|\log W|^p<\infty$ if and only if $\sum_{2}^{\infty} p_j j(\log j)^{p+1}<\infty$.*

For a proof we refer the reader to Athreya (1971a).

Complements and Problems IV

1. It is reasonable to conjecture that for continuous lifetime distributions, the necessary and sufficient condition for non-explosion in the age-dependent case is the same as in the Markov case.

2. Prove the sufficiency of each of the conditions (i)–(iv) for direct R-integrability listed after theorem 4.1. (See Feller Vol. II.)

3. Let $\hat{G}(\theta)=\int_0^\infty e^{-\theta t}\,dG(t)$, and let $\theta_0=\inf\{\theta\colon \hat{G}(\theta)<\infty\}$. If $\hat{G}(\theta_0)=\infty$, then show that the Malthusian parameter $\alpha(\gamma,G)$ exists for all $\gamma>0$. If $\hat{G}(\theta_0)<\infty$, then $\alpha(\gamma,G)$ exists if and only if $\gamma\geqslant\hat{G}(\theta_0)^{-1}$.

4. Investigate the situation in problem 3 for a distribution G satisfying $1-G(t)\sim t^{-2}e^{-t}$.

5. Show that when the Malthusian parameter α for (γ,G) exists, then $e^{-\alpha t}[1-G(t)]$ is directly R-integrable.

6. Prove lemmas 4.5 and 4.6. The details are given in Chistyakov (1964).

7. Prove theorems 5.3A and 5.3B. Hint for 5.3B: Show that

$$\frac{\mu(t)}{1-G(t)}=\sum m^n\left[\frac{1-G_{n+1}(t)}{1-G(t)}-\frac{1-G_n(t)}{1-G(t)}\right],$$

and appeal to lemma 4.7 and the dominated convergence theorem.

8. Fill in the steps in the proof of (5.7). See Harris (1963) for an outline of the proof.

9. To prove the Vinogradov converse (p. 147), show first that $G_\alpha(\infty)<\infty$ for the given α.

10. In the discrete case $\lim((f_n(s)-q)/\gamma^n)=Q(s)$ satisfies the equation (I.11.10), and this fact was exploited to derive a spectral decomposition for $f_n(s)$. Can these ideas be extended to $\lim((F(s,t)-q)/e^{-\alpha t})$ for the age-dependent case when the Malthusian parameter exists?

11. Carry out the "standard" iteration argument called for in the derivation of (8.2). Also prove (8.3) using the steps suggested.

12. *Open Problem:* Deduce the limit theorem for the supercritical, age-dependent case by using the inverse function of $Q(s)$ (defined in section 7) in a similar fashion to that of section I.11.

13. *Open Problem:* Find the relation between the limit distribution of $Z(t)/\mu(t)$ and that of ζ_n/m^n, where ζ_n is the embedded Galton-Watson process.

14. *Open Problem:* Lamperti (1967a) has shown that in the discrete time case, W is in the domain of a stable law if and only if $\{p_j\}$ is in the same domain. Is there an analogous result for the age-dependent case?

15. *Open Problem:* Prove the existence of a solution of equation (11.3) in the class C defined there, without the hypothesis $\sum p_j j\log j<\infty$.

Possible approach: Approximate $\{p_j; j\geqslant 0\}$ by a sequence of distributions $\{p_j^{(n)}; j\geqslant 0\}$, which satisfy the moment condition for each n. Use a compactness argument.

16. *Non-degeneracy:* In problem 10 of chapter I, the reader was asked to show that if $f(s)$ was not linear, then the conditioned limit distribution $\{b_j\}$ for the sub-critical process was not degenerate. A similar result is to be expected in the age-dependent case when the Malthusian parameter exists. (When it does not exist then we have seen that such a result is false. See Chover, Ney, Wainger (1972)).

To prove that the limiting generating function is not linear: Referring to expression (7.10), it is sufficient to show that

$$\int_0^\infty e^{-\alpha t}\{q-f[F(s,t)]-\gamma(q-F(s,t))\}\,dt$$

is not linear in s. If $f''(1)<\infty$, then $F''(s,t)$ exists for $0\leqslant s\leqslant 1$, and it becomes sufficient to show that the second derivative of the above, namely

$$\int_0^\infty e^{-\alpha t}\{F''(s,t)(m-f'[F(s,t)])-F'^2(s,t)f''[F(s,t)]\}\,dt \qquad (*)$$

is not identically 0. But as $s\to 1$

$$\int_0^\infty e^{-\alpha t}F''(s,t)(m-f'[F(s,t)])\,dt\to 0,$$

while $\int_0^\infty e^{-\alpha t} F'^2(s,t) f''[F(s,t)]\,dt \to f''(1) \int_0^\infty e^{-\alpha t} \mu^2(t)\,dt$, which is positive. Hence $(*) < 0$.

17. *Open Problems:* A strengthening of theorem 11.2 would be to show that (2) implies that $Z(t)/c' e^{\alpha t}$ actually converges almost surely to a random variable W having the same distribution as W of theorem 1, i.e., $E(e^{-uW}) \equiv \varphi(u)$. In Chapters I and III martingale arguments yielded this result. We conjecture that in the age-dependent case an appropriate martingale argument would also work. This is supported by the following heuristics, borrowed from Harris (1963). Let: $Z(x, y, t)$ = the number of particles living at time t and of age exceeding y, for a branching process starting with one particle of age x at $t=0$,

$$M(x,y,t) = EZ(x,y,t),$$

$$V(x) = e^{\alpha x}(1-G(x))^{-1} \int_x^\infty e^{-\alpha u}\,dG(u),$$

$$A(x) = \left(\int_0^x e^{-\alpha t}(1-G(t))\,dt\right)\left(\int_0^\infty e^{-\alpha t}(1-G(t))\,dt\right)^{-1},$$

$$V_t = \sum_{i=1}^{Z(t)} V(x_i),$$

where $x_1, x_2, \ldots, x_{Z(t)}$ are the ages of the $Z(t)$ particles in the system at time t.

Under some mild regularity assumptions it can be shown that $M(x,y,t)$ satisfies an integral equation similar to (1.1), and that $V(y)$ is a right eigenfunction for $M(x,y,t)$ with eigenvalue $e^{\alpha t}$. That is

$$\int_0^\infty V(y) M(x, dy, t) = e^{\alpha t} V(x).$$

This implies that the process $\{V_t e^{-\alpha t}; t \geqslant 0\}$ is a martingale, and since it is non-negative, $\lim_{t\to\infty} V_t e^{-\alpha t} = W'$ exists almost surely. Now look at $V_t(Z(t))^{-1}$ on the set of non-extinction. If $A_x(y,t) = Z(x,y,t)\,(Z(t))^{-1}$ is the age distribution at time t for a process starting from one particle of age x at $t=0$, then a reasonable conjecture is that almost surely (or at least in probability) on the set of non-extinction, $A_x(y,t)$ converges to $A(y)$ as $t \to \infty$, where $A(y)$ is as defined above and thus independent of x. Harris refers to $A(\cdot)$ as the stationary age distribution. Our guess is that this convergence depends solely on the fact that on the set of non-extinction $Z(t) \to \infty$, and does not need any extra assumption like $\sum p_j j \log j < \infty$. On the basis of this convergence it would follow that since

$$Z(t) e^{-\alpha t} = (V_t(Z(t))^{-1})^{-1} V_t e^{-\alpha t} = \left(\int_0^\infty V(y)\,dA_x(y,t)\right)^{-1} V_t e^{-\alpha t},$$

$\lim_{t\to\infty} Z(t) e^{-\alpha t}$ exists almost surely (or at least in probability) and is equal to

$$\left(\int_0^\infty V(y)\,dA(y)\right)^{-1} W'.$$

In order to generalize Seneta's result of Chapter I, it may be useful to find out if there exists a function $C(t)$ such that $V_t(C(t))^{-1}$ converges to a non-degenerate limit law, and then use the convergence of the age distribution.

18. *Conditioned limit laws* (open problem). Study the age-dependent analogs of the limit laws in sections I.14 and I.15, e. g.:

(a) $$\lim P\{Z(t)=j \mid t+\tau<T<\infty\}$$

where $T=$ the time of extinction of the process, and the limit is taken either first on t and then on τ, or in the reverse order;

(b) $$\lim_{t\to\infty} P\{Z(ct)\leqslant k(t) \mid Z(t)>0\}, \qquad (0<c<1).$$

(This is primarily of interest when $m=1$.) Is there a limiting process here, as there was for the Galton-Watson process?

19. If $G(0+)>0$, then the process $Z(t)$ is equivalent to a process $\hat{Z}(t)$ with $\hat{G}(0+)=0$ and generating function $\hat{f}(s)$. Determine $\hat{G}$ and $\hat{f}$ in terms of G and f.

20. In the case of the Galton-Watson process we established the geometric convergence rate of $f_n(s)$ to q by studying $\gamma^{-n}(d/ds)\,f_n(s)$ (see section I.11). In the age-dependent case one can follow an analogous procedure for proving that the rate of convergence of $F(s,t)$ to q is exponential when the Malthusian parameter exists (theorem IV.7.1).

Show that $\partial F(s,t)/\partial s$ satisfies the equation

$$\frac{\partial F(s,t)}{\partial s} = 1-G(t)+\int_0^t f'[F(s,t-y)]\frac{\partial F(s,t-y)}{\partial s}\,dG(y),$$

and use this to prove the convergence of $e^{-\alpha t}(\partial F(s,t)/\partial s)$ along the lines of the proof of theorem IV.7.1.

21. (Doney, 1971). Let $N(t)=$ total number of particles up to time t. By setting up an integral equation for the joint generating function $F(s_1,s_2,t)$ for $(Z(t),N(t))$, and using the methods of section 11, show that if $1<m<\infty$ and $\sum(j\log j)\,p_j<\infty$ then $(Z(t)/EZ(t), N(t)/EN(t))$ converges in distribution to (W,W).

22. Give a heuristic derivation of (5.3) and (5.5) by reasoning similarly to the motivating discussion for (1.1).

23. To prove (5.7) one uses the following fact, whose proof we leave to the reader (or see Feller, Vol. II, Section XI.6):

If $K(x)$ is a defective distribution on $[0,\infty)$ with $\lambda=K(\infty)<1$, and if $\xi(t)$ is bounded, measurable, and $\lim_{t\to\infty}\xi(t)=\xi_0$ exists, then the solution $x(t)$ of $x=\xi+(x*K)$ satisfies $x(t)\to\xi_0/(1-\lambda)$ as $t\to\infty$.

Chapter V

Multi-Type Branching Processes

1. Introduction and Definitions

The processes we have studied till now have all consisted of indistinguishable particles. After starting with fixed unit lifetimes for the Galton-Watson process, we considered exponential, and then arbitrary lifetime distributions.

Another natural direction for generalization is to allow a number of distinguishable particles having different probabilistic behavior, and it is to such processes that we now turn. We again start with the discrete time version resulting from fixed unit lifetimes; and later go into the other lifetime distributions.

Throughout this chapter we consider a finite number (p) of particle types. Such processes arise in a variety of biological and physical applications, where, e. g., they could represent genetic types in an animal population, mutant types in a bacterial population, electrons, photons, nucleons, etc. in a cosmic ray cascade, and so on. It is also possible to construct processes with a countable number or a continuum of types, or even with an abstract space as a "type-space". We shall see some examples of "continuous type" processes in chapter VI, where we discuss cascades and branching random walks. We shall not, however, go into the general theory of branching process on abstract type-spaces. For work in this direction, see, for example, chapter III of Harris (1963), T. Mullikin (1963), or M. Jirina (1958).

We will prove the analogs of the three main limit laws for the Galton-Watson process. The discrete case is treated carefully and in considerable detail. The continuous time cases are covered more quickly and without detailed proofs, since the techniques become increasingly repetitious re-combinations of earlier cases. The reader is referred to the recent book of C. Mode (1971) on multi-type branching processes, where numerous models which we will not consider here are treated. This reference also contains a fairly detailed treatment of some age-dependent multi-type processes.

In section 8 we shall see some further aspects of the structure of the supercritical multi-type process in terms of the asymptotics of functionals of the process. We also look at the relation between branching processes and some classical urn schemes.

To define the particle production of a p-type process, we need p generating functions, each in p variables. The ith generating function, $f^{(i)}$, will determine the distribution of the number of offspring of various types to be produced by a type i particle. Thus we let

$$f^{(i)}(s_1,\ldots,s_p)=\sum_{j_1,\ldots,j_p\geq 0} p^{(i)}(j_1,\ldots,j_p)\,s_1^{j_1}\cdot\cdots\cdot s_p^{j_p}, \tag{1}$$

$$0\leq s_\alpha\leq 1, \qquad \alpha=1,\ldots,p,$$

where $p^{(i)}(j_1,\ldots,j_p)=$ the probability that a type i parent produces j_1 particles of type 1, j_2 of type 2, …, j_p of type p.

Notation. Throughout the chapter, we adopt the following conventions.

(A–1) $\mathscr{E}_p$ = p-dimensional Euclidean space

(A–2) $\mathscr{E}_p^+$ = the p-dimensional nonnegative orthant $=\{(x_1,\ldots,x_p): x_i\geq 0,\ i=1,\ldots,p\}$.

(A–3) Points in $\mathscr{E}_p$ are denoted by heavy type: $\mathbf{x}=(x_1,\ldots,x_p)$.

(A–4) $\mathbf{u}\leq\mathbf{v}$ means $u_\alpha\leq v_\alpha$, $\alpha=1,\ldots,p$, while $\mathbf{u}<\mathbf{v}$ means $u_\alpha\leq v_\alpha$ for all α and $u_\alpha<v_\alpha$ for at least one α.

(A–5) $\mathbf{0}=(0,\ldots,0)$, $\mathbf{1}=(1,\ldots,1)$

(A–6) $\mathscr{C}_p$ is the unit cube in $\mathscr{E}_p=\{\mathbf{x}:\mathbf{0}\leq\mathbf{x}<\mathbf{1}\}$.

(A–7) $e_i=(0\ldots0,1,0\ldots0)$, with the 1 in the ith component.

(A–8) $\mathscr{R}_p$ is the p-dimensional lattice space, i.e. the set of all points of $\mathscr{E}_p$ with integer coordinates.

(A–9) $\mathscr{R}_p^+=\{\mathbf{x}\in\mathscr{R}_p: x_i\geq 0,\ i=1,\ldots,p\}$.

(A–10) When there is no possible confusion, we will drop the subscripts p on $\mathscr{E}_p$, $\mathscr{E}_p^+$, $\mathscr{C}_p$, $\mathscr{R}_p$, $\mathscr{R}_p^+$.

(A–11) The vector of absolute values is

$$|\mathbf{x}|=(|x_1|,\ldots,|x_p|).$$

(A–12) The sup norm is

$$\|\mathbf{x}\|=\max(|x_1|,\ldots,|x_p|).$$

(A–13) We use the product notation

$$\mathbf{x}^{\mathbf{y}}=\prod_{i=1}^{p} x_i^{y_i}.$$

For a matrix $\mathbf{M}$, the sup norm is

$$\|\mathbf{M}\|=\max(|m_{ij}|,\ i,j=1,\ldots,p). \tag{A–14}$$

(A–15) Symbols which are traditionally used for integers will be used for points in $\mathscr{R}_p$. Thus we will write

$$\boldsymbol{i}=(i_1,\ldots,i_p)$$

without explicitly saying $\boldsymbol{i}\in\mathscr{R}_p$. This usage will always be clear from the context.

For the particle production probability and generation functions we write

(B–1) $$\boldsymbol{p}(\boldsymbol{j})=(p^{(1)}(\boldsymbol{j}),\ldots,p^{(p)}(\boldsymbol{j})),$$

and

(B–2) $$\boldsymbol{f}(\boldsymbol{s})=(f^{(1)}(\boldsymbol{s}),\ldots,f^{(p)}(\boldsymbol{s})).$$

Thus the coordinates of $\boldsymbol{p}$ and $\boldsymbol{f}$ are denoted by superscripts rather than subscripts (as in A–3). (This will not cause any confusion, and is consistent with the notation in (C–1) below, denoting a quantity associated with a process initiated by a single particle of the type indicated in the superscript.)

We may now write (1) in the form

$$\boldsymbol{f}(\boldsymbol{s})=\sum_{\boldsymbol{j}\in\mathscr{R}_p}\boldsymbol{p}(\boldsymbol{j})\boldsymbol{s}^{\boldsymbol{j}},\qquad \boldsymbol{s}\in\mathscr{C}_p\,. \tag{2}$$

The total datum of the problem is the function $\boldsymbol{p}(\boldsymbol{j})$, or equivalently $\boldsymbol{f}(\boldsymbol{s})$.

Definition. *A multi (p -type Galton-Watson process is a Markov chain* $\{\boldsymbol{Z}_n;\, n=0,1,2,\ldots\}$ *on* $\mathscr{R}_p^+$, *with transition function*

$$\begin{aligned} P(\boldsymbol{i},\boldsymbol{j})&=P\{\boldsymbol{Z}_{n+1}=\boldsymbol{j}\mid \boldsymbol{Z}_n=\boldsymbol{i}\},\qquad \boldsymbol{i},\boldsymbol{j}\in\mathscr{R}_p^{+} \\ &=\textit{coefficient of } \boldsymbol{s}^{\boldsymbol{j}} \textit{ in } [\boldsymbol{f}(\boldsymbol{s})]^{\boldsymbol{i}}. \end{aligned} \tag{3}$$

(C–1) When the process is initiated in state $\boldsymbol{i}$, we will denote it by $\boldsymbol{Z}_n^{(\boldsymbol{i})}$. In terms of coordinates, we write

$$\boldsymbol{Z}_n=(Z_{n1},\ldots,Z_{np}),\quad \text{or}\quad \boldsymbol{Z}_n^{(\boldsymbol{i})}=(Z_{n1}^{(\boldsymbol{i})},\ldots,Z_{np}^{(\boldsymbol{i})}).$$

Thus $Z_{nj}^{(\boldsymbol{i})}=$ the number of type j particles in the nth generation for a process with $\boldsymbol{Z}_0=\boldsymbol{i}$. When we do not want to specify an initial state we leave off the superscript.

(C–2) The generating function of $\boldsymbol{Z}_n^{(\boldsymbol{i})}$ will be denoted by $\boldsymbol{f}_n^{(\boldsymbol{i})}(\boldsymbol{s})$.

(C–3) We also write

$$\boldsymbol{Z}_n^{(\boldsymbol{e}_i)}=\boldsymbol{Z}_n^{(i)},\quad \text{and}\quad \boldsymbol{f}_n^{(\boldsymbol{e}_i)}(\boldsymbol{s})=\boldsymbol{f}_n^{(i)}(\boldsymbol{s}).$$

Generating Functions will play their usual important role. Using by now familiar arguments one can show that

$$\boldsymbol{f}_{k+n}(\boldsymbol{s})=\boldsymbol{f}_k[\boldsymbol{f}_n(\boldsymbol{s})],\qquad \boldsymbol{s}\in\mathscr{C}_p\,. \tag{4}$$

Probabilistic Setting. Again we will have occasion to make reference to the basic probability space $(\Omega, \mathbb{F}, P_j)$ for the process. Here Ω is the space of trees, $\mathbb{F}$ is generated by the cylinder sets of Ω, and P_j is the probability measure on $(\Omega, \mathbb{F})$ when the process is initiated with $\boldsymbol{Z}_0 = \boldsymbol{e}_j$.

The expectation operator for P_j will be denoted by E_j. Similar remarks hold about the construction of the above space as in sections I.1 and III.2. We also have again the

Additive Property. The process $\{\boldsymbol{Z}_n^{(\mathbf{i})}; n=0,1,2,\ldots\}$ is the sum of $i_1+\cdots+i_p$ independent process, i_j of which are initiated in state $\boldsymbol{e}_j$, $j=1,\ldots,p$.

Assumption: Nonsingularity. If $\boldsymbol{f}(s)=\boldsymbol{M}s'$, where $\boldsymbol{M}$ is a $p \times p$ matrix of non-negative elements, then we say $\{\boldsymbol{Z}_n\}$ is a *singular* process. In this case each particle has exactly one offspring, and hence the branching process will be equivalent to an ordinary finite Markov chain. We assume nonsingularity throughout.

2. Moments and the Frobenius Theorem

Let $m_{ij} = E Z_{1j}^{(\mathbf{i})}$ = the expected number of type j offspring of a single type i particle in one generation. We always *assume that m_{ij} exists*, and define the *mean matrix*

$$\boldsymbol{M} = \{m_{ij};\, i,j=1,\ldots,p\}.$$

Clearly

$$m_{ij} = \frac{\partial f^{(i)}}{\partial s_j}(\mathbf{1}),$$

and

$$E Z_{nj}^{(\mathbf{i})} = \frac{\partial f_n^{(\mathbf{i})}}{\partial s_j}(\mathbf{1}). \tag{1}$$

Using this fact and the chain rule for differentiation, we get

$$E[\boldsymbol{Z}_n | \boldsymbol{Z}_0] = \boldsymbol{Z}_0 \boldsymbol{M}^n. \tag{2}$$

Let $m_{ij}^{(n)}$ = the (i,j)th element of $\boldsymbol{M}^n$.

Assumption. There is an n such that $m_{ij}^{(n)} > 0$ for all $i,j=1,\ldots,p$. A matrix $\boldsymbol{M}$ having this property is called *strictly positive.* If $\boldsymbol{M}$ is strictly positive, then the process $\{\boldsymbol{Z}_n;\, n=0,1,2,\ldots\}$ is called *positive regular.*

Higher moments, when they exist, can again be defined in terms of derivatives of $\boldsymbol{f}_n(\boldsymbol{s})$ at $\mathbf{1}$. We use the notation

$$q_n^{(\alpha)}(i,j) = E\{Z_{ni}^{(\alpha)} Z_{nj}^{(\alpha)} - \delta_{ij} Z_{ni}^{(\alpha)}\} \qquad (3)$$
$$= \frac{\partial^2 f_n^{(\alpha)}}{\partial s_i \partial s_j}(\mathbf{1}), \qquad i,j,\alpha = 1, \ldots, \not p;$$

and define the matrix

$$\boldsymbol{Q}_n^{(\alpha)} = \{q_n^{(\alpha)}(i,j)\,;\, i,j = 1, \ldots, \not p\}, \qquad (4)$$

the vector of matrices $\mathbb{Q}_n = (\boldsymbol{Q}_n^{(1)}, \ldots, \boldsymbol{Q}_n^{(\not p)})$, the quadratic form

$$\boldsymbol{Q}_n^{(\alpha)}[\boldsymbol{s}] = \tfrac{1}{2} \sum_{i=1}^{\not p} \sum_{j=1}^{\not p} s_i q_n^{(\alpha)}(i,j) s_j, \qquad (5)$$

and the vectors of quadratic forms

$$\mathbb{Q}_n[\boldsymbol{s}] = (\boldsymbol{Q}_n^{(1)}[\boldsymbol{s}], \ldots, \boldsymbol{Q}_n^{(\not p)}[\boldsymbol{s}]). \qquad (6)$$
$$\mathbb{Q}[\boldsymbol{s}] \equiv \mathbb{Q}_1[\boldsymbol{s}].$$

By elementary manipulation of $\boldsymbol{f}_n(\boldsymbol{s})$ one can derive an expression for $\boldsymbol{Q}_n^{(\alpha)}$ in terms of $\mathbb{Q}$ and $\boldsymbol{M}$, which is the multidimensional analog of (I.2.2). For example, a formula for $E(\boldsymbol{Z}_{ni}\boldsymbol{Z}_{nj})$ can be found on page 37 of Harris (1963). These expressions are rather bulky, and since we shall not be using them explicitly, we will not derive them here. We will, however, derive the analogous results in the continuous time case in section 7.3.

We shall have frequent occasion to apply the *Frobenius theorem* for positive matrices, which we quote here. For a proof of this theorem, see Karlin (1966).

Theorem 1. *A strictly positive matrix* $\boldsymbol{M}$ *has a maximal eigenvalue* ρ *which is positive, simple, and has associated positive right and left eigenvectors* $\boldsymbol{u}$ *and* $\boldsymbol{v}$. *If these are normalized so that* $(\boldsymbol{u} \cdot \boldsymbol{v}) = 1$ *and* $(\boldsymbol{u} \cdot \mathbf{1}) = 1$, *then one can write*

$$\boldsymbol{M}^n = \rho^n \boldsymbol{P} + \boldsymbol{R}^n,$$

where, $\boldsymbol{P}$ is the matrix whose (i,j)th entry is $u_i v_j$, and where

(i) $\boldsymbol{PR} = \boldsymbol{RP} = \mathbf{0}$, *and*

(ii) $r_{ij}^{(n)} \leqslant c\rho_0^n$, $i,j = 1, \ldots, \not p$, *for some* $c < \infty$, *and* $0 < \rho_0 < \rho$ *(*$r_{ij}^{(n)}$ *being the* (i,j) th *entry of* $\boldsymbol{R}^n$*).*

Assumption. From now on $\boldsymbol{M}$ will always denote the mean matrix and will be assumed finite, while $\boldsymbol{u}$, $\boldsymbol{v}$ and ρ will be as given in the theorem above (and with the same normalizations).

3. Extinction Probability and Transience

The role of the crucial criticality parameter (analogous to the mean m in one dimension) is now played by the maximal eigenvalue ρ of $\boldsymbol{M}$, as can be seen in theorem 2 below. The state $(0, \ldots, 0)$ is an absorbing state, and the process is transient except in the singular case. The proof of this fact is not entirely trivial, however, since we now have the complication that there may be states from which absorption is impossible and others from which it is possible. The idea of the proof is to show that in the former case the process must blow up. The main results are given below. Since complete proofs are given in Harris ((1963), sections II.6 and II.7), we shall not repeat them here.

Theorem 1. *If $\{\boldsymbol{Z}_n\}$ is positive regular and nonsingular, then*

$$P\{\boldsymbol{Z}_n=\boldsymbol{j} \text{ infinitely often}\}=0$$

for any $\boldsymbol{j}\neq\boldsymbol{0}, \boldsymbol{j}\in\mathscr{R}_p$.

Harris points out with the illustration $f^{(1)}(s_1,s_2)=s_1 s_2$, $f^{(2)}(s_1,s_2)=1$, that nonsingularity is *not* sufficient for the transience of the nonzero states, since in this example the state $(1,1)$ is recurrent.

Let $q^{(i)}=$ the probability of eventual extinction of the process initiated with a single particle of type i $(=P\{\boldsymbol{Z}_n^{(i)}=\boldsymbol{0}$ for some $n\})$, and let $\boldsymbol{q}=(q^{(1)}, \ldots, q^{(p)})\geqslant\boldsymbol{0}$.

Theorem 2. *Assume $\{\boldsymbol{Z}_n\}$ is positive regular and nonsingular, and let ρ be the maximum eigenvalue of $\boldsymbol{M}$.*

(i) *If $\rho\leqslant 1$ then $\boldsymbol{q}=\boldsymbol{1}$. If $\rho>1$, then $\boldsymbol{q}<\boldsymbol{1}$;*

(ii) $\lim_{n\to\infty} \boldsymbol{f}_n(\boldsymbol{s})=\boldsymbol{q}$ *for all $\boldsymbol{s}\in\mathscr{C}_p$;*

(iii) *The only solution of $\boldsymbol{f}(\boldsymbol{s})=\boldsymbol{s}$ in $\mathscr{C}_p$* is $\boldsymbol{q}$.

We shall call the process supercritical, critical or subcritical accordind as $\rho>$, $=$ or <1.

4. Limit Theorems for the Subcritical Case

The multitype analog of the Yaglom theorem of Section I.8 in its sharpest form was proved by A. Joffe and F. Spitzer (1967). Earlier forms under stronger moment hypotheses were due to T. Harris (1963) and M. Jirina (1957). The Joffe-Spitzer results are contained in theorems 1 and 2 below.

Theorem 1. *If $\rho<1$, then*

$$\frac{\boldsymbol{v}\cdot[\boldsymbol{1}-\boldsymbol{f}_n(\boldsymbol{s})]}{\rho^n}\downarrow Q(\boldsymbol{s})\geqslant 0 \quad \text{as } n\to\infty,\ \boldsymbol{s}\in\mathscr{C}_p, \tag{1}$$

where $Q(\cdot)$ *is nonincreasing and* >0 *if and only if* $E\|Z_1\|\log\|Z_1\|<\infty$;

$$\lim_{n\to\infty}\frac{\mathbf{1}-f_n(s)}{\rho^n}=Q(s)\,u\,; \tag{2}$$

and

$$\lim_{n\to\infty}\rho^{-n}P\{Z_n\neq\mathbf{0}\,|\,Z_0=i\}=Q(\mathbf{0})\,(i\cdot u). \tag{3}$$

Theorem 2. *If* $\rho<1$ *then*

$$\lim_{n\to\infty}P\{Z_n=j\,|\,Z_0=i,\,Z_n\neq\mathbf{0}\}=b(j) \tag{4}$$

exists, is independent of $\mathbf{i}$, *and is a probability measure on* $\mathscr{R}^+$. *Furthermore*

$$\sum j\,b(j)<\infty \quad \text{if and only if } E\|Z_1\|\log\|Z_1\|<\infty. \tag{5}$$

When $Q(s)>0$, then (2), (3) and (4) are particularly easy to prove; and in turn, when the second moments of Z_1 exist, then the fact that $Q(s)>0$ is easily established. Since the idea of the proof is already illustrated in this setting, and since we gave complete proofs in the one-dimensional case (Sections I.8 and I.11), we will here give details only for the case $E\|Z_1\|^2<\infty$; and refer the reader to the paper of Joffe and Spitzer for the proofs in the general case (which involve considerably more technical difficulties).

The main part of the following lemma is an analog for matrices of the well known criterion for the convergence of products of numbers. (For the proof see the above reference.)

Lemma 1. *Let* P *be a strictly positive matrix with maximal eigenvalue* 1 *and associated right and left eigenvectors* u *and* v; *the latter normalized so that* $u\cdot v=1$ *and* $u\cdot\mathbf{1}=1$. *Let* A_n, $\mathbf{0}\leqslant A_n\leqslant P$ *be a given sequence of matrices, define*

$$B_n=\prod_{k=0}^{n}(P-A_k),$$

and let x *be a fixed vector such that* $x\geqslant\mathbf{0}$ *and* $B_n x\neq\mathbf{0}$ *for all* $n\geqslant 1$. *Then*

(i) $\lim_{n\to\infty}A_n=\mathbf{0}$ *implies* $\lim_{n\to\infty}B_n x/(v\cdot(B_n x))=u$, *the convergence being uniform for* $x\geqslant 0$, $B_n x\neq 0$;

(ii) $\lim_{n\to\infty}B_n x=x_0$ *always exists;*

(iii) $x_0\neq\mathbf{0}$ *if and only if* $\sum A_k<\infty$.

By the integral form of the remainder term in the multivariate Taylor expansion of $f(s)$, one can show that there is a non-negative matrix valued function $A(s)$ such that

$$\mathbf{1}-f(s)=(\mathbf{1}-s)\,(M'-A(s)). \tag{6}$$

(Compare with the function $A(s)$ in I.10.)

If $E\|\boldsymbol{Z}_1\|^2<\infty$, then

$$\boldsymbol{A}(\boldsymbol{s})=O(\|\boldsymbol{1}-\boldsymbol{s}\|). \tag{7}$$

We can now easily get the weak forms of theorems 1 and 2.

Theorem 3. *If $\rho<1$ and $E\|\boldsymbol{Z}_1\|^2<\infty$, then as $n\to\infty$*

$$\rho^{-n}\boldsymbol{v}\cdot[\boldsymbol{1}-\boldsymbol{f}_n(\boldsymbol{s})]\downarrow Q(\boldsymbol{s})>0, \qquad \boldsymbol{s}\in\mathscr{C}_p, \tag{8}$$

where $Q(\cdot)$ is nonincreasing; $Q(\boldsymbol{1}-)=0$. Furthermore

$$\lim_{n\to\infty}\rho^{-n}[\boldsymbol{1}-\boldsymbol{f}_n(\boldsymbol{s})]=Q(\boldsymbol{s})\boldsymbol{u}. \tag{9}$$

Proof. Since by (6)

$$\rho^{-(n+1)}\boldsymbol{v}\cdot[\boldsymbol{1}-\boldsymbol{f}_{n+1}(\boldsymbol{s})]=\rho^{-n}\boldsymbol{v}\cdot(\boldsymbol{1}-\boldsymbol{f}_n(\boldsymbol{s}))-\rho^{-(n+1)}(\boldsymbol{1}-\boldsymbol{f}_n(\boldsymbol{s}))\boldsymbol{A}(\boldsymbol{f}_n(\boldsymbol{s})),$$

we see that $\rho^{-n}\boldsymbol{v}\cdot(\boldsymbol{1}-\boldsymbol{f}_n(\boldsymbol{s}))$ is decreasing in n. This proves that the limit in (8) exists. Call it $Q(\boldsymbol{s})$.

Iterating (6) yields

$$\rho^{-n}[\boldsymbol{1}-\boldsymbol{f}_n(\boldsymbol{s})]=(\boldsymbol{1}-\boldsymbol{s})\prod_{i=0}^{n-1}\left\{\frac{\boldsymbol{M}'}{\rho}-\frac{\boldsymbol{A}(\boldsymbol{f}_i(\boldsymbol{s}))}{\rho}\right\}. \tag{10}$$

Hence lemma 1 (ii) implies that

$$\lim_{n\to\infty}\rho^{-n}[\boldsymbol{1}-\boldsymbol{f}_n(\boldsymbol{s})]\equiv \boldsymbol{V}(\boldsymbol{s}) \quad \text{exists.} \tag{11}$$

By (7) $\boldsymbol{A}[\boldsymbol{f}_i(\boldsymbol{s})]=O(\|\boldsymbol{1}-\boldsymbol{f}_i(\boldsymbol{s})\|)$, which by (11) $=O(\rho^i)$. Hence $\sum \boldsymbol{A}[\boldsymbol{f}_i(\boldsymbol{s})]<\infty$, and by lemma 1 (iii) we conclude that $\boldsymbol{V}(\boldsymbol{s})\neq\boldsymbol{0}$.

Finally lemma 1 (i) applied to (10) implies that

$$\lim_{n\to\infty}\frac{\boldsymbol{1}-\boldsymbol{f}_n(\boldsymbol{s})}{\boldsymbol{v}\cdot[\boldsymbol{1}-\boldsymbol{f}_n(\boldsymbol{s})]}=\boldsymbol{u} \tag{12}$$

and thus $\boldsymbol{V}(\boldsymbol{s})=Q(\boldsymbol{s})\boldsymbol{u}\neq\boldsymbol{0}$. □

Theorem 4. *If $\rho<1$ and $E\|\boldsymbol{Z}_1\|^2<\infty$, then*

$$\lim_{n\to\infty}\rho^{-n}P\{\boldsymbol{Z}_n\neq 0\,|\,\boldsymbol{Z}_0=\boldsymbol{i}\}=Q(\boldsymbol{0})\,(\boldsymbol{i}\cdot\boldsymbol{u}), \tag{13}$$

and

$$\lim_{n\to\infty}P\{\boldsymbol{Z}_n=\boldsymbol{j}\,|\,\boldsymbol{Z}_0=\boldsymbol{i},\boldsymbol{Z}_n\neq\boldsymbol{0}\}=v(\boldsymbol{j}) \tag{14}$$

exists, is independent of $\boldsymbol{i}$ *and is a probability measure on $\mathscr{R}^+$.*

Proof. Clearly

$$P\{\boldsymbol{Z}_n\neq\boldsymbol{0}\,|\,\boldsymbol{Z}_0=\boldsymbol{i}\}=1-\left\{\boldsymbol{1}-\rho^n\frac{\boldsymbol{1}-\boldsymbol{f}_n(\boldsymbol{0})}{\rho^n}\right\}^{\boldsymbol{i}},$$

which by (9)

$$=Q(\boldsymbol{0})\,(\boldsymbol{i}\cdot\boldsymbol{u})\,\rho^n+O(\rho^{2n}).$$

This implies (13).

Let

$$\begin{aligned} g_n(\boldsymbol{i}, \boldsymbol{s}) &= \sum_{\boldsymbol{j} \in \mathscr{R}^+} P\{\boldsymbol{Z}_n = \boldsymbol{j} \mid \boldsymbol{Z}_0 = \boldsymbol{i}, \boldsymbol{Z}_n \neq \boldsymbol{0}\} \boldsymbol{s}^{\boldsymbol{j}} \\ &= 1 - \frac{1 - [\boldsymbol{f}_n(\boldsymbol{s})]^{\boldsymbol{i}}}{1 - [\boldsymbol{f}_n(\boldsymbol{0})]^{\boldsymbol{i}}} = 1 - \frac{1 - [\boldsymbol{1} - \{\boldsymbol{1} - \boldsymbol{f}_n(\boldsymbol{s})\}]^{\boldsymbol{i}}}{1 - [\boldsymbol{1} - \{\boldsymbol{1} - \boldsymbol{f}_n(\boldsymbol{0})\}]^{\boldsymbol{i}}}. \end{aligned} \tag{15}$$

We want to show that

$$\lim_{n \to \infty} g_n(\boldsymbol{i}, \boldsymbol{s}) = g(\boldsymbol{s}) \tag{16}$$

exists, is independent of $\boldsymbol{i}$, and is a probability generating function.

But by (15) and (9)

$$\lim_{n \to \infty} 1 - g_n(\boldsymbol{i}, \boldsymbol{s}) = \lim_{n \to \infty} \frac{\boldsymbol{i} \cdot [\boldsymbol{1} - \boldsymbol{f}_n(\boldsymbol{s})] \rho^{-n}}{\boldsymbol{i} \cdot [\boldsymbol{1} - \boldsymbol{f}_n(\boldsymbol{0})] \rho^{-n}} = \frac{\boldsymbol{i} \cdot \boldsymbol{u} Q(\boldsymbol{s})}{\boldsymbol{i} \cdot \boldsymbol{u} Q(\boldsymbol{0})} \equiv 1 - g(\boldsymbol{s}).$$

Since $Q(\boldsymbol{1}-)=0$ (theorem 3), $g(\boldsymbol{1}-)=1$, and $g(\boldsymbol{s})$ is a generating function. □

5. Limit Theorems for the Critical Case

The asymptotic behavior of the multi-type critical process offers no new surprises. As in the one-dimensional case, we condition on non-extinction, and normalize by dividing the process by the generation number. The limit law is again exponential. As in the super-critical case (to be considered in the next section) the limit measure is concentrated on the ray $r\boldsymbol{v}$, $r>0$. The projection of the process $\boldsymbol{Z}_n/n$ (conditioned on $\boldsymbol{Z}_n \neq 0$) on a subspace orthogonal to $\boldsymbol{v}$ goes to $\boldsymbol{0}$ in distribution, but in this case dividing by $\sqrt{n}$ instead of n leads to a nondegenerate limit law (see Ney (1967)).

The exponential limit law was first proved by Mullikin (1963) under a third moment assumption. The present approach, assuming a second moment, is due to Joffe and Spitzer (1967). The key step in the proof is an extension of the basic lemma I.9.1, with which we begin.

Multitype Basic Lemma. *If* $\rho=1$ *and* $E\|\boldsymbol{Z}_1\|^2<\infty$, *then*

$$\lim_{n \to \infty} \frac{1}{n} \left[\frac{1}{\boldsymbol{v} \cdot [\boldsymbol{1} - \boldsymbol{f}_n(\boldsymbol{s})]} - \frac{1}{\boldsymbol{v} \cdot [\boldsymbol{1} - \boldsymbol{s}]} \right] = \boldsymbol{v} \cdot \mathbb{Q}[\boldsymbol{s}]$$

uniformly for $\boldsymbol{s} \in \mathscr{C}_p$, *where* $\mathbb{Q}$ *is as defined in* (2.6).

If $E(\|\boldsymbol{Z}_1\|^2)<\infty$, then one can expand $\boldsymbol{f}(\boldsymbol{s})$ up to second order terms in the form

$$\boldsymbol{1} - \boldsymbol{f}(\boldsymbol{s}) = (\boldsymbol{1} - \boldsymbol{s}) M' - \mathbb{Q}[\boldsymbol{1} - \boldsymbol{s}] + \mathbb{B}_{\boldsymbol{s}}[\boldsymbol{1} - \boldsymbol{s}], \tag{1}$$

where $\mathbb{B}_{\boldsymbol{s}}[\boldsymbol{t}]$ $(\boldsymbol{s}, \boldsymbol{t} \in \mathscr{C}_p)$ is a vector valued function, which can be given explicitly in terms of the integral form of the remainder for the multivariate Taylor expansion. (See formula (4.34) of Joffe and Spitzer.) From this expression, one sees that

(i) $\mathbf{0} \leqslant \mathbb{B}_s[t] \leqslant \mathbb{Q}[t]$;
(ii) $\mathbb{B}_s[t]$ is nonincreasing in s; (2)
(iii) $\lim_{s \to 1} \mathbb{B}_s[t] = \mathbf{0}$.

Taking inner products with $\boldsymbol{v}$ and using the fact that $\boldsymbol{v} = \boldsymbol{v}\boldsymbol{M}$, (1) implies

$$\boldsymbol{v} \cdot [\boldsymbol{f}(\boldsymbol{s}) - \boldsymbol{s}] = \boldsymbol{v} \cdot \mathbb{Q}[\mathbf{1} - \boldsymbol{s}] - \boldsymbol{v} \cdot \mathbb{B}_s[\mathbf{1} - \boldsymbol{s}]. \tag{3}$$

Proceeding as in section I.9, we let

$$a(s) \equiv \boldsymbol{v} \cdot \mathbb{Q}\left[\frac{\mathbf{1} - \boldsymbol{s}}{\boldsymbol{v} \cdot (\mathbf{1} - \boldsymbol{s})}\right]$$

$$\varepsilon(s) \equiv \boldsymbol{v} \cdot \mathbb{B}_s\left[\frac{\mathbf{1} - \boldsymbol{s}}{\boldsymbol{v} \cdot (\mathbf{1} - \boldsymbol{s})}\right]$$

$$\delta(s) \equiv \frac{1}{\boldsymbol{v} \cdot (\mathbf{1} - \boldsymbol{s})} + a(s) - \frac{1}{\boldsymbol{v} \cdot [\mathbf{1} - \boldsymbol{f}(\boldsymbol{s})]},$$

and get from (3) that

$$\delta(s) = \frac{\varepsilon(s) - a(s)\boldsymbol{v} \cdot (\mathbf{1} - \boldsymbol{s})[a(s) - \varepsilon(s)]}{1 - \boldsymbol{v} \cdot (\mathbf{1} - \boldsymbol{s})[a(s) - \varepsilon(s)]}. \tag{4}$$

By (2.i), $\varepsilon(s) \leqslant a(s)$, and thus

$$-(a(s))^2 \boldsymbol{v} \cdot (\mathbf{1} - \boldsymbol{s}) \leqslant \varepsilon(s) - a(s)\boldsymbol{v} \cdot (\mathbf{1} - \boldsymbol{s})[a(s) - \varepsilon(s)] \leqslant \delta(s) \leqslant \varepsilon(s). \tag{5}$$

Replacing $\boldsymbol{s}$ by $\boldsymbol{f}_n(\boldsymbol{s})$ and adding over n, we get

$$\frac{1}{n}\sum_{k=0}^{n-1} \delta[\boldsymbol{f}_k(\boldsymbol{s})] = \frac{1}{n}\left[\frac{1}{\boldsymbol{v} \cdot (\mathbf{1} - \boldsymbol{s})} - \frac{1}{\boldsymbol{v} \cdot (\mathbf{1} - \boldsymbol{f}_n(\boldsymbol{s}))}\right] + \frac{1}{n}\sum_{k=0}^{n-1} a[\boldsymbol{f}_k(\boldsymbol{s})]. \tag{6}$$

Now by (5) and (2.ii)

$$\begin{aligned} &-\boldsymbol{v} \cdot [\mathbf{1} - \boldsymbol{f}_n(\boldsymbol{s})]\left\{\boldsymbol{v} \cdot \mathbb{Q}\left[\frac{\mathbf{1} - \boldsymbol{f}_n(\boldsymbol{s})}{\boldsymbol{v} \cdot [\mathbf{1} - \boldsymbol{f}_n(\boldsymbol{s})]}\right]\right\}^2 \\ &\leqslant \delta[\boldsymbol{f}_n(\boldsymbol{s})] \leqslant \boldsymbol{v} \cdot \mathbb{B}_{\boldsymbol{f}_n(\mathbf{0})}\left[\frac{\mathbf{1} - \boldsymbol{f}_n(\boldsymbol{s})}{\boldsymbol{v} \cdot [\mathbf{1} - \boldsymbol{f}_n(\boldsymbol{s})]}\right]. \end{aligned} \tag{7}$$

Applying (2) we see that both sides of (7) go to zero uniformly for $\boldsymbol{s} \in \mathscr{C}_p$ as $n \to \infty$, and hence so does the left side of (6). Finally, we easily verify that (4.12) also holds when $\rho = 1$, and hence we see that

$$\frac{1}{n}\sum_{k=0}^{n-1} a[\boldsymbol{f}_k(\boldsymbol{s})] \to \boldsymbol{v} \cdot \mathbb{Q}(\boldsymbol{u}).$$

Thus (6) implies the lemma. □

Corollary. *If* $\rho=1$, *and* $E\|\boldsymbol{Z}_1\|^2<\infty$, *then*

$$\lim_{n\to\infty} nP\{\boldsymbol{Z}_n\neq 0\,|\,\boldsymbol{Z}_0=\boldsymbol{i}\}=\frac{\boldsymbol{i}\cdot\boldsymbol{u}}{\boldsymbol{v}\cdot\mathbb{Q}[\boldsymbol{u}]}. \tag{8}$$

Proof. We have remarked that (4.12) also holds when $\rho=1$. From this fact and the basic lemma, we get that

$$nP\{\boldsymbol{Z}_n\neq 0\,|\,\boldsymbol{Z}_0=\boldsymbol{e}_j\}\to\frac{\boldsymbol{u}_j}{\boldsymbol{v}\cdot\mathbb{Q}[\boldsymbol{u}]},$$

from which the corollary follows. □

Finally, we state the main result

Theorem 1. *If* $\rho=1$ *and* $E\|\boldsymbol{Z}_1\|^2<\infty$, *and if* $\boldsymbol{w}\cdot\boldsymbol{v}>0$, *then* $[\boldsymbol{Z}_n\cdot\boldsymbol{w}]/n$, *conditioned on* $\boldsymbol{Z}_n\neq 0$, *converges in distribution to the random variable with density*

$$f(x)=\frac{1}{\gamma_1}e^{-\frac{x}{\gamma_1}},\qquad x\geqslant 0, \tag{9}$$

where

$$\gamma_1=\frac{\boldsymbol{v}\cdot\boldsymbol{w}}{\boldsymbol{v}\cdot\mathbb{Q}[\boldsymbol{u}]}.$$

Proof. Define the conditioned random vector

$$\boldsymbol{W}_n(\boldsymbol{i})=\left\{\frac{1}{\boldsymbol{v}\cdot\mathbb{Q}[\boldsymbol{u}]}\,\frac{\boldsymbol{Z}_n}{n\boldsymbol{v}}\,\middle|\,\boldsymbol{Z}_n\neq\boldsymbol{0},\boldsymbol{Z}_0=\boldsymbol{i}\right\},$$

where

$$\frac{\boldsymbol{Z}_n}{\boldsymbol{v}}\equiv\left(\frac{Z_{n1}}{v_1},\dots,\frac{Z_{np}}{v_p}\right).$$

Then

$$Ee^{-\boldsymbol{W}_n\cdot\boldsymbol{\lambda}}=\frac{[\boldsymbol{f}_n(\boldsymbol{s}_n)-\boldsymbol{f}_n(\boldsymbol{0})]^{\boldsymbol{i}}}{[\boldsymbol{1}-\boldsymbol{f}_n(\boldsymbol{0})]^{\boldsymbol{i}}}, \tag{10}$$

where $\boldsymbol{s}_n=(s_{n1},\dots,s_{np})$, and

$$s_{nj}=\exp\{-\lambda_j(n\boldsymbol{v}\cdot\mathbb{Q}[\boldsymbol{u}]v_j)^{-1}\}.$$

Application of the basic lemma to (10) then implies

$$\lim_{n\to\infty}E(e^{-(\boldsymbol{W}_n\cdot\boldsymbol{\lambda})})=\frac{1}{1+(\boldsymbol{\lambda}\cdot\boldsymbol{1})}, \tag{11}$$

which is the Laplace transform of the exponential density with unit parameter, concentrated on the ray $r\boldsymbol{1}$, $r\geqslant 0$. A change of variable then yields the theorem. The details are very similar to the one-dimensional case and can be found in Joffe-Spitzer (1967), and Mullikin (1963). □

If the vector $\boldsymbol{w}$ in the theorem is orthogonal to $\boldsymbol{v}$, then the limit law degenerates, i. e., $\gamma_1=0$, and $[\boldsymbol{Z}_n\cdot\boldsymbol{w}]/n$ (conditioned on $\boldsymbol{Z}_n\neq\boldsymbol{0}$) goes to 0 in distribution. The correct normalizing factor for this case turns out to be $\sqrt{n}$, and the limit law Laplacian. (See Ney (1967).) Namely,

Theorem 2. *If* $\rho=1$, $E\|\boldsymbol{Z}_1\|^2<\infty$, *and* $\boldsymbol{w}\cdot\boldsymbol{v}=0$, *then* $(\boldsymbol{Z}_n\cdot\boldsymbol{w})/\sqrt{n}$, *conditioned on* $\boldsymbol{Z}_n\neq 0$, *converges in distribution to the random variable with density*

$$f_2(x)=\frac{1}{2\gamma_2}e^{-\frac{|x|}{\gamma_2}},\qquad -\infty<x<\infty,\qquad \gamma_2>0. \tag{12}$$

6. The Supercritical Case and Geometric Growth

Our assumptions throughout this section are supercriticality ($\rho>1$), nonsingularity, and positive regularity. We shall also be making frequent reference to the condition

$$E\{Z_{1j}^{(i)}\log Z_{1j}^{(i)}\}<\infty\quad\text{for all } 1\leqslant i,j\leqslant p. \tag{*}$$

Theorem 1. *If* $\rho>1$, *and* $\{\boldsymbol{Z}_n\}$ *is nonsingular and positive regular, then*

$$\lim_{n\to\infty}\left(\frac{\boldsymbol{Z}_n}{\rho^n}\right)=\boldsymbol{v}W\ \text{ a.s.}, \tag{1}$$

where W *is a nonnegative random variable such that*

$$P\{W>0\}>0\quad\textit{if and only if (*) holds}. \tag{2}$$

Theorem 2. *Assume that* (*) *holds, and let* W *be the limit random variable in* (1). *Then*

(i) $E_i(W)=u_i$, $\quad i=1,\dots,p$;
(ii) $P_i(W=0)=q^{(i)}=P_i\{\boldsymbol{Z}_n=0$ for some $n\}$, $\quad i=1,\dots,p$;
(iii) *The transforms* $\varphi_i(\alpha)\equiv E_i e^{-\alpha W}$, $i=1,\dots,p$ *satisfy*

$$\boldsymbol{\phi}(\alpha)=\boldsymbol{f}\left[\boldsymbol{\phi}\left(\frac{\alpha}{\rho}\right)\right], \tag{3}$$

where

$$\boldsymbol{\phi}(\alpha)=(\varphi_1(\alpha),\dots,\varphi_p(\alpha));$$

(iv) *There exist strictly positive functions* $w_i(x)$ *on* $(0,\infty)$, *such that for any* $0<a<b<\infty$

$$P_i\{a<W\leqslant b\}=\int_a^b w_i(x)\,dx.$$

These two theorems were first obtained by H. Kesten and B. Stigum (1966). Previously, T. Harris (1963) had established the a.s. convergence of $\boldsymbol{Z}_n/\rho^n$ to $\boldsymbol{v}W$ under the assumption that all second moments of $\boldsymbol{Z}_n$ exist. It should be remarked that such extra moment assumptions considerably simplify the proofs, and that the really difficult part of Theorem 1 is the necessary and sufficient condition for the non-degeneracy of W.

We will take yet a third approach (based on Athreya (1970)) which we feel is somewhat more direct than that of Kesten and Stigum. This is developed in the next three theorems.

The first of these (theorem 3 below) demonstrates the one-dimensional character of the limit law—namely that the proportions of particles of various types approach the corresponding ratios of the components of the eigenvector $\boldsymbol{v}$ (which of course are nonrandom quantities). Only the *magnitude* of $\lim \boldsymbol{Z}_n \rho^{-n}$ is random.

Theorem 3. *Let* $\mathsf{A}=\{\omega\colon \boldsymbol{Z}_n(\omega)\to\infty \text{ as } n\to\infty\}$. *Then, for any* $\varepsilon>0$

$$\lim_{n\to\infty} P\left\{\omega\colon \omega\in\mathsf{A}, \left\| \frac{\boldsymbol{Z}_n(\omega)}{\boldsymbol{u}\cdot\boldsymbol{Z}_n(\omega)} - \boldsymbol{v}\right\| > \varepsilon\right\} = 0. \tag{4}$$

Remark 1. Since the convergence in (4) is "in probability", we will be able to conclude the convergence in theorem 1 only "in probability". The required a.s. convergence in (4) has recently been proved by T. Kurtz (unpublished) under (*) but we will only give the proof of the weaker result here.

Remark 2. The salient feature of theorem 3 is that nothing like (*) need be assumed. This should be contrasted with inferring (4) by assuming (*) to ensure $W\neq 0$, and then applying theorem 1.

The main part of the convergence is a consequence of

Theorem 4. *Let* $W_n=(\boldsymbol{u}\cdot\boldsymbol{Z}_n)\rho^{-n}$ *and* $\mathbb{F}_n=$ *the σ-algebra generated by* $\{\boldsymbol{Z}_i; i=1,\ldots,n\}$. *Then* $\{(W_n,\mathbb{F}_n); n\geqslant 0\}$ *is a nonnegative martingale, and hence* $\lim_{n\to\infty} W_n$ *exists* a.s. (*This limit is the* W *in* (1).)

Finally the nondegeneracy of W and the functional equation part of theorem 2 is contained in

Theorem 5. *Let* $$\mathsf{C}=\Big\{\boldsymbol{\psi}(x)=(\psi_1(x),\ldots,\psi_p(x));\ \psi_i(x)=\int_0^\infty e^{-tx}\,dF_i(t),\ F_i \text{ a distribution such that } F_i(0+)<1,\ \int_0^\infty t\,dF_i(t)<\infty,$$
$$0<\lim_{x\downarrow 0} x^{-1}[1-\psi_i(x)]<\infty,\ i=1,\ldots,p\Big\},$$

and let $\boldsymbol{\phi}(x)$ *be as in theorem* 2 (iii). *Then* $\boldsymbol{\phi}$ *satisfies* (3) *and*

(i) $\boldsymbol{\phi} \in \mathsf{C}$ *if and only if* (*) *holds; and*

(ii) *If* $\boldsymbol{\phi} \in \mathsf{C}$ *then* $-\phi_i'(0) = v_i$ *and* $P_i(W=0) = q^{(i)}$, $i = 1, \ldots, p$.

Theorems 3, 4 and 5 imply all of theorems 1 and 2 except for the a.s. convergence in (1) (instead of which they establish convergence in probability), and except for the absolute continuity in theorem 2 (iv). The latter can be proved along similar lines to theorem I.10.4.

We thus confine ourselves to the proofs of theorems 3, 4 and 5. Starting with theorem 4, we observe that it is a special case of

Theorem 4′. *Let* ξ *be any right eigenvector of* $\boldsymbol{M}$, *and* λ *be the corresponding eigenvalue. Then*

$$\{Y_n\} = \{(\xi \cdot \boldsymbol{Z}_n)\lambda^{-n},\ \mathbb{F}_n;\ n \geq 0\}$$

is a (complex valued) martingale.

Proof. By the additive property of the process (see section 1)

$$E\{Y_n | \mathbb{F}_0\} = \lambda^{-n}(\boldsymbol{M}^n \xi) \cdot \boldsymbol{Z}_0 = \xi \cdot \boldsymbol{Z}_0 = Y_0 .$$

Hence, by the Markov property and the stationarity of the transition probabilities

$$E\{Y_{n+k} | \mathbb{F}_k\} = \frac{1}{\lambda^k} E\left\{\frac{\xi \cdot \boldsymbol{Z}_{n+k}}{\lambda^n}\middle| \mathbb{F}_k\right\} = \lambda^{-k} \xi \cdot \boldsymbol{Z}_k = Y_k . \qquad \square$$

Turning to the proof of theorem 3, we shall make use of the following lemma, which is in the same spirit as the Frobenius theorem. (See Karlin (1966) for a proof.)

Lemma 1. *Let* $\mathsf{K} = \{\boldsymbol{x} = (x_1, \ldots, x_p);\ x_i > 0,\ \boldsymbol{x} \cdot \boldsymbol{u} = 1\}$. *Then*

$$\lim_{n \to \infty} \sup_{\boldsymbol{x} \in \mathsf{K}} \|\boldsymbol{x} \boldsymbol{M}^n \rho^{-n} - \boldsymbol{v}\| = 0 .$$

Proof of theorem 3. Recall that $Z_{ni}^{(j)}$ = the number of type i particles in the nth generation of a process initiated with a single type j particle. Let $Z_{mi}^{(j)l}(n)$, denote the number of type i offspring at time $m+n$, of the l'th type j particle of the n'th generation. That these are well defined random variables follows for the same reasons as in the one-dimensional case (see section I.1). Then by the additive property

$$Z_{n+m,i}^{(k)} = \sum_{j=1}^{p} \sum_{l=1}^{Z_{nj}^{(k)}} Z_{mi}^{(j)l}(n) . \tag{5}$$

Let

$$\boldsymbol{X}_n \equiv \boldsymbol{Z}_n (\boldsymbol{u} \cdot \boldsymbol{Z}_n)^{-1} , \tag{6}$$

i.e.,

$$\frac{X_{nj}}{Z_{nj}} = (\boldsymbol{u}\cdot\boldsymbol{Z}_n)^{-1}, \qquad j=1,\dots,p .$$

Then

$$\begin{aligned}\boldsymbol{Z}_{n+m}^{(k)} &= \sum_{j=1}^{p}\sum_{l=1}^{Z_{nj}^{(k)}} \boldsymbol{Z}_m^{(j)l}(n)\\ &= \sum_{j=1}^{p}\sum_{l=1}^{Z_{nj}^{(k)}} (\boldsymbol{Z}_m^{(j)l}(n)-\boldsymbol{e}_j\boldsymbol{M}^m)+\sum_{j=1}^{p} Z_{nj}^{(k)}\boldsymbol{e}_j\boldsymbol{M}^m .\end{aligned} \tag{7}$$

(Observe that $E\boldsymbol{Z}_m^{(j)l}(n)=\boldsymbol{e}_j\boldsymbol{M}^m$). From (7) we can compute $(\boldsymbol{u}\cdot\boldsymbol{Z}_{n+m}^{(k)})$ and get

$$\boldsymbol{X}_{n+m}^{(k)} \equiv \frac{\boldsymbol{Z}_{n+m}^{(k)}}{(\boldsymbol{u}\cdot\boldsymbol{Z}_{n+m}^{(k)})} = \frac{\sum_{j=1}^{p} Z_{nj}^{(k)}\boldsymbol{e}_j\boldsymbol{M}^m + \sum_{j=1}^{p}\sum_{l=1}^{Z_{nj}^{(k)}}[\boldsymbol{Z}_m^{(j)l}(n)-\boldsymbol{e}_j\boldsymbol{M}^m]}{\rho^m(\boldsymbol{u}\cdot\boldsymbol{Z}_n^{(k)})+\sum_{j=1}^{p}\sum_{l=1}^{Z_{nj}^{(k)}}[\boldsymbol{u}\cdot\boldsymbol{Z}_m^{(j)l}(n)-\rho^m u_j]} . \tag{8}$$

Suppressing the superscript k, and letting

$$r_{nm}=\sum_{j=1}^{p}\left(\frac{X_{nj}}{Z_{nj}}\right)\sum_{l=1}^{Z_{nj}}\rho^{-m}[\boldsymbol{u}\cdot\boldsymbol{Z}_m^{(j)l}(n)-\rho^m u_j],$$

and

$$\boldsymbol{\alpha}_{nm}=\sum_{j=1}^{p}\left(\frac{X_{nj}}{Z_{nj}}\right)\sum_{l=1}^{Z_{nj}}\rho^{-m}[\boldsymbol{Z}_m^{(j)l}(n)-e_j\boldsymbol{M}^m],$$

we can write

$$\boldsymbol{X}_{n+m}-\boldsymbol{v}=\{(\boldsymbol{X}_n\boldsymbol{M}^m\rho^{-m}-\boldsymbol{v})-\boldsymbol{v}r_{nm}+\boldsymbol{\alpha}_{nm}\}\{1+r_{nm}\}^{-1}. \tag{9}$$

Let $\varepsilon>0$ be arbitrary. Then by lemma 1, there is an m_0 such that for all $m\geqslant m_0$

$$\sup_{\boldsymbol{x}\in K}\|\boldsymbol{x}\boldsymbol{M}^m\rho^{-m}-\boldsymbol{v}\|\leqslant\varepsilon.$$

Fix an $m\geqslant m_0$. Then by the weak law of large numbers, it follows that for any $\eta>0$

$$\left.\begin{aligned}&\lim_{n\to\infty} P\{\omega\colon \boldsymbol{Z}_n(\omega)\to\infty,\ |r_{nm}|>\eta\}=0\\ &\lim_{n\to\infty} P\{\omega\colon \boldsymbol{Z}_n(\omega)\to\infty,\ \|\boldsymbol{\alpha}_{nm}\|>\eta\}=0\end{aligned}\right\}. \tag{10}$$

From (9) we have, for any $\varepsilon>0$, $\eta>0$, that

$$\begin{aligned}&P\{\omega\colon \boldsymbol{Z}_n(\omega)\to\infty,\ \|\boldsymbol{X}_{n+m}-\boldsymbol{v}\|\leqslant(\varepsilon+\eta+\|\boldsymbol{v}\|\eta)(1-\eta)^{-1}\}\\ &\geqslant 1-P\{\omega\colon \boldsymbol{Z}_n(\omega)\to\infty,\ |r_{nm}|>\eta\}-P\{\omega\colon \boldsymbol{Z}_n(\omega)\to\infty,\ \|\boldsymbol{\alpha}_{nm}\|>\eta\}.\end{aligned}$$

This, with (10), implies that for any $\varepsilon>0$, $\eta>0$

$$\limsup_{N\to\infty} P\{\omega\colon \omega\in \mathsf{A},\ \|X_N-v\|>(\varepsilon+\eta+\|v\|\eta)(1-\eta)^{-1}\}=0,$$

which proves theorem 3. $\square$

Before getting into the proof of theorem 5, let us make a few observations.

If $g(s)$, $s\in\mathscr{C}_p$ is the generating function of any random variable X taking values in $\mathscr{R}_p^+$, and having finite mean vector m, then one can define the function $A(s)$ by

$$g(\mathbf{1}-s)=1-s\cdot m+A(s) \tag{11}$$

Then

$$\frac{\partial A}{\partial s_j}\geqslant 0, \qquad s\in\mathscr{C}_p,$$

and hence

$$s\geqslant t \quad \text{implies} \quad A(s)\geqslant A(t), \qquad s,t\in\mathscr{C}_p. \tag{12}$$

Note that $\bar{g}(s)\equiv g(s\mathbf{1})$, where $0\leqslant s\leqslant 1$, is the generating function of the nonnegative integer valued random variable $X\cdot\mathbf{1}$ with mean $\bar{m}=m\cdot\mathbf{1}$. Hence, defining

$$\bar{A}(s)\equiv\bar{m}-\frac{1-\bar{g}(1-s)}{s}=\frac{A(s\mathbf{1})}{s}>0, \qquad \bar{A}(0)=0,$$

we see by corollary I.10.2 that

$$\bar{A}(s) \text{ is nondecreasing and continuous on } [0,1]; \tag{13}$$

and for any $0<c$, $r<1$,

$$\sum \bar{A}(c r^n)<\infty \quad \text{and} \quad \lim_{c\downarrow 0}\sum \bar{A}(c r^n)=0 \tag{14}$$

if and only if

$$E[X\cdot\mathbf{1}]\log[X\cdot\mathbf{1}]<\infty. \tag{**}$$

Proof of theorem 5. That ϕ satisfies (3) follows from a continuity argument exactly analogous to the one-dimensional case. We shall first prove that

<u>$\phi\in\mathsf{C}\Rightarrow(*)$:</u>

Suppose $\phi\in\mathsf{C}$ and (*) is false. Define $\theta(x)$ by

$$\theta(x)=\frac{1}{x}\sum_{i=1}^{p} v_i[1-\phi_i(x)].$$

Replacing g by $f^{(i)}$ in (11), and letting $A^{(i)}$ and $\bar{A}^{(i)}$ be the functions corresponding to A and $\bar{A}$, we get from (3)

$$\theta(x)=\theta(x\rho^{-1})\left\{1-\frac{\sum_{i=1}^{p} v_i A^{(i)}\left(\mathbf{1}-\boldsymbol{\phi}\left(\frac{x}{\rho}\right)\right)}{\rho\sum_{i=1}^{p} v_i\left[1-\phi_i\left(\frac{x}{\rho}\right)\right]}\right\}. \tag{15}$$

Since $\boldsymbol{\phi}\in \mathsf{C}$, there exist constants $0<c<\infty$, $\delta>0$, such that for $x\leqslant\delta$ we have

$$1-\phi_j(x)\geqslant c\,x \quad \text{for all } j.$$

This, together with (12) implies

$$A^{(i)}(\mathbf{1}-\boldsymbol{\phi}(x\rho^{-1}))\geqslant A^{(i)}(c\,x\rho^{-1}\mathbf{1})=c'\,x\bar{A}^{(i)}(c'\,x), \tag{16}$$

where $c'=c\rho^{-1}$.

Also for $x>0$, we have the inequality $(1-\phi_i(x))/x\leqslant E(W|\mathbf{Z}_0=\boldsymbol{e}_i)$. Thus $\sum_{i=1}^{p} v_i[1-\phi_i(x\rho^{-1})]\leqslant x\rho^{-1}\sum_{i=1}^{p} v_i E(W|\mathbf{Z}_0=\boldsymbol{e}_i)$, which by Fatou's lemma and the facts that $W_n\to W$ a.s. and is a martingale, is

$$\leqslant x\rho^{-1}\sum_{i=1}^{p} v_i E(W_0|\mathbf{Z}_0=\boldsymbol{e}_i)=x\rho^{-1} \quad \left(\text{since } \sum_{i=1}^{p} u_i v_i=1\right).$$

This inequality, with (16), implies that for $x\leqslant\delta$

$$\begin{aligned}\rho^{-1}\left[\sum_{i=1}^{p} v_i[1-\phi_i(x\rho^{-1})]^{-1}\right]&\left[\sum_{i=1}^{p} v_i A^{(i)}(\mathbf{1}-\boldsymbol{\phi}(x\rho^{-1}))\right]\\ &\geqslant x^{-1}c'\,x\sum_{i=1}^{p} v_i\bar{A}^{(i)}(c'\,x).\end{aligned} \tag{17}$$

Using the inequality $1-x\leqslant e^{-x}$ for $x\geqslant 0$, we conclude from (15)—(17) that for $0\leqslant x\leqslant\delta$

$$0<c\leqslant\theta(x)\leqslant\theta(x\rho^{-1})\exp\left\{-c'\sum_{i=1}^{p} v_i\bar{A}^{(i)}(c'\,x)\right\}. \tag{18}$$

(The constant c is not necessarily the same one each time.)

Iteration of (18) yields (for $x\leqslant\delta$)

$$0<c\leqslant\theta(x)\leqslant\exp\left\{-c'\sum_{i=1}^{p} v_i\sum_{n=0}^{\infty}\bar{A}^{(i)}(c'\,x\rho^{-n})\right\} \tag{19}$$

where we have used the fact that $\lim_{x\to 0}\sum(v_i[1-\phi_i(x)]/x)=\sum v_i u_i=1$.

If (*) is false, then there exists a pair (i_0,j_0) such that

$$E(Z_{1\,j_0}^{(i_0)}\log Z_{1\,j_0}^{(i_0)})=\infty\,,$$

and hence

$$E[(\mathbf{Z}_1^{(i_0)}\cdot\mathbf{1})(\log(\mathbf{Z}_1^{(i_0)}\cdot\mathbf{1}))]=\infty.$$

Now, by (14)

$$\sum_{n=0}^{\infty}\bar{A}^{(i_0)}(c'x\rho^{-n})=\infty. \tag{20}$$

Clearly, (19) and (20) yield the needed contradiction, namely,

$$0<c\leqslant\theta(x)\leqslant 0.$$

Corollary 1. (*) *false implies* $W=0$ a.s.

Proof. By the "only if" part above, we know that $\boldsymbol{\phi}\notin\mathsf{C}$. But by Fatou's Lemma we know that $\lim_{x\downarrow 0}(1-\phi_i(x))x^{-1}\leqslant u_i<\infty$ for $1\leqslant i\leqslant p$. Thus

$$E(W|\mathbf{Z}_0=e_i)=0 \quad \text{for every } i.$$

We continue with the proof of Theorem 5 by proving that

(*) $\Rightarrow\boldsymbol{\phi}\in\mathsf{C}$.

For $x>0$, let

$$\left.\begin{aligned}&\phi_{ni}(x)=E_i(e^{-xW_n}),\\&\boldsymbol{\phi}_n(x)=(\phi_{n1}(x),\phi_{n2}(x),\ldots,\phi_{np}(x)),\\&\theta_n(x)=x^{-1}\sum_{i=1}^{p}v_i[1-\phi_{ni}(x)].\end{aligned}\right\} \tag{21}$$

Define the values of the above functions at 0 by right continuity.

Using the basic branching property, we get for $x>0$ $\boldsymbol{\phi}_n(x)=\boldsymbol{f}(\boldsymbol{\phi}_{n-1}(x/\rho))$, which in turn yields

$$0<1-\theta_n(x)=[1-\theta_{n-1}(x\rho^{-1})]+\sum_{i=1}^{p}v_i x^{-1}A^{(i)}(\mathbf{1}-\boldsymbol{\phi}_{n-1}(x\rho^{-1})). \tag{22}$$

The martingale property yields

$$x^{-1}(1-\phi_{ni}(x))\leqslant E(W_n|\mathbf{Z}_0=e_i)=u_i\leqslant\|\boldsymbol{u}\|,$$

and thus by (12)

$$x^{-1}A^{(i)}(\mathbf{1}-\boldsymbol{\phi}_{n-1}(x\rho^{-1}))\leqslant\bar{u}\bar{A}^{(i)}(\bar{u}x), \quad \text{where } \bar{u}=\|\boldsymbol{u}\|\rho^{-1}.$$

Hence from (22), we have

$$0<1-\theta_n(x)\leqslant 1-\theta_{n-1}(x\rho^{-1})+\sum_{i=1}^{p}v_i\bar{u}\bar{A}^{(i)}(\bar{u}x),$$

which on iteration yields

$$0<1-\theta_n(x)\leqslant 1-\theta_0(x\rho^{-n})+\sum_{i=1}^{p} v_i\bar{u}\sum_{r=0}^{n-1}\bar{A}^{(i)}(\bar{u}x\rho^{-r}). \tag{23}$$

It is easily seen that (*) implies

$$E[(\mathbf{Z}^{(i)}\cdot\mathbf{1})\log(\mathbf{Z}^{(i)}\cdot\mathbf{1})]<\infty$$

for all i.

Now appeal to corollary I.10.2, and in (23) let $n\to\infty$ first and then $x\downarrow 0$, to get $\lim_{x\downarrow 0}[1-\theta(x)]=0$.
But since
$\sum_{i=1}^{p} u_i v_i=1$, and $x^{-1}[1-\varphi_i(x)]\leqslant u_i$, it follows that

$$\lim_{x\downarrow 0}\{u_i-x^{-1}[1-\phi_i(x)]\}=0, \tag{24}$$

thus proving that $\boldsymbol{\phi}\in\mathsf{C}$ as well as that $-\phi_i'(0)=v_i$ for every i. This proves part (i).

Finally, on letting $u\to\infty$ in (3), we see that if $\bar{q}^{(i)}=P_i(W=0)$ $=\lim_{u\to\infty}\phi_i(u)$, then $\bar{\boldsymbol{q}}=(\bar{q}^{(1)},\bar{q}^{(2)},\ldots,\bar{q}^{(p)})$ satisfies

$$\bar{\boldsymbol{q}}=\boldsymbol{f}(\bar{\boldsymbol{q}}). \tag{25}$$

Since $\boldsymbol{\phi}\in\mathsf{C}$ implies $\bar{q}^{(i)}<1$ for all i, we may conclude from (25) that $\bar{\boldsymbol{q}}$ has to coincide with the extinction probability vector $\boldsymbol{q}$.

The proof of theorem 5 is now complete. □

7. The Continuous Time, Multitype Markov Case

7.1 Definitions and Preliminaries

When particles have random lifetimes, then there is a natural analog for multitype processes of the continuous and age-dependent processes of chapters III and IV. If the particle lifetime distributions are exponential we get a continuous time Markov process with discrete state space $\mathscr{R}_p^+$. The construction and limit theory of these processes naturally utilizes a blend of the techniques of chapter III and the previous sections of this chapter. The detailed development, in large part, thus involves arguments which are by now quite familiar. *We will therefore limit ourselves to an outline sketch of the theory.*

Let $Z_j(t)$ = the number of type j particles existing at time t, and set

$$\boldsymbol{Z}(t)=(Z_1(t),\ldots,Z_p(t)), \qquad t\geqslant 0.$$

Definition. A stochastic process $\{\boldsymbol{Z}(t,\omega);t\geqslant 0\}$ on a probability space $(\Omega,\mathbb{F},P)$ is called a *p-dimensional continuous time Markov branching process if:*

(i) Its state space is $\mathscr{R}_p^+$;
(ii) It is a stationary strong Markov process with respect to the fields

$$\mathbb{F}_t=\sigma\{\boldsymbol{Z}(s,\omega);\, s\leqslant t\};$$

(iii) The transition probabilities $P(\boldsymbol{i},\boldsymbol{j};t)$ satisfy

$$\sum_{\boldsymbol{j}\in\mathscr{R}_p^+} P(\boldsymbol{i},\boldsymbol{j};t)\boldsymbol{s}^{\boldsymbol{j}} = \prod_{k=1}^{p}\Big[\sum_{\boldsymbol{j}\in\mathscr{R}_p^+} P(\boldsymbol{e}_k,\boldsymbol{j};t)\boldsymbol{s}^{\boldsymbol{j}}\Big]^{i_k};$$

for all $\boldsymbol{i}\in\mathscr{R}_p^+$ and $\boldsymbol{s}\in\mathscr{C}_p$.

Notation. Note that as in the discrete case $\{\boldsymbol{Z}^{(\boldsymbol{j})}(t,\omega)\}$ represents a process initiated in state $\boldsymbol{j}$; and when $\boldsymbol{j}=\boldsymbol{e}_i$ we write $\boldsymbol{Z}^{(\boldsymbol{e}_i)}(t,\omega)\equiv\boldsymbol{Z}^{(i)}(t,\omega)$.

Construction of the Process. The transition functions are determined by the infinitesimal parameters

$$\boldsymbol{a}=(a_1,\ldots,a_p)\in\mathscr{E}_p^+,$$
$$\boldsymbol{0}\leqslant \boldsymbol{p}(\boldsymbol{j})=(p^{(1)}(\boldsymbol{j}),\ldots,p^{(p)}(\boldsymbol{j})),\quad \sum_{\boldsymbol{j}\in\mathscr{R}^+} p^{(i)}(\boldsymbol{j})=1,$$

as solutions of the Kolmogorov equations. For notational simplicity we express these directly in their generating function form. Let

$$\boldsymbol{f}(\boldsymbol{s})=(f^{(1)}(\boldsymbol{s}),\ldots,f^{(p)}(\boldsymbol{s})),$$

where

$$f^{(i)}(\boldsymbol{s})=\sum_{\boldsymbol{j}\in\mathscr{R}^+} p^{(i)}(\boldsymbol{j})\boldsymbol{s}^{\boldsymbol{j}}.$$

Let

$$u^{(i)}(\boldsymbol{s})=a_i[f^{(i)}(\boldsymbol{s})-s_i],$$

and

$$F(\boldsymbol{i},\boldsymbol{s};t)=\sum_{\boldsymbol{j}\in\mathscr{R}^+} P(\boldsymbol{i},\boldsymbol{j};t)\boldsymbol{s}^{\boldsymbol{j}};$$

and write

$$F(\boldsymbol{e}_i,\boldsymbol{s};t)=F(i,\boldsymbol{s};t),$$
$$\boldsymbol{F}(\boldsymbol{s};t)=(F(1,\boldsymbol{s};t)),\ldots,F(p,\boldsymbol{s};t)),$$
$$\boldsymbol{u}(\boldsymbol{s})=(u^{(1)}(\boldsymbol{s}),\ldots,u^{(p)}(\boldsymbol{s})).$$

The function $\boldsymbol{u}(\boldsymbol{s})$ is called the infinitesimal generating function. The Kolmogorov equations are

$$\frac{\partial}{\partial t} F(\boldsymbol{i}, \boldsymbol{s}; t) = \sum_{j=1}^{p} u^{(j)}(\boldsymbol{s}) \frac{\partial}{\partial s_j} F(\boldsymbol{i}, \boldsymbol{s}; t) \qquad \text{(forward equation)}, \tag{1}$$

and

$$\frac{\partial}{\partial t} F(i, \boldsymbol{s}; t) = u^{(i)}[\boldsymbol{F}(\boldsymbol{s}; t)], \qquad \text{(backward equation)}. \tag{2}$$

$$i = 1, \ldots, p.$$

The minimal process is described as follows (and yields the appropriate probabilistic interpretation of the infinitesimal parameters): If a particle of type i is alive at a particular time, then its additional life length is exponentially distributed with parameter a_i. Upon its death it produces offspring of the p-types according to the distribution $p^{(i)}(\boldsymbol{j})$. Particles live and produce independently of each other, and of the past.

In order to guarantee that a.s. there cannot be infinitely many particles produced in a finite time, we shall assume throughout that all the means exist, namely

$$\left. \frac{\partial f^{(i)}(\boldsymbol{s})}{\partial s_j} \right|_{s=1} < \infty \quad \text{for all } i, j. \tag{3}$$

This is only a sufficient condition and is not necessary. A necessary and sufficient condition for nonexplosion in finite time can be derived from the more general work of T. Savits (1969).

Under this hypothesis we are also guaranteed that equations (1) and (2), subject to the boundary condition

$$P(\boldsymbol{i}, \boldsymbol{j}; 0+) = \delta(\boldsymbol{i}, \boldsymbol{j}) = \begin{cases} 1 & \text{if } \boldsymbol{i} = \boldsymbol{j}, \\ 0 & \text{otherwise}, \end{cases} \tag{4}$$

have a unique solution which is the generating function of $P(\boldsymbol{i}, \boldsymbol{j}; t)$, with $\sum_{\boldsymbol{j}} P(\boldsymbol{i}, \boldsymbol{j}; t) = 1$; and which corresponds to the minimal process.

These considerations are entirely analogous to the first three sections of chapter III.

There is also an analogue of the Harris construction of a probability space whose sample points are family trees. (See the discussion at the end of section III.2.) We thus again obtain the familiar additive property:

For any $t, u \geqslant 0$, there exist measurable functions $\boldsymbol{Z}_t^{(i),\alpha}(u, \omega)$ for $\alpha = 1, \ldots, Z_i(t, \omega)$, $i = 1, \ldots, p$, such that a.s.

$$\boldsymbol{Z}(t+u, \omega) = \sum_{i=1}^{p} \sum_{\alpha=1}^{Z_i(t,\omega)} \boldsymbol{Z}_t^{(i),\alpha}(u, \omega),$$

where $\{\boldsymbol{Z}_t^{(i),\alpha}(u, \omega); u \geqslant 0\}$ are conditionally (given $\boldsymbol{Z}(t, \omega)$) independent stochastic processes equivalent to $\boldsymbol{Z}^{(i)}(\cdot, \omega)$, $i = 1, \ldots, p$, respectively.

7.2 Means

It can be shown that (3) implies that

$$m_{ij}(t) = E(Z_j^{(i)}(t)) < \infty. \tag{5}$$

This is proved analogously to the one-dimensional case (see complements).

The mean matrix of the multitype process is the $p \times p$ matrix

$$\boldsymbol{M}(t) = \{m_{ij}(t); i,j=1,\dots,p\}. \tag{6}$$

From (2), one derives the semigroup property

$$\boldsymbol{M}(t+u) = \boldsymbol{M}(t)\boldsymbol{M}(u), \qquad t,u \geqslant 0, \tag{7}$$

and the continuity condition

$$\lim_{t\to 0} \boldsymbol{M}(t) = \boldsymbol{I}. \tag{8}$$

It is well known that (7) and (8) imply the existence of a matrix $\boldsymbol{A}$, called the infinitesimal generator of the semigroup $\{\boldsymbol{M}(t); t \geqslant 0\}$, such that

$$\boldsymbol{M}(t) \equiv \exp(\boldsymbol{A}\,t) = \sum_0^\infty \frac{t^r \boldsymbol{A}^r}{r\,!},$$

the series converging in the strong (norm) sense. We can identify the matrix $\boldsymbol{A} = ((a_{ij}))$ as

$$a_{ij} = a_i b_{ij}, \qquad b_{ij} = \left.\frac{\partial f^{(i)}(\boldsymbol{s})}{\partial s_j}\right|_{\boldsymbol{s}=\boldsymbol{1}} - \delta_{ij}. \tag{9}$$

As in the discrete case, we make the basic assumption of *positive regularity*, namely that there exists a $t_0, 0 < t_0 < \infty$ such that

$$m_{ij}(t_0) > 0 \quad \text{for all } i, j. \tag{10}$$

The Perron-Frobenius theory of positive matrices is again used to deduce the existence of a strictly positive eigenvalue $\rho_1(t_0)$ of $\boldsymbol{M}(t_0)$, such that (i) any other eigenvalue $\rho(t_0)$ of $\boldsymbol{M}(t_0)$ satisfies $|\rho(t_0)| < \rho_1(t_0)$, and (ii) the algebraic and geometric multiplicities of $\rho_1(t_0)$ are both unity. Since $\boldsymbol{M}(t) = \exp(\boldsymbol{A}\,t)$ the eigenvalues of $\boldsymbol{M}(t)$ are given by $e^{\lambda_i t}$, $i=1,2,\dots,p$, where $\lambda_1, \lambda_2, \dots, \lambda_p$ are the eigenvalues of $\boldsymbol{A}$; and $\boldsymbol{M}(t)$ and $\boldsymbol{A}$ have the same eigenvectors. This implies that we can arrange the eigenvalues of $\boldsymbol{A}$ as

$$\lambda_1 > \operatorname{Re}\lambda_2 \geqslant \operatorname{Re}\lambda_3 \geqslant \cdots \geqslant \operatorname{Re}\lambda_p, \tag{11}$$

and determine the left and right eigenvectors $\boldsymbol{v}$ and $\boldsymbol{u}$ of λ_1 with all coordinates strictly positive, and such that

$$\boldsymbol{u} \cdot \boldsymbol{v} = 1, \qquad \boldsymbol{u} \cdot \mathbf{1} = 1. \tag{12}$$

The process is *supercritical*, *critical*, or *subcritical* according as λ_1, $>$, $=$, or <0.

7.3 Second Moments

Let us suppose that

$$\left. \frac{\partial^2 f^{(i)}(\boldsymbol{s})}{\partial s_j \partial s_k} \right|_{\boldsymbol{s}=\mathbf{1}} < \infty \quad \text{for all } i, j \text{ and } k. \tag{13}$$

As in the case of the mean, (13) implies that

$$d_{ij}^{(r)}(t) \equiv E(Z_i^{(r)}(t) Z_j^{(r)}(t)) < \infty \quad \text{for all } i, j, r, t. \tag{14}$$

Applying (2), yields the matrix differential equation

$$\frac{d}{dt}(\boldsymbol{D}^{(r)}(t)) = \boldsymbol{A}' \boldsymbol{D}^{(r)}(t) + \boldsymbol{D}^{(r)}(t) \boldsymbol{A} + \sum_{i=1}^{p} m_{ri}(t) \boldsymbol{B}^{(i)}, \tag{15}$$

where $\boldsymbol{D}^{(r)}(\mathrm{t})$ is the matrix $((d_{ij}^{(r)}(t)))$,

$$\boldsymbol{A}' = \text{transpose of } \boldsymbol{A},$$

$$\boldsymbol{B}^{(i)} = ((b_{jk}^{(i)})),$$

$$b_{jk}^{(i)} = a_i \left[\left. \frac{\partial^2 f^{(i)}(\boldsymbol{s})}{\partial s_j \partial s_k} \right|_{\boldsymbol{s}=\mathbf{1}} + \delta_{jk} - b_{ij} \delta_{ik} - b_{ik} \delta_{ij} - \delta_{ij} \delta_{ik} \right],$$

with a_{ij} and b_{ij} having the same meaning as in (9).

It is known (see Coddington and Levinson (1955)) that the unique solution of (15) is

$$\boldsymbol{D}^{(r)}(t) = (\boldsymbol{M}(t))' \boldsymbol{D}^{(r)}(0) (\boldsymbol{M}(t)) + \int_0^t (\boldsymbol{M}(t-\tau))' \left(\sum_{i=1}^{p} m_{ri}(\tau) \boldsymbol{B}^{(i)} \right) (\boldsymbol{M}(t-\tau)) \, d\tau. \tag{16}$$

Remark. The analogous discrete time results mentioned in section 2 are derived similarly using difference instead of differential equations.

7.4 Growth Rates of the First Two Moments

From the Perron-Frobenius theorem, we can determine the growth rate of $\boldsymbol{M}(t)$. In fact, by that theorem,

$$\lim_{t \to \infty} \boldsymbol{M}(t) e^{-\lambda_1 t} = \boldsymbol{P} \equiv ((u_i v_j)). \tag{17}$$

We shall see that this fact in turn enables us to completely determine the growth rate of the second moments as well. We consider the cases $\lambda_1 <, =, > 0$ in that order.

(i) $\lambda_1 < 0$

From (16), we get

$$\begin{aligned} \boldsymbol{D}^{(r)}(t)e^{-\lambda_1 t} &= (\boldsymbol{M}(t))' \boldsymbol{D}^{(r)}(0) \boldsymbol{M}(t) e^{-\lambda_1 t} \\ &+ e^{\lambda_1 t} \int_0^t (\boldsymbol{M}(t-\tau)e^{-\lambda_1(t-\tau)})' \left(\sum_{i=1}^{p} m_{ri}(\tau) e^{-\lambda_1 \tau} \boldsymbol{B}^{(i)} \right) (\boldsymbol{M}(t-\tau)e^{-\lambda_1(t-\tau)}) e^{-\lambda_1 \tau} d\tau. \end{aligned} \tag{18}$$

Then making the change of variable $t-\tau=u$, applying (17) and the dominated convergence theorem, we have

$$\lim_{t\to\infty} \boldsymbol{D}^{(r)}(t)e^{-\lambda_1 t} = u_r \int_0^\infty (\boldsymbol{M}(u)e^{-\lambda_1 u})' \left(\sum_{i=1}^{p} v_i \boldsymbol{B}^{(i)} \right) (\boldsymbol{M}(u)e^{-\lambda_1 u}) e^{\lambda_1 u} du, \tag{19}$$

and thus $\boldsymbol{D}^{(r)}(t)$ decays exponentially fast.

(ii) $\lambda_1 = 0$

Multiplying both sides of (16) by t^{-1} and using the fact that $\boldsymbol{M}(t) \to \boldsymbol{P}$ we have

$$\lim_{t\to\infty} \boldsymbol{D}^{(r)}(t)t^{-1} = u_r \sum_{i=1}^{p} v_i (\boldsymbol{P}' \boldsymbol{B}^{(i)} \boldsymbol{P}) \tag{20}$$

and thus $\boldsymbol{D}^{(r)}(t)$ grows like t.

(iii) $\lambda_1 > 0$

Multiplying both sides of (16) by $e^{-2\lambda_1 t}$ and again using dominated convergence, we have

$$\lim_{t\to\infty} \boldsymbol{D}^{(r)}(t)e^{-2\lambda_1 t} = \boldsymbol{P}' \boldsymbol{D}^{(r)}(0) \boldsymbol{P} + \int_0^\infty \boldsymbol{P}' \left(\sum_{i=1}^{p} m_{ri}(\tau) e^{-\lambda_1 \tau} \boldsymbol{B}^{(i)} \right) \boldsymbol{P} e^{-\lambda_1 \tau} d\tau, \tag{21}$$

and thus $\boldsymbol{D}^{(r)}(t)$ blows up exponentially fast. The rates in all the cases are exact since one can show that the right sides of (19), (20) and (21) are strictly positive.

In the next section we will study the asymptotic behavior of functionals of the process, for example $\xi \cdot \boldsymbol{Z}(t)$, where ξ is an eigenvector of $\boldsymbol{A}$.

For this purpose we will need estimates of the growth of $V_r(t) \equiv E|\xi \cdot \boldsymbol{Z}^{(r)}(t)|^2$. For any positive vector ξ the answer is easily obtained from (19), (20), and (21). However, if ξ is orthogonal to $\boldsymbol{u}$ or $\boldsymbol{v}$, which is the case if ξ is an eigenvector with eigenvalue $\lambda \neq \lambda_1$, then the above estimates on $D^{(r)}(t)$ only yield an upper bound (lower bound if $\lambda_1<0$) for the growth of $V_r(t)=\bar{\xi} D^{(r)}(t)\xi'$ (where the bar stands for conjugation). However, using (16) directly and arguing as for $D^{(r)}(t)$, we arrive at the following result. (The proof is just like (19)–(21), and is left as an exercise.)

Theorem 1. *Let ξ be a right eigenvector of A with eigenvalue $\lambda \neq \lambda_1$. Let $a=\operatorname{Re}\lambda$. Then, under* (13), *the following hold:*

(i) *If $2a<\lambda_1$, then*

$$\lim_{t\to\infty} V_r(t)e^{-\lambda_1 t} = u_r c(\lambda_1-2a)^{-1}. \tag{22}$$

(ii) *If $2a=\lambda_1$, then*

$$\lim_{t\to\infty} V_r(t)e^{-\lambda_1 t}t^{-1} = u_r c\,. \tag{23}$$

(iii) *If $2a>\lambda_1$ then*

$$\lim_{t\to\infty} V_r(t)e^{-2at} = |\xi|_r^2 + \sum_{i=1}^{p} c_i \int_0^\infty m_{ri}(\tau)e^{-\lambda_1\tau}e^{-(2a-\lambda_1)\tau}d\tau\,, \tag{24}$$

where

$$c_i = v_i \bar{\xi} B^{(i)}\xi', \qquad c=\sum_{i=1}^{p} c_i, \text{ and } B^{(i)} \text{ is defined in (15).}$$

Remarks. 1. In view of (11), we see that if $\lambda_1 \leqslant 0$ then the cases (ii) and (iii) of Theorem 1 do not arise since $2a<a<\lambda_1$ for any $a<\lambda_1\leqslant 0$.

2. At the cost of some repetition, we wish to remind the reader that the above estimates work just as well in discrete time where instead of an integral in the right side of (16), one has a sum.

7.5 Limit Theorems

There are analogs of the results of sections III.6 and III.7. The results on skeleton processes and the use of the Kingman theorem go through exactly as before, and yield continuous time analogs of theorems in sections 4 and 5 of this chapter.

In the supercritical case there are problems of almost sure convergence as well, which cannot be dealt with by the skeleton technique. We therefore limit ourselves to some remarks on this case.

To begin with, the extinction probability vector $q=(q^{(1)}, q^{(2)},\dots,q^{(p)})$ where $q^{(r)}=P_r\{\omega: Z^{(r)}(t,\omega)=\mathbf{0}$ for some $t\}$ is the unique solution of the equation

$$u(s)=\mathbf{0} \tag{25}$$

for $s\in\mathscr{C}_p$. This can be shown directly as in chapter III. It implies that $P_r\{\omega: Z(t,\omega)\to\infty$ as $t\to\infty\}=1-q^{(r)}>0$ (which is why the process is called *supercritical*). Noting that $Z(n\delta)$ is a discrete time Galton-Watson process for each $\delta>0$ leads us to expect the following analog of results of section 6.

Theorem 2. *Let $\boldsymbol{Z}(t,\omega)$ be positive regular and nonsingular. Then*

$$\lim_{t\to\infty} \boldsymbol{Z}(t,\omega)e^{-\lambda_1 t} = W(\omega)\boldsymbol{v} \tag{26}$$

exists a.s., *where $W\equiv W(\omega)$ is a nonnegative numerical random variable. Furthermore*

$$P_r\{\omega: W(\omega)>0\}>0 \quad \textit{for some } r \textit{ if and only if}$$

$$E\xi_{ij}\log\xi_{ij}<\infty \quad \textit{for all } i \textit{ and } j, \tag{*}$$

where $(\xi_{i1},\xi_{i2},\dots,\xi_{ip})$ is a random variable with p.g.f. $f^{(i)}(\boldsymbol{s})$.

Under (*)

$$P_r\{\omega: W(\omega)>0\}=q^{(r)}, \qquad E_r(W)=u_r, \tag{27}$$

and $\varphi_r(x)\equiv E_r(\exp(-xW))$ satisfies the equation

$$\varphi_r(x)=a_r\int_0^\infty f^{(r)}(\boldsymbol{\phi}(xe^{-\lambda_1 y}))e^{-a_r y}\,dy \tag{28}$$

for $r=1,2,\dots,p$, where $\boldsymbol{\phi}(x)=(\varphi_1(x),\varphi_2(x),\dots,\varphi_p(x))$.

There are two approaches we can suggest to the proof of this result.

Method 1. Show first that (*) holds if and only if $EZ_j^{(i)}(\delta)\log Z_j^{(i)}(\delta)<\infty$ for all i,j and $\delta>0$. This yields the existence of a set A_δ in $\mathbb{F}$ and a random variable $W_\delta(\omega)$ such that $P(\mathsf{A}_\delta)=1$, and such that for ω in A_δ $\boldsymbol{Z}(n\delta,\omega)e^{-\lambda_1 n\delta}\to W_\delta(\omega)\boldsymbol{v}$. By Theorem 8.1 (see the next section) $\{\boldsymbol{u}\cdot\boldsymbol{Z}(t,\omega)e^{-\lambda_1 t}, \mathbb{F}_t, t\geqslant 0\}$ is a martingale, and being nonnegative, converges a.s. to a random variable $W(\omega)$. It is immediate that $W(\omega)=W_\delta(\omega)$ a.s. for all $\delta>0$. It is also clear that this W has all the distribution properties asserted. The only thing that needs to be done is to go from the existence (for each $\delta>0$) of a set A_δ on which the skeleton $Z(n\delta,\omega)e^{-\lambda_1 n\delta}$ converges to $W(\omega)$ a.s., to the existence of a set A on which $Z(t,\omega)e^{-\lambda_1 t}$ converges to $W(\omega)$ a.s. Unfortunately, the Kingman type argument used in chapter III will fail here since A_δ varies with δ. The proof is nontrivial and we refer the reader to Athreya (1967) for details.

Method 2. Show first that the limit $W(\omega)$ of the nonnegative martingale $\{\boldsymbol{u}\cdot\boldsymbol{Z}(t,\omega)e^{-\lambda_1 t}, \mathbb{F}_t; t\geqslant 0\}$ has all the distributional properties asserted by directly imitating the proof in section 6. Next, show that $(\boldsymbol{Z}(t,\omega))(\boldsymbol{u}\cdot\boldsymbol{Z}(t,\omega))^{-1}$ converges a.s. on the set of nonextinction to $\boldsymbol{u}$. This approach has not been carried out.

7.6 Split Times

As in chapter III, the times at which particles die in a MCMBP (Multi-type continuous time Markov branching process) will be called the *split times*. We shall denote by $\tau_n(\omega)$ the nth split time for the realization corresponding to ω. A new element enters in the analysis of the τ_n's in the multitype case since they must now be classified according to the type of the particle that split. If $N_i(n)$ denotes the number of splits of particles of type i among the first n splits, the behavior of the proportions $n^{-1}N_i(n)$ is of interest. We shall consider these now and then use them in section 9 in studying the embedding of urn schemes in MCMBP's.

As in chapter III, it is not difficult to see that on the set of extinction, there can be at most a finite number of deaths; while on the set of non-extinction there must be an infinite number, with $\tau_n(\omega)\to\infty$ as $n\to\infty$. The results of chapter III extend to the present case in a natural way. These are contained in Theorems 3 and 4 below.

Theorem 3. *Assume* (*) *of Theorem* 2 *holds. Let* $\mathsf{A}=\{\omega:\boldsymbol{Z}(t,\omega)\to\infty$ *as* $t\to\infty\}$.

(i) *For each* i, $\lim_{n\to\infty} N_i(n)/n$ *exists* a.s. *on* A *and equals*

$$p_i\equiv(a_i v_i)\left(\sum_{j=1}^{p} a_j v_j\right)^{-1}.$$

(ii) $\lim_{n\to\infty} n e^{-\lambda_1\tau_n}=\mu^{-1}W$ a.s., *where* $\mu=\sum_{i=1}^{p} p_i(\lambda_1+a_i)u_i a_i^{-1}$.

Let $\mathscr{F}_n$ be the σ-algebra associated with τ_n, which is an increasing sequence of stopping times. We shall assume for simplicity that all extinction probabilities $q^{(i)}$ are zero. In this case for almost all ω, the number of splits $N(\omega)$ will be infinite. Set

$$Y_n(\omega)=\tau_n(\omega)-\sum_{j=0}^{n-1} X(\tau_j(\omega),\omega)\,, \tag{29}$$

where

$$X(t,\omega)=\boldsymbol{a}\cdot\boldsymbol{Z}(t,\omega)\,.$$

Theorem 4.

(i) $\{Y_n,\mathscr{F}_n;n=1,2,\ldots\}$ *is a uniformly square integrable martingale.*

(ii) $\tau_n(\omega)-(1/\lambda_1)\log n$ *converges to a finite limit on a set of positive probability, and hence* $P(W(\omega)>0)>0$ *if and only if the hypothesis* (*) *of Theorem* 2 *holds.*

The proof of Theorem 4 follows very much along the lines of the one-dimensional case, and we refer to Athreya-Karlin (1967) for details. Turning to the proof of Theorem 3, we need the following variant of the strong law for martingales (due to P. Levy (1937), pp. 250).

Lemma 1. *Let* δ_j, $j=1,2,\ldots$ *be a sequence of* $0-1$ *valued random variables, on some probability space* $(\Omega,\mathbb{F},P)$, *measurable with respect to* *a sequence* $\mathscr{F}_j, j=1,2,\ldots$ *of increasing sub* σ*-algebras of* $\mathbb{F}$. *Define* $p_n=E(\delta_n|\mathscr{F}_{n-1})$, $n=1,2,\ldots$. *Then*

$$\frac{1}{n}\sum_{j=1}^{n}(\delta_j-p_j)\to 0 \qquad \text{a.s.} \tag{30}$$

Proof of Theorem 3. Fix i and define

$$\delta_j^{(i)}=\begin{cases}1 & \text{if the } j\text{th split is of a type } i \text{ particle,}\\ 0 & \text{otherwise}.\end{cases}$$

Clearly, $N_i(n)=\sum_{j=1}^{n}\delta_j^{(i)}$.

Consider the family of random variables and σ-algebras $\{\delta_j^{(i)}, \mathscr{F}_j, j=1,2,\ldots\}$ where $\mathscr{F}_n$ is as in Theorem 4. From the nature of the process and by virtue of the strong Markov property we have

$$p_j^{(i)}=E(\delta_j^{(i)}|\mathscr{F}_{j-1})=\frac{a_i Z_i(\tau_{j-1})}{X(\tau_{j-1})}. \tag{31}$$

Since all the $q^{(i)}$'s are zero, $\tau_n(\omega)\to\infty$ as $n\to\infty$ a.s., and combining this with Theorem 2 we have under (*)

$$\lim_{j\to\infty} p_j^{(i)}=\lim_{j\to\infty}\left\{\frac{a_i Z_i(\tau_{j-1})e^{-\lambda_1\tau_{j-1}}}{X(\tau_{j-1})e^{-\lambda_1\tau_{j-1}}}\right\}=\frac{a_i v_i}{\boldsymbol{a}\cdot\boldsymbol{v}}\equiv p_i \qquad \text{a.s.}$$

Now, appealing to Lemma 1 yields

$$\lim_{n\to\infty}\frac{1}{n}\sum_{j=0}^{n}\delta_j^{(i)}=\lim_{n\to\infty}\frac{1}{n}\sum_{j=0}^{n}p_j^{(i)}=p_i \qquad \text{a.s.}$$

This establishes (i).

Turning to (ii), we note first that

$$\lim_{n\to\infty}\boldsymbol{u}\cdot\boldsymbol{Z}(\tau_n)e^{-\lambda_1\tau_n}=W \qquad \text{a.s.}$$

Thus it suffices to show that

$$\lim_{n\to\infty}\frac{\boldsymbol{u}\cdot\boldsymbol{Z}(\tau_n)}{n}=\mu. \tag{32}$$

This is easily done by the strong law of large numbers, by expressing $\boldsymbol{u}\cdot\boldsymbol{Z}(\tau_n)$ as

$$\boldsymbol{u}\cdot\boldsymbol{Z}(\tau_n)=\sum_{j=1}^{p}\sum_{r=1}^{N_j(n)}\eta_r^{(j)}, \tag{33}$$

where

$$\eta_r^{(j)}=\sum_{i=1}^{p}u_i[Z_i(\tau_t^{(j)}+0)-Z_i(\tau_r^{(j)}-0)],$$

and $\tau_r^{(j)}$ is the rth split of a particle of type j.

It is easily checked that for every j, $\{\eta_r^{(j)}, r=1,2,\ldots\}$ is a sequence of independently identically distributed random variables with mean $(\lambda_1+a_j)u_j a_j^{-1}$. Now, dividing both sides of (33) by n, letting $n\to\infty$, and using (i), we get (32). □

We suspect that Theorem 3 is true without the hypothesis (*). Indeed, we can show by using arguments similar to those in Theorem 4.3, that $(\boldsymbol{Z}(t))(\boldsymbol{u}\cdot\boldsymbol{Z}(t))^{-1}$ converges in probability to $\boldsymbol{u}$. It would be interesting to find out if this still holds when $t\to\infty$ through a sequence of stopping times such as τ_n. This would yield the convergence in probability of $p_j^{(i)}$, as $j\to\infty$ and we could then conclude that Theorem 3 holds with convergence in probability replacing convergence a.s. This problem is open.

8. Linear Functionals of Supercritical Processes

Let $\boldsymbol{Z}(t)$ be the multi-type Markov process of the last section. (We carry over all notation.) If ξ is any vector not orthogonal to $\boldsymbol{v}$, then the limiting behavior of $\xi\cdot\boldsymbol{Z}(t)$ is easily determined from results about $\boldsymbol{Z}(t)$ which we have just discussed. However, if $\xi\cdot\boldsymbol{v}=0$, new methods are required. In the present section we treat only the case when ξ is a right eigenvector of $\boldsymbol{A}$. The case of a general vector makes use of the Jordan canonical form of $\boldsymbol{A}$, and is more complicated. It has been treated by Kesten and Stigum (1966a, b) for the discrete case, and Athreya (1969a, b) for the continuous case. The following results for the eigenvalue case were also first obtained in these references.

At the outset, we have a martingale theorem of by now familiar form.

Theorem 1. *Let ξ be any right eigenvector of $\boldsymbol{A}$ and λ be the corresponding eigenvalue. Then*

$$\{Y(t,\omega)=\boldsymbol{\xi}\cdot\boldsymbol{Z}(t,\omega)e^{-\lambda t};\ \mathbb{F}_t;\ t\geqslant 0\}$$

is a martingale.

Proof. Exactly the same as the discrete case (theorem 6.4′). □

In section 7 we saw that the behavior of $E|\xi\cdot\boldsymbol{Z}(t)|^2$ depended on whether $2\operatorname{Re}\lambda-\lambda_1$ was >0, $=0$, or <0; and hence it is not surprising that this trichotomy arises in the limit theory of $\xi\cdot\boldsymbol{Z}(t)$. The easy case arises when $2\operatorname{Re}\lambda>\lambda_1$, since then by theorem 7.1 (iii)

$$\sup_t E_r|Y(t)|^2<\infty, \qquad r=1,\ldots,p.$$

From this we get at once that

Theorem 2. *If* ξ *is a right eigenvector of* $\boldsymbol{A}$ *with eigenvalue* $\lambda \neq \lambda_1$, *if* $2\operatorname{Re}\lambda > \lambda_1$, *and if the second moments of* $\boldsymbol{f}$ *exist* (see (7.13)), *then there is a random variable* $Y(\omega)$ *such that*

$$Y(t,\omega) \to Y(\omega) \quad \text{a.s. } \textit{and in mean square.}$$

Observe again that one can have $2\operatorname{Re}\lambda > \lambda_1$ only when $\lambda_1 > 0$. The assumption of supercriticality is made throughout this section.

When $2\operatorname{Re}\lambda < \lambda_1$, then under appropriate normalization, $\boldsymbol{\xi}\cdot\boldsymbol{Z}(t)$ converges in law to a mixture of normal distributions. More significantly, there exists a random normalizing factor under which the limit distribution is a single normal. This leads to an asymptotic independence.

Theorem 3. *Assume that* $2\operatorname{Re}\lambda < \lambda_1$. *Let* $\boldsymbol{\xi} = \xi' + \sqrt{-1}\ \xi''$ *where* ξ' *and* ξ'' *are both real vectors. Let* l_1 *and* l_2 *be two arbitrary real numbers,* $Y(t) = (l_1\xi' + l_2\xi'')\cdot\boldsymbol{Z}(t)$ *and* $Y_1(t) = \boldsymbol{u}\cdot\boldsymbol{Z}(t)$. *Then*

$$\lim_{t\to\infty} P\left\{0 < x_1 \leqslant W \leqslant x_2 < \infty,\ \frac{Y(t)}{\sqrt{Y_1(t)}} \leqslant y \,\middle|\, \boldsymbol{Z}(0)\right\} \tag{1}$$
$$= P\{0 < x_1 \leqslant W \leqslant x_2 | \boldsymbol{Z}(0)\}\, \Phi(y/\sigma)$$

where

$$\sigma^2 = \lim_{t\to\infty} \sigma^2(t)$$
$$\sigma^2(t) = \sum_{i=1}^{p} u_i \sigma_i^2(t)$$
$$\sigma_i^2(t) = e^{-\lambda_1 t} E(|Y(t)|^2 | \boldsymbol{Z}(0) = e_i)$$
$$\Phi(x) = (2\pi)^{-\frac{1}{2}} \int_{-\infty}^{x} e^{-\frac{t^2}{2}} dt\,.$$

Remarks. 1. Using the estimates given in (7.22) it can be shown that σ^2 is >0 for $\boldsymbol{\xi} \neq \boldsymbol{0}$.

2. On the ω set $\{\omega: 0 < x_1 \leqslant W(\omega) < x_2 < \infty\}$, $\boldsymbol{Z}(t,\omega)$ never vanishes, and since $u_i > 0$ for all i the random variable $Y(t)/\sqrt{Y_1(t)}$ is well defined on this set.

3. Theorem 3 implies the asymptotic independence of W and $Y(t)/\sqrt{Y_1(t)}$ and we can deduce easily the following:

Corollary 1. *Under the conditions of Theorem 3*

$$\lim_{t\to\infty} P\left\{0<x_1<W\leqslant x_2<\infty,\ Y(t)e^{-\frac{\lambda_1 t}{2}}\leqslant x \,\middle|\, \boldsymbol{Z}(0)\right\}$$
$$=\int_{x_1}^{x_2} \Phi\left(\frac{x}{\sigma\sqrt{y}}\right) d_y P(W\leqslant y|\boldsymbol{Z}(0)).$$

It is in this form that Kesten and Stigum stated their result for the discrete time case.

4. By the so-called Cramer-Wold device (see theorem 7.7 of Billingsley (1968)) the arbitrariness of l_1 and l_2 in Theorem 3 implies that the joint distribution of

$$\left(\frac{\xi'\cdot\boldsymbol{Z}(t)}{\sqrt{Y_1(t)}},\ \frac{\xi''\cdot\boldsymbol{Z}(t)}{\sqrt{Y_2(t)}}\right)$$

on the set $\{\omega\colon W(\omega)>0\}$ approaches the bivariate normal.

We now embark on the *proof of Theorem 3*. From the additive property mentioned, in part 1 of section 7, we can write (suppressing ω)

$$Y(t+s)=\sum_{i=1}^{p}\sum_{j=1}^{Z_i(t)}\eta_{ij}(s)+\tilde{Y}(t,s). \tag{2}$$

where

$$\eta_{ij}(s)=l_1\operatorname{Re}(\xi\cdot\boldsymbol{Z}_t^{(i),j}(s)-e^{\lambda s}\xi_i)+l_2\operatorname{Im}(\xi\cdot\boldsymbol{Z}_t^{(i),j}(s)-e^{\lambda s}\xi_i)$$

and

$$\tilde{Y}(t,s)=l_1\operatorname{Re}(e^{\lambda s}\xi\cdot\boldsymbol{Z}(t))+l_2\operatorname{Im}(e^{\lambda s}\xi\cdot\boldsymbol{Z}(t)).$$

Consider a point ω in the set $\mathsf{A}\equiv\{\omega\colon 0<x_1\leqslant W(\omega)\leqslant x_2<\infty\}$. For this ω, $Y_1(t,\omega)$ never vanishes. Dividing both sides of (2) by $\sqrt{Y_1(t+s)}$ we get

$$\frac{Y(t+s)}{\sqrt{Y_1(t+s)}}=\sum_{i=1}^{p}\sqrt{v_i}\frac{1}{\sqrt{Z_i(t)}}\sum_{j=1}^{Z_i(t)}\eta_{ij}(s)e^{-\frac{\lambda_1 s}{2}} \tag{3}$$
$$+\sum_{i=1}^{p}\left(\sqrt{\frac{Z_i(t)e^{-\lambda_1 t}}{Y_1(t+s)e^{-\lambda_1(t+s)}}}-\sqrt{v_i}\right)\frac{1}{\sqrt{Z_i(t)}}\cdot\left(\sum_{j=1}^{Z_i(t)}\eta_{ij}(s)e^{-\frac{\lambda_1 s}{2}}\right)$$
$$+\frac{\tilde{Y}(t,s)}{\sqrt{Y_1(t+s)}}\equiv A_1(t,s)+A_2(t,s)+A_3(t,s)\qquad\text{(say)}.$$

Here is the basic idea of the proof. Using the estimate on $E|\tilde{Y}(t,s)|^2$ developed in (7.22) we first choose s large enough to make A_3 small in probability, uniformly in t. Next, with this large but fixed s we show (using the central limit theorem) that as $t\to\infty, A_1$ converges to the desired normal distribution. Finally, for this fixed s, we show, using the a.s. convergence of $\boldsymbol{Z}(t)e^{-\lambda_1 t}$, that $A_2 \to 0$ in probability as $t \to \infty$.

We shall carry this out in a series of lemmas. Let $0 < x_1 < x_2 < \infty$ and $I = [x_1, x_2]$. Fix this I.

Lemma 1. *Given* $\eta > 0$, $\delta > 0$ *we can find a* $s_0 = s(\eta, \delta)$ *such that* $s \geq s_0$ *implies*

$$\sup_t P\{W \in I, |A_3(t,s)| > \eta\} \leq \delta .$$

Proof. We start with the obvious inequality

$$\begin{aligned} &P\{W \in I, |A_3(t,s)| > \eta\} \\ &\leq P\{W \in I, |A_3(t,s)| > \eta, |Y_1(t+s)e^{-\lambda_1(t+s)} - W| \leq \varepsilon\} \\ &\quad + P\{W \in I, |Y_1(t+s)e^{-\lambda_1(t+s)} - W| > \varepsilon\} \end{aligned}$$

where $\varepsilon = x_1/2$. But $Y_1(t)e^{-\lambda_1 t} \to W$ a.s. and hence, given $\delta > 0$, we can find an $s_0' = s(\delta)$ such that for $s > s_0'$

$$\sup_t P\{|Y_1(t+s)e^{-\lambda_1(t+s)} - W| > \varepsilon\} < \frac{\delta}{2} .$$

Let $a = \operatorname{Re} \lambda$.

Next observe that

$$\begin{aligned} &P\{W \in I, |A_3(t,s)| > \eta, |Y_1(t+s)e^{-\lambda_1(t+s)} - W| \leq \varepsilon\} \\ &\leq P\left\{\left|\tilde{Y}(t,s)e^{-\lambda_1 \frac{(t+s)}{2}}\right| > \eta\left(\frac{x_1}{2}\right)^{\frac{1}{2}}\right\} \\ &\leq 2\left(\frac{E|\tilde{Y}(t,s)|^2}{\eta^2 x_1}\right) e^{-\lambda_1(t+s)} \\ &\leq 2\frac{(l_1^2 + l_2^2)}{\eta^2 x_1} (E\{|\boldsymbol{\xi} \cdot \boldsymbol{Z}(t)|^2\} e^{-\lambda_1 t}) e^{(2a - \lambda_1)s} . \end{aligned}$$

By section 7.4 we know that $\sup_t E\{|\xi \cdot \boldsymbol{Z}(t)|^2\} e^{-\lambda_1 t} \equiv K < \infty$ and since $2a < \lambda_1$, we can choose $s_0''(\delta, \eta)$ large enough so that

$$2\frac{(l_1^2 + l_2^2)}{\eta^2 x_1} K e^{(2a - \lambda_1)s_0''} < \frac{\delta}{2} .$$

It is clear that $s_0 = \max(s_0', s_0'')$ is the desired value. □

Lemma 2. *Fix* s. *Then for any* $\eta > 0$

$$\lim_{t \to \infty} P\{W \in I, |A_2(t,s,y)| > \eta\} = 0 .$$

Proof. Write

$$A_2(t,s)=\sum_{i=1}^{p} C_i(t,s)\,D_i(t,s)\,,$$

where

$$C_i=\left(\sqrt{\frac{Z_i(t)\,e^{-\lambda_1 t}}{Y_1(t+s)\,e^{-\lambda_1(t+s)}}}-\sqrt{v_i}\right)$$

$$D_i=\frac{1}{\sqrt{Z_i(t)}}\sum_{j=1}^{Z_i(t)}\eta_{ij}(s)\,e^{-\frac{\lambda_1 s}{2}}.$$

It suffices to show that

$$\lim_{t\to\infty} P\{W\in I,\ |C_i D_i|>\eta\}=0 \quad \text{for each } i\,.$$

Now

$$P\{W\in I,\ |C_i D_i|>\eta\}\leqslant P\{W\in I,\ |C_i D_i|>\eta,\ |D_i|\leqslant M\} + P\{W\in I,\ |C_i D_i|>\eta,\ |D_i|>M\}\,.$$

Since $\eta_{ij}(s)e^{-\lambda_1 s/2}$ are conditionally (given $Z_i(t)$) independent with mean zero and finite variance, we get

$$E|D_i|^2=E\left|\eta_{ij}(s)\,e^{-\frac{\lambda_1 s}{2}}\right|^2\leqslant K \qquad \text{(as in lemma 1)}\,.$$

Given an $\varepsilon>0$ choose M large enough so that $KM^{-2}<\varepsilon$. Fix this M. Trivially

$$P(W\in I,\ |C_i D_i|>\eta,\ |D_i|\leqslant M\}\leqslant P\{W\in I,\ |C_i|>\eta M^{-1}),$$

and since $I=[x_1,x_2]$ with $0<x_1$ we have, using Theorem 7.2

$$\lim_{t\to\infty} P(W\in I,\ |C_i|>\eta M^{-1})=0\,.$$

Thus

$$\limsup P(W\in I,\ |C_i D_i|>\eta)\leqslant\varepsilon\,,$$

where ε is arbitrary. □

Combining the above two lemmas yields the following important

Corollary 2. *For every* $\eta>0,\ \delta>0$ *there exists an* $s_0=s_0(\eta,\delta)$ *such that for any real* y

$$\begin{aligned}
&\liminf_t P(W\in I,\ A_1(t,s_0)\leqslant y-\eta)-\delta\\
&\leqslant \liminf_t P\left\{W\in I,\ \frac{Y(t+s_0)}{\sqrt{Y_1(t+s_0)}}\leqslant y\right\}\\
&\leqslant \limsup_t P\left\{W\in I,\ \frac{Y(t+s_0)}{\sqrt{Y_1(t+s_0)}}\leqslant y\right\}\\
&\leqslant \limsup_t P\{W\in I,\ A_1(t,s_0)\leqslant y+\eta\}+\delta\,.
\end{aligned}$$

Again referring to Theorem 7.2 we can easily assert the following:

Lemma 3. *Fix* $\eta < x_1/2$. *Then*

$$\lim_{t\to\infty} P(\{W\in I\}\triangle\{d(Y_1(t)e^{-\lambda_1 t}, I)>\eta\})=0,$$

where $\triangle$ *means symmetric difference, and*

$$d(x, I)=\inf_{y\in I}|x-y|\,.$$

A simple variant of the central limit theorem yields

Lemma 4. *Fix* s_0, y, *and* $\eta>0$. *Then*

$$\lim_{t\to\infty}\Big|P\{d(Y_1(t)e^{-\lambda_1 t}, I)<\eta,\ A_1(t,s_0)\leqslant y\} - P\{d(Y_1(t)e^{-\lambda_1 t}, I)<\eta\}\,\Phi\left(\frac{y}{\sigma(s_0)}\right)\Big|=0\,,$$

where $\sigma^2(s_0)$, *and* Φ *are as in theorem 1.*

Now let us put all the pieces together. Let $\varepsilon>0$ be arbitrary and y be fixed. Choose η such that

$$\Phi\left(\frac{y+\eta}{\sigma}\right)-\Phi\left(\frac{y-\eta}{\sigma}\right)<\frac{\varepsilon}{2}\,.$$

Pick s_1 such that $s\geqslant s_1$ implies

$$\left|\Phi\left(\frac{y+r\eta}{\sigma(s)}\right)-\Phi\left(\frac{y+r\eta}{\sigma}\right)\right|<\frac{\varepsilon}{2}\quad\text{for } r=\pm 1\,.$$

Let $\delta=\varepsilon/2$, $s^*=\max(s_0(\eta,\delta), s_1)$, where $s_0(\eta,\delta)$ is as in lemma 1. By corollary 2, and lemmas 3 and 4 we then have

$$\liminf P\left(W\in I, \frac{Y(t)}{\sqrt{Y_1(t)}}\leqslant y\right)\geqslant P(W\in I)\,\Phi\left(\frac{y-\eta}{\sigma(s^*)}\right)-\frac{\varepsilon}{2}\,,$$

and

$$\limsup P\left(W\in I, \frac{Y(t)}{\sqrt{Y_1(t)}}\leqslant y\right)\leqslant P(W\in I)\,\Phi\left(\frac{y+\eta}{\sigma(s^*)}\right)+\frac{\varepsilon}{2}\,.$$

Letting $\varepsilon\downarrow 0$ forces $\eta\downarrow 0$, $s^*\to\infty$, and we get our result. □

We now turn to the case $2\,\mathrm{Re}\,\lambda=\lambda_1$. The main result here is

Theorem 4.

$$\lim_{t\to\infty} P\left\{0<x_1<W\leqslant x_2<\infty, \frac{Y(t)}{\sqrt{Y_1(t)\log Y_1(t)}}\leqslant x\,\middle|\,\boldsymbol{Z}(0)\right\}$$
$$=P\{0<x_1\leqslant W\leqslant x_2<\infty|\boldsymbol{Z}(0)\}\,\Phi\left(\frac{x}{\sigma}\right)$$

where the notation is as in theorem 3 *except that* $\sigma_i^2(t)$ *is now given by*

$$\sigma_i^2(t)=t^{-1}e^{-\lambda_1 t}E(|Y(t)|^2|\boldsymbol{Z}(0)=e_i)\,.$$

The remarks that followed theorem 3 apply equally here. Of course, $Y(t)/\sqrt{Y_1(t)\log Y_1(t)}$ and not $Y(t)/\sqrt{Y_1(t)}$ is asymptotically independent of W, and we have

Corollary 2. *Under the conditions of theorem* 4

$$\lim_{t\to\infty} P\left\{0<x_1\leqslant W\leqslant x_2<\infty, \frac{Y(t)}{\sqrt{t e^{\lambda_1 t}}}\leqslant x\,\middle|\,\boldsymbol{Z}(0)\right\}$$
$$=\int_{x_1}^{x_2}\Phi\left(\frac{x}{\sigma\sqrt{y}}\right)d_y P\{W\leqslant y|\boldsymbol{Z}(0)\}\,.$$

The approach here is similar to the case $2\,\mathrm{Re}\,\lambda<\lambda_1$ but with one very important difference. We still use an equation similar to (3) but in the limiting process, we can no longer choose and fix a large s, and then let $t\to\infty$, but must let t and s go to ∞ simultaneously in such a way that $t/s\to 0$. This leads to a triangular array with a random number in each row.

We first outline the key steps in the proof. Just as (3) before, we here obtain

$$\frac{Y(t+s)}{\sqrt{Y_1(t+s)\log Y_1(t+s)}}=\sqrt{\frac{s}{\log Y_1(t+s)}}\sum_{i=1}^{p}\sqrt{v_i}\,\frac{1}{\sqrt{Z_i(t)}}\sum_{j=1}^{Z_i(t)}(s\,e^{\lambda_1 s})^{-\frac{1}{2}}\eta_{ij}(s)$$
$$+\sqrt{\frac{s}{\log Y_1(t+s)}}\sum_{i=1}^{p}\left(\sqrt{\frac{Z_i(t)e^{-\lambda_1 t}}{Y_1(t+s)e^{-\lambda_1(t+s)}}}-\sqrt{v_i}\right)\frac{1}{\sqrt{Z_i(t)}}\sum_{j=1}^{Z_i(t)}(s\,e^{\lambda_1 s})^{-\frac{1}{2}}\eta_{ij}(s)$$
$$+\frac{\tilde{Y}(t,s)}{\sqrt{Y_1(t+s)\log Y_1(t+s)}}\equiv A_1+A_2+A_3\,. \tag{4}$$

If we tried to follow the same route now, our first difficulty would be the analog of lemma 1. In fact, it does not hold. This is because

$$E|\boldsymbol{\xi}\cdot\boldsymbol{Z}(t)|^2\sim\text{constant}\cdot e^{-\lambda_1 t}t^{-1}.$$

and so for each fixed s

$$\lim_{t\to\infty}(E|\tilde{\boldsymbol{Y}}(t,s)|^2)e^{-\lambda_1(t+s)}(t+s)^{-1} \quad \text{exists and is } >0\,.$$

However, if we let t depend on s, say $t=\varepsilon s$, then A_3 can be made small in probability by choosing ε properly, and then letting $s\to\infty$. Slutsky's theorem (see Loeve (1963)) shows that A_2 goes to zero in distribution since

$$\sqrt{\frac{Z_i(t)e^{-\lambda_1 t}}{Y_1(t+s)e^{-\lambda_1(t+s)}}}\to\sqrt{v_i} \quad\text{and}\quad \sqrt{\frac{s}{\log Y_1(t+s)}}\to\frac{1}{\sqrt{\lambda_1}} \quad \text{a.s. on } \{W>0\};$$

Finally if we show that for each i

$$X_i(s)=\frac{1}{\sqrt{Z_i(t)}}\sum_{j=1}^{Z_i(t)}(s e^{\lambda_1 s})^{-\frac{1}{2}}\eta_{ij}(s)\to N(0,\sigma_i^2) \tag{5}$$

as $s\to\infty$, $\varepsilon s=t$, then we are done.

Thus it remains only to establish (5).

In view of the additive property $X_i(s)$ has the same distribution as

$$X(s)=\frac{1}{\sqrt{v(t)}}\sum_{j=1}^{v(t)}\eta_j(s) \quad \text{with } t=\varepsilon s\,, \tag{6}$$

where (i) for each s $\{\eta_j(s):j=1,2\ldots\}$ is a sequence of independent random variables all distributed as $(s e^{\lambda_1 s})^{-\frac{1}{2}}\eta_{ij}(s)$, (ii) the stochastic process $\{v(t);t\geqslant 0\}$ is independent of $\{\eta_j(s);j=1,2\ldots\}$ for all s, and has the same distribution as the stochastic process $\{Z_i(t);t\geqslant 0\}$.

We shall use the Lindeberg-Feller criterion for triangular arrays of infinitesimal random variables where the number in each row is non-random. We consider the distribution of $X(s)$ conditioned on the entire realization of the process $\{v(t);t\geqslant 0\}$, and show that for almost all realizations of the $v(t)$ process such that $\lim_{t\to\infty} v(t)e^{-\lambda_1 t}$ exists and is >0, this distribution converges to $N(0,\sigma_i^2)$. Since the limit is independent of the conditioning event, the unconditional limit distribution is also the same. We note that on the set $\{\omega: 0<x_1\leqslant W(\omega)\leqslant x_2\}$, $\lim_{t\to\infty} Z_i(t,\omega)e^{-\lambda_1 t}$ $\geqslant x_1>0$ a.s., and so we may assume that for almost all realizations, $\lim_{t\to\infty} v(t)e^{-\lambda_1 t}\geqslant x_1/2>0$. Appealing to the Lindeberg-Feller criterion we need to check only that for each $0<c<\infty$,

$$\lim_{s\to\infty}\int_{|x|>ce^{\lambda_1\varepsilon s}} x^2\,dF(x,s)=0\,, \tag{7}$$

where $F(x,s)$ is the distribution function of $\eta_i(s)$. Since $|\eta_{ij}(s)|^2$ $<(l_1^2+l_2^2)|\boldsymbol{\xi}\cdot\boldsymbol{Z}^{(ij)}(s)-e^{\lambda s}\xi_i|^2$, (7) is a corollary of the following

Lemma 5 (Athreya 1971b). *Let* $1<\theta<\infty$ *and*

$$\bar{M}(s)=E_i\{|R(s)|^2;|R(s)|>\theta^s\}, \tag{8}$$

where E_i *stands for expectation with initial condition* $\boldsymbol{Z}(0)=e_i$ *and* $R(s)=(\boldsymbol{\xi}\cdot\boldsymbol{Z}(s)-e^{\lambda s}\xi_i)(e^{\lambda s}\sqrt{s})^{-1}$. *Then* $\lim_{s\to\infty}\bar{M}(s)=0$.

Proof. We shall assume that s is an integer. The proof can be easily extended to the case when s is not an integer. Write $M_n=\bar{M}(n)$. We claim $\bar{M}_n=nM_n$ satisfies the recurrence relation

$$\bar{M}_{n+1}\leqslant\bar{M}_n+C_n, \tag{9}$$

where

$$C_n=C_{n1}+C_{n2},$$
$$C_{n1}=2E\{|S_n|^2;\ \mathsf{A}_n\cup\mathsf{A}_{n+1}\},$$
$$C_{n2}=2nE\{|Y_n|^2;\ \mathsf{A}_n^c\,\mathsf{A}_{n+1}\},$$
$$Y_n=R(n),$$
$$S_n=\left(Y_{n+1}-Y_n\sqrt{\frac{n}{n+1}}\right)(\sqrt{(n+1)})e^{\lambda}, \tag{10}$$
$$\mathsf{A}_n=\{|Y_n|>\theta^n\}.$$

Clearly,

$$E\{|Y_{n+1}|^2;\ \mathsf{A}_{n+1}\}=E\{|Y_{n+1}|^2;\ \mathsf{A}_n\}+E\{|Y_{n+1}|^2;\ \mathsf{A}_n^c\,\mathsf{A}_{n+1}\}. \tag{11}$$

Using the martingale property of $\xi\cdot\boldsymbol{Z}(s)e^{-\lambda s}$ we see that

$$\begin{aligned}E\{|Y_{n+1}|^2;\ \mathsf{A}_n\}&=E\left\{\left|((\sqrt{n+1})e^{\lambda})^{-1}S_n+Y_n\sqrt{\frac{n}{n+1}}\right|^2;\ \mathsf{A}_n\right\}\\&=\frac{1}{(n+1)e^{\lambda_1}}E\{|S_n|^2;\ \mathsf{A}_n\}+\frac{n}{(n+1)}E\{|Y_n|^2;\ \mathsf{A}_n\}\end{aligned} \tag{12}$$

(using the orthogonality of S_n and $Y_n\chi_{\mathsf{A}_n}$). Next, from (10) we note that

$$|Y_{n+1}|^2\leqslant2\left[((n+1)e^{\lambda_1})^{-1}|S_n|^2+|Y_n|^2\frac{n}{(n+1)}\right],$$

and hence

$$\begin{aligned}E\{|Y_{n+1}|^2;\ \mathsf{A}_n^c\,\mathsf{A}_{n+1}\}\leqslant{}&2[(n+1)e^{\lambda_1}]^{-1}E\{|S_n|^2;\ \mathsf{A}_n^c\,\mathsf{A}_{n+1}\}\\&+\frac{2n}{(n+1)}E\{|Y_n|^2;\ \mathsf{A}_n^c\,\mathsf{A}_{n+1}\}.\end{aligned} \tag{13}$$

Multiplying both sides of (11) by $(n+1)$ and using (12) and (13), we get (9). Iterating (9) yields

$$\bar{M}_{n+1}\leqslant\sum_{j=1}^{n}C_j+\bar{M}_1, \tag{14}$$

or equivalently

$$M_{n+1} \leqslant \frac{1}{(n+1)} \left[\sum_{j=1}^{n} C_j + \bar{M}_1 \right]. \tag{15}$$

It suffices to show that $\lim_{n\to\infty} C_n = 0$.

Let us first look at $\lim_{n\to\infty} C_{n1}$. If we show that the sequence $\{|S_n|^2\}$ is uniformly integrable we may conclude that $\lim C_{n1}=0$, since $P(A_n) \leqslant E|Y_n|^2/\theta^{2n} \to 0$ as $\sup_n E|Y_n|^2 < \infty$. To establish the uniform integrability of $\{|S_n|^2\}$, note that S_n can be written as

$$S_n = \left[\sum_{r=1}^{p} \sum_{r=1}^{Z_n^r} (\xi \cdot \boldsymbol{Z}^{(r)j}(1) - e^{\lambda} \xi_r) \right] (e^{-\lambda n}),$$

where $\boldsymbol{Z}^{(r)j}(1)$ denotes the offspring vector in one unit of time, of the jth particle of type r in the nth generation. Again by the additive property we know that on the set $\{\boldsymbol{Z}_n \to \infty\}$, $|S_n|$ converges in law to $|S|$, where $S \equiv a \sum_{r=1}^{p} \sqrt{v_r}\, W N_r$. Here $(N_1, N_2, \ldots, N_p)$ are independent random variables, N_r being complex normal with mean 0 and covariance same as that of $\xi \cdot \boldsymbol{Z}^{(r)}(1) - e^{\lambda} \xi_r{}^2$; and W is independent of $N_1, N_2, \ldots N_p)$, with the same distribution as $\lim_{t\to\infty} \boldsymbol{Z}(t) e^{-\lambda_1 t}$ when $\boldsymbol{Z}(0) = e_i$. Also $E|S_n|^2 = E\left(\sum_{r=1}^{p} e^{-\lambda_1 n} Z_n^r \sigma_r^2 \right)$, and hence $E|S_n|^2 \to E|S|^2$. The facts that $|S_n| \to |S|$ in law, and $E|S_n|^2 \to E|S|^2$, imply the uniform integrability of $\{|S_n|^2\}$. (See Chung (1966), Chapter 4.)

Now consider C_{n2}. It is easily seen that for large n

$$|Y_n| < \theta^n, \quad |Y_{n+1}| > \theta^{n+1} \;\Rightarrow\; |S_n| > \sqrt{(n+1)}\, \theta^n (\theta - 1).$$

If we set $\bar{\mathsf{A}}_{n+1} = \{|S_n| > \sqrt{(n+1)}\, \theta^n (\theta-1)\}$ then for large enough n

$$\mathsf{A}_{n+1}\, \mathsf{A}_n^c \subset \bar{\mathsf{A}}_{n+1}\, \mathsf{A}_n^c .$$

Thus

$$E\{|Y_n|^2 \chi_{\mathsf{A}_{n+1} \mathsf{A}_n^c}\} \leqslant E\{|Y_n|^2 \chi_{\bar{\mathsf{A}}_{n+1} \mathsf{A}_n^c}\} \leqslant E[|Y_n|^2 \chi_{\mathsf{A}_n^c} E(\chi_{\bar{\mathsf{A}}_{n+1}} | \mathbb{F}_n)], \tag{16}$$

where $\mathbb{F}_t$ is the σ algebra generated by $\boldsymbol{Z}(s)$ for $s \leqslant t$. Now using Chebychev's inequality we get the bound

$$E(\chi_{\mathsf{A}_{n+1}} | \mathbb{F}_n) \leqslant \frac{E(|S_n|^2 | \mathscr{F}_n)}{(n+1)\, \theta^{2n} (\theta-1)^2} \qquad \text{a.s.}$$

But $E(|S_n|^2 | \mathbb{F}_n) = \sum_{r=1}^{p} e^{-\lambda_1 n} Z_n^r \sigma_r^2$. Substituting this in (16) yields

$$nE\{|Y_n|^2 \chi_{A_{n+1}} \chi_{A_n^c}\} \leqslant E\left\{\frac{|Y_n|^2 \chi_{A_n^c}}{\theta^{2n}(\theta-1)^2} \sum_{r=1}^{p} e^{-\lambda_1 n} Z_n^r \sigma_r^2\right\} \leqslant CE\left\{\frac{|Y_n|^2 \chi_{A_n^c}}{\theta^{2n}} W_n\right\},$$

where $W_n = e^{-\lambda_1 n} \sum_{r=1}^{p} u_r Z_{nr}$, and C is an appropriate constant. Now $\{W_n\}$ is a martingale sequence and under our second moment assumptions is uniformly integrable. Also the sequence $|Y_n|^2 \chi_{A_n^c}/\theta^{2n}$ is bounded by one, and $E(|Y_n|^2 \chi_{A_n^c}/\theta^{2n}) \leqslant E|Y_n|^2/\theta^{2n} \to 0$ since $E|Y_n|^2$ is bounded in n. Therefore, by a slight variant of the Lebesgue dominated convergence theorem we may conclude that

$$\lim_{n\to\infty} E\left\{\frac{|Y_n|^2 \chi_{A_n^c}}{\theta^{2n}} W_n\right\} = 0,$$

and hence that $\lim_{n\to\infty} C_{n2} = 0$. We have thus shown that $\lim_{n\to\infty} C_n = 0$. This proves lemma 5 and hence theorem 4. □

9. Embedding of Urn Schemes into Continuous Time Markov Branching Processes

9.1 Introduction & Polya's Urn

We shall now explore some connections between urn schemes and branching processes. Before we define a general urn scheme let us examine the familiar example of Polya's urn. We start with an urn containing W_0 white and B_0 black balls. A *draw* is effected as follows: (i) choose a ball at random from the urn; (ii) observe its color, return the ball to the urn; (iii) add α balls of the color drawn. Let (W_n, B_n) denote the composition of the urn after n successive draws. The stochastic process $\{(W_n, B_n); n=0,1,2,\ldots\}$ is called Polya's urn scheme with parameters W_0, B_0 and α. This process has applications in the study of contagious diseases (see Feller Vol. I (1968); also Blackwell and Kendall (1964), who computed the Martin boundary for this process). We shall now see that Polya's scheme can be embedded in a continuous time Markov branching process. Let us start with a one dimensional Markov branching process $\{X(t); t \geqslant 0\}$ with initial size $X(0) = W_0 + B_0$. Assume the infinitesimal generating function to be of the form $u(s) = [s^{\alpha+1} - s]$. Thus the life times of the particles are all unit exponential (exponentially distributed with mean one) and on death each creates $(\alpha+1)$ new particles. Divide the process at the start into two groups one consisting of W_0 particles and the other consisting of B_0 particles. Let τ_n, $n \geqslant 1$ denote the successive times at which splits or deaths occur in the whole collection. *If* $(X_1(t), X_2(t))$ *denotes the vector of population sizes in the two groups at time t then* $\{(X_1(\tau_n), X_2(\tau_n)); n=0,1,2,\ldots\}$ *is a stochastic process having the same distribution as Polya's urn scheme with parameters* W_0, B_0 *and* α. The proof of this is a special case of a more general theorem to be proved shortly. A very quick corollary to this embedding is the following

Theorem 1. *Let* $\{(W_n, B_n); n=0,1,2,\ldots\}$ *be a Polya's urn with parameters* W_0, B_0 *and* α. *Then there exists a random variable Y with a Beta* $(W_0/\alpha, B_0/\alpha)$ *distribution, such that*

$$\lim_{n\to\infty}\left(\frac{W_n}{W_n+B_n}\right)=Y \qquad \text{a.s.} \tag{1}$$

Proof. From the limit theory of chapter III we know that

$$\lim_{t\to\infty} X(t)e^{-\alpha t}=W \qquad \text{exists a.s.,}$$

where $X(t)$ is the branching processes mentioned in the above embedding. We can in this case explicitly solve the functional equation which $\varphi(u)=E(e^{-uW}|X(0)=1)$ satisfies, and identify the distribution of W given $X(0)=1$ to be Gamma with parameters $1/\alpha$ and $1/\alpha$. Furthermore using the additive property of the branching process we note that there exist independent random variables $W^{(j)}, j=1,2,\ldots, W_0+B_0$, all having $\Gamma(1/\alpha, 1/\alpha)$ distribution, such that

and
$$\begin{cases} \lim\limits_{t\to\infty} X_1(t)e^{-\alpha t}=\sum\limits_{j=1}^{W_0} W^{(j)} \qquad \text{a.s.,} \\ \lim\limits_{t\to\infty} X_2(t)e^{-\alpha t}=\sum\limits_{j=W_0+1}^{W_0+B_0} W^{(j)}. \end{cases} \tag{2}$$

From the results on split times in Chapter III we know that $\tau_n\to\infty$ a.s. It is immediate that

$$\lim_{n\to\infty}\left(\frac{X_1(\tau_n)}{X_1(\tau_n)+X_2(\tau_n)}\right)=\lim_{n\to\infty}\left(\frac{X_1(\tau_n)e^{-\alpha\tau_n}}{(X_1(\tau_n)+X_2(\tau_n))e^{-\alpha\tau_n}}\right)=\frac{\sum\limits_{j=1}^{W_0} W^{(j)}}{\sum\limits_{j=1}^{(W_0+B_0)} W^{(j)}}.$$

It is well known that the limit random variable above has the Beta distribution with parameters W_0/α, B_0/α. □

For a converse of theorem I, see Athreya (1969e).

9.2 The General Embedding Theorem

The following urn scheme is a generalization of Polya's urn scheme. An urn has balls of p different colors. We start with Y_{0i} balls of color i $(i=1,2,\ldots,p)$. A *draw* is effected as follows: (i) choose a ball at random from the urn; (ii) observe its color C and return the ball to the urn, and (iii) if $C=i$ add a random number R_{ij} of balls of color j $(j=1,2,\ldots,p)$. Let $\boldsymbol{Y}_n=(Y_{n1},Y_{n2},\ldots,Y_{np})$ denote the composition of the urn after n consecutive draws, where Y_{ni} is the number of balls of color i. The stochastic process $\{\boldsymbol{Y}_n: n=0,1,2,\ldots\}$ on the p dimensional integer lattice will be called a generalized Polya's urn scheme (GPU), and denoted by $GP(\boldsymbol{Y}_0, f_1(\boldsymbol{s}),\ldots,f_p(\boldsymbol{s}))$, where $f_i(\boldsymbol{s})$ is the p dimensional probability generating function of the vector $\boldsymbol{R}_i=(R_{i1},R_{i2},\ldots,R_{ip})$. A special case of this model is B. Friedman's urn, for which $p=2$, and $\boldsymbol{R}_1$ and $\boldsymbol{R}_2$ have degenerate distributions given by $P(R_{11}=\alpha+1, R_{12}=\beta)=1$ and $P(R_{21}=\beta, R_{22}=\alpha+1)=1$. D. Freedman (1965) developed a number of limit theorems for Friedman's urn scheme. We shall prove these, and their generalizations to GPU by embedding in a multitype continuous time Markov branching process. The embedding is accomplished as follows.

Let $\{\boldsymbol{Y}_n: n=0,1,2,\ldots\}$ be a $GP(\boldsymbol{Y}_0, f_1(\boldsymbol{s}),\ldots,f_p(\boldsymbol{s}))$ process. Let $\{\boldsymbol{Z}(t); t\geqslant 0\}$ be a p-type MCMBP of the type discussed in the last two sections. Assume that (i) the life times of particles of all types are unit exponentials; and that (ii) an ith type particle creates, on death, new particles of all types according to the p.g.f. $s_i f_i(\boldsymbol{s})$. Let $\boldsymbol{Z}(0)=\boldsymbol{Y}_0$, and $\{\tau_n: n=0,1,2,\ldots; \tau_0=0\}$ denote the *split times* for this process. Then,

Theorem 2. *The stochastic processes* $\{\boldsymbol{Y}_n; n=0,1,2,\ldots\}$ *and* $\{\boldsymbol{Z}(\tau_n); n=0,1,2,\ldots\}$ *are equivalent.*

Proof. We first note that both the stochastic processes are discrete time, discrete state space Markov chains with stationary transition probabilities. Thus if we show that they have the same transition probabilities then we are done since $\boldsymbol{Z}(0)=\boldsymbol{Y}_0$. We need to look just at the first step. We start the branching process with $Z_i(0)=Y_{0i}$ particles of type i for $i=1,2,\ldots,p$. The life times of all the particles are independently distributed unit exponentials. Hence the probability that the first split is of a type i particle is simply

$$\frac{Z_i(0)}{\sum_{i=1}^{p} Z_i(0)}=\frac{Y_{0i}}{\sum_{i=1}^{p} Y_{0i}}.$$

After this split the parent particle dies but creates a random number $R_{ij}+\delta_{ij}$ of new particles of type j, where $(R_{ij}; j=1,2,\ldots,p)$ has $f_i(\boldsymbol{s})$

as its p.g.f. Now the process starts all over again since the remaining life time of all other particles is a fresh exponential. This shows that if $Y_0 = Z(0)$, then Y_1 has the same distribution as $Z(\tau_1)$. Thus the transition mechanisms of both the processes are identical. (Note that we have implicitly assumed that the process $Z(t)$ is right continuous in t with probability one.) □

9.3 Some Quick Applications and Almost Sure Convergence

We shall now exploit the embedding theorem to obtain results about urn schemes from branching processes. Let $\{Y_n; n=0,1,2,\ldots\}$ be a $GP(Y_0, f_1(s),\ldots,f_p(s))$ and $\{Z(t); t\geqslant 0\}$ be a branching process associated with the urn scheme in the manner described earlier.

Assume that

(i) The distribution of $\{R_{ij}; j=1,2,\ldots,p\}$ is such that $P(R_{ij}=0$ for all $j)=0$, i.e. that after each draw the composition of the urn does in fact change. This will make the split times coincide with the discontinuities of the sample paths $\{Z(t); t\geqslant 0\}$. It also assures us that the process is non-singular and supercritical with extinction probability zero.

(ii) The $f_i(s)$ are all such that there is no explosion in the branching process in finite time. Recall that a sufficient condition for this is that all the means

$$m_{ij} = \left|\frac{\partial f_i(s)}{\partial s_j}\right|_{s=1}$$

are finite. It then follows that $\tau_n \to \infty$ a.s. (see section 7).

(iii) The matrix $M = \{(m_{ij})\}$ is strictly positive. This implies that $\{Z(t); t\geqslant 0\}$ is positive regular.

(iv) The log moment hypothesis (*) of Section 7.5, namely that

$$E R_{ij} \log R_{ij} < \infty \quad \text{for all } i,j \tag{*}$$

is satisfied.

It is now immediate from theorems 7.2 and 7.3, and the fact that $\tau_n \to \infty$ a.s., that for any $GP(Y_0, f_1, f_2, \ldots, f_p)$ satisfying (*)

$$n^{-1} Y_n \to v\,\mu \qquad \text{a.s.} \tag{3}$$

This result is worth some comment. It implies that for each i the proportion of balls of the ith color

$$\frac{Y_{ni}}{\sum_{i=1}^{p} Y_{ni}} \to v_i \qquad \text{a.s.} \tag{4}$$

A special case of this is the Friedman's urn that we mentioned earlier. Here $u_1 = u_2 = \frac{1}{2}$.

This must be contrasted with the Polya's urn case. There the associated branching process is nonsingular but *not positively regular*. Thus theorem 7.2 is not applicable. However, the process decomposes into (p) independent identical one dimensional branching processes. In this case the proportion $Y_{n1}/(Y_{n1}+Y_{n2})$ will converge to a random variable rather than a constant as in (4).

Let us return to the positively regular case, and to the question of the limiting behavior of linear functionals on a GPU. If ξ is an arbitrary vector such that $\xi \cdot \boldsymbol{v} \neq 0$ then (3) gives an immediate answer, namely

$$n^{-1}\,\xi \cdot \boldsymbol{Y}_n \to (\boldsymbol{v} \cdot \xi)\,\mu \qquad \text{a.s.}$$

But if ξ is a right eigenvector with eigenvalue $\lambda \neq \lambda_1$, then $\xi \cdot \boldsymbol{v} = 0$, and we can combine theorem 8.1 and theorem 7.3 to obtain (using the same notation as in section 7):

Theorem 3. *Let* $2\,\mathrm{Re}\,\lambda > \lambda_1$ *and* $\beta = \lambda/\lambda_1$. *Then* $n^{-\beta}\,\xi \cdot \boldsymbol{Y}_n$ *converges* a.s. *to a random variable.*

The case of a general η satisfying $\eta \cdot \boldsymbol{v} = 0$ is discussed in Athreya (1969b and 1971b).

9.4 Convergence in Law

Going from the almost sure limit behavior of $\xi \cdot \boldsymbol{Z}(t)$ to that of $\xi \cdot \boldsymbol{Z}(\tau_n)$ was easy, but in the case of law behavior the transition is much harder. That is, knowing that $\xi \cdot \boldsymbol{Z}(t)$ (appropriately normalized) converges in law (as $t \to \infty$) to a random variable X, it is nontrivial to prove a corresponding convergence in law for $\xi \cdot \boldsymbol{Z}(\tau_n)$. Needless to say, one of the reasons for attempting the transition is that we can use the embedding to get a result for the GP urn schemes.

This raises a general problem concerning Markov processes and associated embedded chains. Let $\{X(t);\, t \geqslant 0\}$ be a continuous time Markov process with a discrete state space, and let $\{\tau_n;\, n = 1, 2, \ldots\}$ denote the jump times. Suppose $P\{\tau_n \to \infty\} = 1$ and $Y(t) = f(X(t))$ is a functional on the process such that $F(t,x) = P\{Y(t) \leqslant x\}$ converges to some distribution function, say, $F(x)$ as $t \to \infty$. Since $Y(t) = Y(\tau_n)$ for $\tau_n \leqslant t < \tau_{n+1}$, $n = 1, 2, \ldots$, (assuming right continuity of $Y(t)$) it is tempting to conjecture that $F_n(x) = P\{Y(\tau_n) \leqslant x\}$ converges in distribution to F. Of course, such a conjecture is false without some kind of aperiodicity on the embedded chain $X(\tau_n)$. As a counterexample consider the telegraph signal process. Here $X(t)$ alternates from $+1$ to -1 with the sojourn times in either state having unit exponential distribution. It is easy to check that

$$\lim_{t\to\infty} P\{X(t)=1\,|\,X(0)=1\}=\tfrac{1}{2},$$
$$\lim_{n\to\infty} P\{X(\tau_{2n})=1\,|\,X(0)=1\}=1,$$
$$\lim_{n\to\infty} P\{X(\tau_{2n+1})=-1\,|\,X(0)=1\}=1.$$

However, if the embedded chain $\{X(\tau_n)\}$ is aperiodic and ergodic then both $X(t)$ and $X(\tau_n)$ have the same limiting distributions as $t\to\infty$ and $n\to\infty$. Of course this will imply the same conclusion for all continuous functionals $g(X(t))$. All other possibilities are open problems.

Returning to the branching process we shall be content with stating the following analog of theorem 8.2. We refer the reader to Athreya and Karlin (1968) for the proof.

Theorem 4. *Assume the hypothesis of theorem 8.2. Then*

$$\lim_{n\to\infty} P\left\{0<x_1\leqslant W\leqslant x_2<\infty, \frac{Y(\tau_n)}{\sqrt{Y_1(\tau_n)}}\leqslant x\,|\,\mathbf{Z}(0)\right\}$$
$$=P\{0<x_1\leqslant W\leqslant x_2<\infty\,|\mathbf{Z}(0)\}\,\Phi\left(\frac{x}{\sigma}\right).$$

By the embedding theorem 2 and the above theorem we get the following result on GP urn schemes.

Theorem 5. *Let* $\{\mathbf{Y}_n; n=0,1,2,\ldots\}$ *be a* $GP(\mathbf{Y}_0, f_1(s),\ldots,f_p(s))$. *Assume that*

$$E R_{ij}^2<\infty \qquad \textit{for all } i,j.$$

Let ξ *be an eigenvector of* M *corresponding to an eigenvalue* λ. *Assume* $2\operatorname{Re}\lambda<\lambda_1$. *Then, as* $n\to\infty$,

$$\frac{\xi\cdot\mathbf{Y}_n}{\sqrt{n}}\xrightarrow{d} N(0,c),$$

where c *is a constant and* $\xrightarrow{d}$ *means convergence in law.*

Theorems 3 and 5 are generalizations of Freedman's result for Friedman's urn. For the third case, $2\operatorname{Re}\lambda=\lambda_1$, we have a log factor in the normalization, as can be seen from theorem 8.2. The translation of this result for the split times, and hence urn schemes, has not been published but is very similar to the case $2\operatorname{Re}\lambda<\lambda_1$. Such a result will clearly subsume Freedman's result (1965).

10. The Multitype Age-Dependent Process

We conclude by observing that here again there are analogs of some of the classical limit laws, and summarize the known results.

The population again consits of p types of particles, whose reproductive behavior is governed by a p-dimensional generating function $\boldsymbol{f}(\boldsymbol{s})$, exactly as defined in section 1. The lifetime of a type i particle is a random variable with distribution $G_i(\cdot), i=1,\ldots,p$. Both particle production and lifetimes are independent of the history of the process prior to the birth of the particle in question.

Again let $\boldsymbol{Z}(t)=(Z_1(t),\ldots,Z_p(t))$ denote the number of particles of the various types existing at time t, and let

$$F^{(i)}(\boldsymbol{s},t)=\sum_{\boldsymbol{j}\in\mathscr{R}_p^+} P\{\boldsymbol{Z}(t)=\boldsymbol{j}\,|\,\boldsymbol{Z}(0)=\boldsymbol{e}_i\}\,\boldsymbol{s}^{\boldsymbol{j}}, \tag{1}$$

$$\boldsymbol{F}(\boldsymbol{s},t)=\big(F_1(\boldsymbol{s},t),\ldots,F_p(\boldsymbol{s},t)\big).$$

Then as in the one-dimensional age-dependent process in chapter IV, one can show that

$$F^{(i)}(\boldsymbol{s},t)=s_i[1-G_i(t)]+\int_0^t f^{(i)}[\boldsymbol{F}(\boldsymbol{s},t-y)]\,dG_i(y), \qquad i=1,\ldots,p. \tag{2}$$

The existence and uniqueness theory of the system of equations in (2) can be developed along the lines of section IV.2.

Let $\mu_{ij}(t)=E[Z_j(t)|\boldsymbol{Z}(0)=\boldsymbol{e}_i]$, and $\boldsymbol{U}(t)=\{\mu_{ij}(t)\}$, $i,j=1,\ldots,p$ be the matrix of means at time t, and let $\boldsymbol{M}$ be the particle production mean matrix associated with $\boldsymbol{f}(\boldsymbol{s})$. If $\|\boldsymbol{M}\|<\infty$, then $\|\boldsymbol{U}(t)\|$ is bounded on finite intervals, and $\boldsymbol{U}(t)$ satisfies the matrix equation

$$\boldsymbol{U}'(t)=\boldsymbol{D}[\boldsymbol{1}-\boldsymbol{G}(t)]+\int_0^t \boldsymbol{U}'(t-y)\boldsymbol{M}'\,d[\boldsymbol{G}(y)],$$

where $\boldsymbol{D}[\boldsymbol{x}]$ is the diagonal matrix with x_1 in the ith place, and $d[\boldsymbol{G}(y)]$ is the diagonal matrix with $dG_i(y)$ in the ith entry. In fact, $U(t)$ is the unique solution bounded on finite intervals.

Here again there is an analog to the concept of a Malthusian parameter α introduced in chapter IV[14]. It can then be shown that there exists a constant matrix $\boldsymbol{C}$ and a matrix $\boldsymbol{B}(t)$ such that

$$\boldsymbol{U}(t)=\boldsymbol{C}e^{\alpha t}+\boldsymbol{B}(t),$$

where $\|\boldsymbol{B}(t)\|=O(e^{\beta t})$ for some $\beta<\alpha$. In the supercritical case, the Malthusian parameter α always exists. The details of the above dis-

---

[14] Let $\hat{\boldsymbol{M}}(\alpha)$ be the matrix whose (i,j) entry is $m_{ij}\int e^{-\alpha t}\,dG_i(t)$. The Malthusian parameter is that number (unique, if it exists) such that the maximal eigenvalue of $\hat{\boldsymbol{M}}(\alpha)$ is 1.

cussion can be found in Mode (1971). In the subcritical case, the parameter may not exist, and the analog of the discussion of chapter IV for this case has not yet been carried out. (Super- or subcritical again mean that the maximal eigenvalue of $\boldsymbol{M}$ is >1 or <1.)

The limit laws for $\boldsymbol{Z}(t)$ take the following form:

Supercritical Case ($\rho>1$). We assume the same regularity conditions on $\boldsymbol{M}$ as in section 2. Let $\alpha>0$ be the Malthusian parameter and $\boldsymbol{W}(t)=e^{-\alpha t}\boldsymbol{Z}(t)$. Assume that all second moment associated $\boldsymbol{f}(\boldsymbol{s})$ exist and that $G_i(t)$, $i=1,\ldots,p$ is absolutely continuous with a density in L_2. Then it has been proved (see Mode (1971)) that $\boldsymbol{W}(t)\to\boldsymbol{v}W$ a.s., $\boldsymbol{v}$ being the left eigenvector of $\boldsymbol{M}$ associated with the maximal eigenvalue, and W being a one-dimensional random variable. Furthermore, the Laplace transforms

$$\varphi_i(\theta)=E\{\exp(-\theta W)|\boldsymbol{Z}(0)=\boldsymbol{e}_i\}, \qquad \theta\geqslant 0,$$

satisfy the equations

$$\varphi_i(\theta)=\int_0^\infty f^{(i)}[\boldsymbol{\phi}(\theta e^{-\alpha t})]\,dG_i(t), \quad i=1,\ldots,p,$$

where $\boldsymbol{\phi}=(\varphi_1,\ldots,\varphi_p)$.

Some of the above results were first announced without proof by Snow (1959a, b). The second moment assumptions on $\boldsymbol{f}$ and the extra regularity conditions on $\boldsymbol{G}$ are surely not necessary. It would be desirable to have a proof of the result under the same logarithmic moment conditions as in the discrete case, and to extend the other parts of theorems 6.1 and 6.2 to the present setting (see complements).

The Subcritical Case ($\rho<1$) (T. Ryan (1968)). If $(*)$ of section 4 is satisfied, if the Malthusian parameter (say α) exists, and if $\int t e^{\alpha t}\,dG_i(t)<\infty$ for $i=1,\ldots,p$, then as $t\to\infty$

(i) $P(\boldsymbol{Z}(t)\neq\boldsymbol{0}|\boldsymbol{Z}(0)=\boldsymbol{e}_i)\sim c_i e^{-\alpha t}$, $c_i>0$;

and

(ii) $P\{\boldsymbol{Z}(t)=\boldsymbol{j}|\boldsymbol{Z}(0)=\boldsymbol{e}_i,\ \boldsymbol{Z}(t)\neq\boldsymbol{0}\}\to b(\boldsymbol{j})$,

where $b(\boldsymbol{j})$ is a p.f. on $\mathscr{R}_p^+-\{\boldsymbol{0}\}$.

The Critical Case ($\rho=1$). The extension from the single to multitype process of the limit law of section IV has not yet been entirely achieved, and leaves us with one of the most challenging unsolved problems in the theory of branching process. The first step in this direction, namely the asymptotic behavior of the extinction probability, has been determined by M. Goldstein (1969, 1971) through an extension of his lemma in section IV.3, and H. Weiner (1970), using an extension of the techniques of Chover, Ney (1968) for the one dimensional case. The result is as follows:

Make the usual assumptions on $\boldsymbol{M}$; assume that all the second moments associated with $\boldsymbol{f}(\boldsymbol{s})$ exist (this is clearly necessary here); and also assume that $t^2[1-G_i(t)]\to 0$ as $t\to\infty, i=1,\ldots,p$. Then

$$\lim_{t\to\infty} t\,P\{\boldsymbol{Z}(t)\neq\boldsymbol{0}\,|\,\boldsymbol{Z}(0)=\boldsymbol{e}_i\} = \left[\frac{\boldsymbol{\mu}\cdot(\boldsymbol{u}\otimes\boldsymbol{v})}{Q[\boldsymbol{u}]}\right]u_i\,, \tag{3}$$

where $\boldsymbol{\mu}=(\mu_1,\ldots,\mu_p)$ is the vector of means of $G_1(t),\ldots,G_p(t)$; $\boldsymbol{u}$ and $\boldsymbol{v}$ are the usual right and left eigenvectors of $\boldsymbol{M}$; $\boldsymbol{u}\otimes\boldsymbol{v}=(u_1 v_1,\ldots,u_p v_p)$ and $\boldsymbol{Q}$ is a second moment quadratic form associated with $\boldsymbol{f}$.

The corresponding exponential limit law for $\{\boldsymbol{Z}(t)/t\,|\,\boldsymbol{Z}(t)\neq\boldsymbol{0}\}$ has been proved by H. Weiner (1970) under the very strong assumption that all moments of $\boldsymbol{f}(\boldsymbol{s})$ exist. He shows that

$$P\left\{\frac{Z_i(t)}{t}\leqslant r\,\middle|\,Z_i(t)>0\right\}$$

converges to an exponential distribution which is independent of i, and has all its mass concentrated along the ray $c(\boldsymbol{u}\otimes\boldsymbol{v})$, $c\geqslant 0$.

To prove this result under a second moment hypotheses is still an open problem.

Complements and Problems V

1. There is an analog of theorem I.11.5 for the multitype case (Joffe-Spitzer, 1967); namely, if $\rho<1$, $\boldsymbol{j}\to\infty$ in such a way that $\boldsymbol{j}\cdot\boldsymbol{u}\to\infty$, and $n\to\infty$ in such a way that $(\boldsymbol{j}\cdot\boldsymbol{u})\rho^n\gamma\to c$, then

$$\lim_{n\to\infty} P\{\boldsymbol{Z}_n=\boldsymbol{i}\,|\,\boldsymbol{Z}_0=\boldsymbol{j}\}=d_{\boldsymbol{i}}\,,$$

where $d_{\boldsymbol{i}}$ is a probability distribution on $\mathscr{R}_p^+$.
It has generating function $e^{-c[1-g(s)]}$ where g is as in (4.16).

2. Let $\boldsymbol{M}_n(\boldsymbol{s})=((m_{ij}^{(n)}(\boldsymbol{s})))$, $1\leqslant i,j\leqslant p$, where

$$m_{ij}^{(n)}(\boldsymbol{s}) = \frac{\partial f_n^{(i)}(\boldsymbol{s})}{\partial s_j}\,.$$

Show that

$$\boldsymbol{M}_n(\boldsymbol{s})=\boldsymbol{M}(\boldsymbol{f}_{n-1}(\boldsymbol{s}))\,\boldsymbol{M}_{n-1}(\boldsymbol{s})\,,$$

where we write $\boldsymbol{M}(\boldsymbol{s})$ for $\boldsymbol{M}_1(\boldsymbol{s})$.

3. Let $\lambda(\boldsymbol{s})$ be the largest eigenvalue of $\boldsymbol{M}(\boldsymbol{s})$. Noting that $\boldsymbol{M}_n(\boldsymbol{q})=\boldsymbol{M}^n(\boldsymbol{q})$, where $\boldsymbol{q}$ is the extinction probability vector, and that the transience of the nonzero states implies $\boldsymbol{M}_n(\boldsymbol{s})\to 0$ for $\boldsymbol{s}<\boldsymbol{1}$, show that $\lambda(\boldsymbol{q})<1$ when $\rho\neq 1$ (Kesten and Stigum (1966 b)).

4. Now use the above two problems and the fact that $\boldsymbol{f}_n(\boldsymbol{s})\to\boldsymbol{q}$ uniformly for $\|\boldsymbol{s}\|\leqslant 1-\delta$, $\delta>0$ to conclude that for $\|\boldsymbol{s}\|\leqslant 1-\delta$ there exists a constant γ in $[0,1)$ such that

$$\sup_{\substack{i,j,n\\ \|\boldsymbol{s}\|\leq 1-\delta}} \gamma^{-n} m_{ij}^{(n)}(\boldsymbol{s})<\infty\,.$$

5. Assume (*) of section 6 holds and let $\varphi_j(it)=E_j(e^{itW})$ be the characteristic function of W (as defined in theorem 6.2). Imitate the methods of chapter *I* to show that $\sup_{\substack{1\le|u|\le\rho\\1\le j\le p}} |\varphi_j(iu)|\equiv 1-\delta$ for some $\delta>0$.

6. Iterating the analog of (6.3) for $\varphi(it)$ show

$$\rho^n\,\varphi_i'(i\rho^n u)=\sum_{j=1}^{p} m_{ij}^{(n)}(\varphi(iu))\,\varphi_i'(iu),$$

and hence that

$$\int|\varphi_i'(iu)|\,du<\infty.$$

Conclude from this, as in chapter I, that W has an absolutely continuous distribution on $(0,\infty)$. (See problem 2 for notation.)

7. Can one improve the result in problem 4 to prove an analog of the geometric convergence of section 11 of chapter I?

8. What about the analog of Seneta's result (theorem I.10.3) for $\boldsymbol{u}\cdot\boldsymbol{Z}_n$ when (*) (section 6) is false?

9. What about the analog of theorem I.10.2 for p-type processes?

10. Show that (7.3)$\Rightarrow$(7.5), i.e., $m_{ij}(t)=E(Z_j^{(i)}(t))<\infty$ for all t, and i,j, if

$$b_{ij}=\left|\frac{\partial f^{(i)}}{\partial s_j}\right|_{s=1}<\infty$$

for all (i,j), by establishing a system of integral inequalities and obtaining a bounded sequence to approximate $m_{ij}(t)$. Show in the process, that under the present hypothesis, with probability one, explosions do not occur. It is harder to show that (7.5)$\Rightarrow$(7.3) also.

11. Show that (7.13)$\Rightarrow$(7.14) by methods similar to the above.

12. *Open problem:* Study the multitype analogs of the conditioned limit laws of sections I.14 and I.15.

13. *Urn Schemes:*
(i) Let $\{(W_n,B_n);\,n=0,1,2,\ldots\}$ be a generalized Polyá urn scheme, where after each draw a random number (with p.g.f. $f(z)$) of balls of the color drawn are added. Show that $W_n/(W_n+B_n)$ converges in distribution to a Gamma distribution with parameters W_0/α and B_0/α if and only if $f(z)=z^\alpha$. This characterizes the classical Polya's urn scheme.
(ii) Generalize the above result to the case of many colors.

14. *Multi-type age-dependent processes:* There are a number of results for single-type age-dependent processes which have not yet been carried over to the multi-type case. These include the asymptotic behavior of the process (moments, generating functions, limit distributions) when the Malthusian parameter does *not* exist. (See sections IV.5, IV.8, IV.10.)

15. Using the discrete time results of section 6, and the idea of the embedded generation process described in section IV.11, show that if $E\,\xi_{ij}\log\xi_{ij}=\infty$ for at least one pair (i,j), then the limit random variable W in the supercritical, age-dependent case (section 10) will be zero a.s.

Chapter VI

Special Processes

There is a long history of applications of branching process models to the physical and biological sciences, and it is likely that much work in the future will be in the direction of developing more realistic (and hence probably more complicated) mathematical models. In this book we make no attempt at a systematic exposition of special processes. Rather we select and describe a few models which have interested us, and which we feel point the direction to some promising problems for the future.

The first few sections of this chapter are devoted to a class of processes called branching random walks and diffusions, which are characterized by the multiplication of particles according to a branching process, and simultaneously their movement in a space according to a random walk or diffusion process.

We start in the next section by describing a simple process of this type, and then discuss some of its applications to special models in the theory of cosmic ray cascades, and to the determination of the distribution of generations in a population. This is followed by a discussion of a branching Brownian motion model.

Some aspects of the analysis of the above models are greatly simplified by recognizing appropriate functions of the processes as martingales. This trick has been used so frequently throughout the book, that it is worthwhile to put these martingale methods in a general setting. This is done in section 4.

In section 5 we discuss processes where particle production distributions do not remain constant from generation to generation, but depend in a (specified) random manner on the environment in which the population is evolving.

All the processes we have studied have been models for the evolution of populations of one or more types of particles. Hence the state spaces have always been the integers, or lattice points of a higher dimensional space. It is of interest to construct processes on continuous state spaces, for which the formal multiplicative property of the discrete process is preserved, and this is done in section 6.

The effect of immigration into a branching population is treated in section 7. In addition to being of obvious physical importance, the limit laws for branching processes with immigration turn out to be related in an interesting manner to the conditioned limit laws of the critical and subcritical Galton-Watson process.

We conclude with another result on the extreme instability of branching processes.

1. A One Dimensional Branching Random Walk

1.1 The Basic Integral Equation

Consider an age-dependent branching process of the type discussed in chapter IV, with particle production generating function $f(s)=\sum p_j s^j$ and lifetime distribution $G(t)$, with $G(0+)=0$. On this process we superimpose the additional structure of particle motion on the line. Consider a parent particle at the point x_0. At its death it splits into k particles with probability p_k, which then move to the random points $x_0+X_1,\ldots,x_0+X_k$. For simplicity we assume that $\{X_n; n\geqslant 1\}$ are independent, identically distributed random variables, and let $\Gamma(\cdot)$ denote their common distribution function. A more general version of this model where the X_i's are dependent and move in a higher dimensional space was treated by Ney (1965a).

Let $Z(x,t|x_0)$ denote the number of particles existing at time t and located in $(-\infty,x]$, given an initial parent at x_0 at time 0. Arguing heuristically as in the motivation of the basic integral equation (IV.1.1) of the age-dependent branching process, we observe that the event

$$\{Z(x,t|x_0)=n\}$$

can occur in the following ways:

(i) the original particle splits at time $y\leqslant t$ into j particles;

(ii) these move to the points $x_1,\ldots,x_j$, respectively;

(iii) in the remaining time $t-y$, the j particles produce a total of n particles in $(-\infty,x]$.

Summing over $y,j,x_1,\ldots,x_j$; taking into account the extra contingencies that if $y>t$ then $Z(x,t|x_0)=1$ when $x_0\leqslant x$ and $=0$ when $x_0>x$, while $Z(x,t|x_0)=0$ when $j=0$ and $y\leqslant t$; and letting

$$p_n(x,t|x_0)=P\{Z(x,t|x_0)=n\},$$

$$D(x)=\begin{cases}1 & \text{if } x\geqslant 0,\\ 0 & \text{if } x<0,\end{cases}$$

$$\delta_{ij}=\text{the Kronecker delta};$$

we get (for $n \geqslant 0$)

$$p_n(x,t|x_0) = [1-G(t)][\delta_{1n} D(x-x_0) + \delta_{0n}(1-D(x-x_0))] + p_0 \delta_{0n} G(t)$$
$$+ \sum_{j=1}^{\infty} p_j \int_0^t G(dy) \int_{-\infty}^{\infty} \Gamma(dx_1 - x_0) \cdots \int_{-\infty}^{\infty} \Gamma(dx_j - x_0) \tag{1}$$
$$\cdot \{p_n(x, t-y|x_1) * \cdots * p_n(x, t-y|x_j)\}.$$

Here $*$ denotes convolution with respect to the subscript n, e.g.

$$p_n(x,t|x_1) * p_n(x,t|x_2) = \sum_{i=0}^{n} p_i(x,t|x_1) p_{n-i}(x,t|x_2).$$

Associated with (1) is an equation for the generating function

$$F(s,x,t|x_0) = \sum_{n=0}^{\infty} p_n(x,t|x_0) s^n, \qquad |s| \leqslant 1, \tag{2}$$

namely

$$F(s,x,t|x_0) = [1-G(t)][sD(x-x_0) + 1 - D(x-x_0)]$$
$$+ \int_0^t f\left[\int_{-\infty}^{\infty} F(s,x,t-y|u) \Gamma(du - x_0)\right] dG(y). \tag{3}$$

Theorem 1. *The set of equations (1) has a unique bounded solution* $\{p_n(x,t|x_0); 0 \leqslant n\}$. *This solution is a probability function* ($p_n \geqslant 0, \sum p_n = 1$). *The generating function F as defined in (2), is the unique bounded solution of (3).*

Thus analytically, equation (3) is our starting point and determines the distribution of $Z(x,t|x_0)$.

The proof is entirely analogous to that of theorem IV.2.1, and we shall omit it.

We now observe by direct substitution that $\{p_n(x-r, t|x_0 - r)\}$ satisfies (1) for any real r. Hence by the uniqueness assertion of theorem 1

$$p_n(x,t|x_0) = p_n(x - x_0, t|0), \qquad n \geqslant 0.$$

Writing

$$Z(x,t|0) = Z(x,t),$$
$$p_n(x,t|0) = p_n(x,t),$$
$$F(s,x,t|0) = F(s,x,t),$$

we can simplify (3) slightly to read

$$F(s,t,x) = [1-G(t)][sD(x) + 1 - D(x)]$$
$$+ \int_0^t f\left[\int_{-\infty}^{\infty} F(s, x-u, t-y) d\Gamma(u)\right] dG(y). \tag{4}$$

This equation plays a role for the branching random walk analogous to (IV.1.1) for the age-dependent branching process. Observe that $Z(\infty, t) = Z(t) =$ total number of particles at time t.

1.2 The Mean Function

The analysis of the moments starts similarly to calculations already performed in chapter IV. Let

$$\mu(x, t) = E\,Z(x, t)\,. \tag{5}$$

Note that since $0 \leqslant \mu(x, t) \leqslant \mu(t)$, this mean always exists. Differentiating through (4) as in section IV.5 we get

$$\mu(x, t) = [1 - G(t)]\,D(x) + m \int_0^t \int_{-\infty}^{\infty} \mu(x - y, t - \tau)\,d\Gamma(y)\,dG(\tau)\,. \tag{6}$$

This is a special case of an equation of the form

$$H(\boldsymbol{x}) = \xi(\boldsymbol{x}) + \gamma \int_{\mathscr{E}_p} H(\boldsymbol{x} - \boldsymbol{y})\,d\lambda(\boldsymbol{y})\,, \tag{7}$$

where $\mathscr{E}_p$ is p-dimensional Euclidean space, $\boldsymbol{x}, \boldsymbol{y}, \in \mathscr{E}_p$, and ξ is a given real-valued function on $\mathscr{E}_p$, and $\boldsymbol{\lambda}(\cdot)$ is a measure on $\mathscr{E}_p$.

Equation (7) is an exact analog for p-dimensions of the renewal equation (3) of section IV.4. Its solution can again be expressed formally as

$$H(\boldsymbol{x}) = \xi(\boldsymbol{x}) * \sum_{n=0}^{\infty} \gamma^n \lambda^{*n}(\boldsymbol{x})\,, \tag{8}$$

where λ^{*n} is the n-fold convolution of λ with itself.

Surprisingly enough, many of the more delicate questions which we answered about the one-dimensional renewal function in section IV.4, are here not yet settled. What is wanted is the asymptotic behavior of $H(\boldsymbol{x})$ as $\boldsymbol{x} \to \infty$; but now of course $\boldsymbol{x}$ can "go to ∞" in different ways, and one would like a complete description of the limiting behavior.

Some aspects of this problem have been studied. When $\gamma = 1$, and the measure λ is concentrated on the lattice points of $\mathscr{E}_p$, a study of $\sum \lambda_n(\boldsymbol{x})$ was carried out by Ney and Spitzer (1965). The construction employed in that paper to describe the asymptotics of $H(\boldsymbol{x})$ as $\boldsymbol{x} \to \infty$ along various rays, (i.e. where $\boldsymbol{x} = \boldsymbol{a}t$, $t \to \infty$ for a fixed vector $\boldsymbol{a}$) also suggests natural analogs of the Malthusian parameter. As in section IV.4 the existence of these parameters would depend on the tail behavior of λ, and this will in turn affect the limit laws for H. Such analyses remain to be carried out.

Thus we are not yet in a position to give a complete description of the asymptotics of $E\,Z(x,t)$, let alone of $Z(x,t)$ itself. We shall, however, treat one aspect of the problem which is of particular interest for branching processes. Before turning to this, let us note the obvious fact that the time and space variables enter into (6) in a symmetric fashion, except that the time distribution is concentrated on $[0, \infty)$. We could equally well talk about two space variables with any distribution in the plane; and of course there are generalizations to higher dimensions. The independence of the time and space variables is also not essential. For a more general result see Ney (1965).

Returning to (6) we shall consider only the case when the Malthusian parameter $\alpha = \alpha(m, G)$, defined by

$$m \int_0^\infty e^{-\alpha t} dG(t) = 1$$

exists. This is the same parameter as in chapter IV, and when $m > 1$ it always exists. We again let

$$G_\alpha(t) = \int_0^t m e^{-\alpha y} dG(y),$$

and let

$$\mu_\alpha = \int_0^\infty t\, dG_\alpha(t), \tag{9.a}$$

$$\sigma_\alpha^2 = \int_0^\infty (t - \mu_\alpha)^2\, dG_\alpha(t), \tag{9.b}$$

$$\nu = \int_{-\infty}^\infty x\, d\Gamma(x), \tag{9.c}$$

$$\theta^2 = \int_{-\infty}^\infty (x - \nu)^2\, d\Gamma(x), \tag{9.d}$$

whenever these moments exist, and let

$$v^2 = \theta^2 \mu_\alpha^{-1} + \sigma_\alpha^2 \nu^2 \mu_\alpha^{-3} \tag{10}$$

and

$$x_t = \nu \mu_\alpha^{-1} t + \gamma v t^{\frac{1}{2}}. \tag{11}$$

Theorem 2. *If the Malthusian parameter and the moments in* (9) *exist, and x_t is given by* (11), *then*

$$\mu(x_t, t) \sim e^{\alpha t} \frac{m-1}{\alpha m \mu_\alpha} \Phi(\gamma) \tag{12}$$

where $\Phi(\cdot)$ is the standard normal distribution. Equivalently, (by theorem IV.5.3A)

$$\lim_{t \to \infty} \frac{\mu(x_t, t)}{\mu(t)} = \Phi(\gamma). \tag{13}$$

Remarks. (i) The result is valid for all $m<\infty$. If $m>1$ then $\alpha<0$ and hence all moments of G_α automatically exist.

(ii) When G_α and/or Γ are in the domain of a stable law, then a similar result holds for a different function x_t (see Athreya-Ney (1971)).

Proof. From (6) we observe that

$$\mu(x,t)=\sum_{n=0}^{\infty} m^n[G^{*n}(t)-G^{*n+1}(t)]\,\Gamma^{*n}(x). \tag{14}$$

Since $\mu(t)=\sum m^n[G^{*n}(t)-G^{*n+1}(t)]$ (by (IV.5.4)), we see that

$$\{(\mu(t))^{-1} m^n[G^{*n}(t)-G^{*n+1}(t)];\ n\geqslant 0\} \tag{15}$$

is a probability function in n for each fixed t; and we denote by N_t a random variable having this distribution. Thus

$$\begin{aligned} P\{N_t\geqslant k\} &= (\mu(t))^{-1}\sum_{n=k}^{\infty} m^n[G^{*n}(t)-G^{*n+1}(t)] \\ &= (\mu(t))^{-1}\sum_{n=0}^{\infty} m^n[G^{*n}(t)-G^{*n+1}(t)]*(m^k G^{*k}(t)) \\ &= m^k\frac{\mu(t)*G^{*k}(t)}{\mu(t)}. \end{aligned} \tag{16}$$

Letting $\mu_\alpha(t)=e^{-\alpha t}\mu(t)$, and multiplying the numerator and denominator in the right side of (16) by $e^{-\alpha t}$ we see that

$$P\{N_t\geqslant k\} = \frac{\mu_\alpha(t)*G_\alpha^{*k}(t)}{\mu_\alpha(t)}. \tag{17}$$

Now let $\hat{S}_n$ be the sum of n independent random variables with distribution G_α. Then

$$G_\alpha^{*n}(t)=P\left\{\frac{\hat{S}_n-n\mu_\alpha}{\sqrt{n}\,\sigma_\alpha}\leqslant\frac{t-n\mu_\alpha}{\sqrt{n}\,\sigma_\alpha}\right\}.$$

and if $(n,t)\to\infty$ in such a way that

$$\frac{(t-n\mu_\alpha)}{\sqrt{n}\,\sigma_\alpha}\to\gamma,$$

i. e. if

$$n_t=\mu_\alpha^{-1}t-\gamma\sigma_\alpha\mu_\alpha^{-\frac{3}{2}}t^{\frac{1}{2}}+o(t), \tag{18}$$

then

$$\lim_{t\to\infty} G_\alpha^{*[n_t]}(t)=\Phi(\gamma).$$

Since $\mu_\alpha(t)\to$constant, we have from (17) that

$$\lim_{t\to\infty} P\{N_t \geqslant n_t\} = \Phi(\gamma). \tag{19}$$

This result was first obtained by M. Samuels (1969) in a somewhat different manner.

Finally from (14) we see that

$$\frac{\mu(x,t)}{\mu(t)} = \sum P\{N_t = n\}\, \Gamma^{*n}(x) = P\{S_{N_t} \leqslant x\}, \tag{20}$$

where S_n is the sum of n independent random variables with distribution Γ, which are also independent of N_t. Hence

$$\lim_{t\to\infty} P\left\{\frac{S_{N_t} - N_t v}{N_t^{\frac{1}{2}}\theta} \leqslant x\right\} = \Phi(x). \tag{21}$$

Writing

$$S_{N_t} = \left(\frac{S_{N_t} - N_t v}{N_t^{\frac{1}{2}}\theta}\right)\left(\frac{N_t \theta^2}{t}\right)^{\frac{1}{2}} t^{\frac{1}{2}} + \left(\frac{N_t - \mu_\alpha^{-1} t}{\sigma_\alpha \mu_\alpha^{-\frac{3}{2}} t^{\frac{1}{2}}}\right)\left(\frac{\sigma_\alpha^2 t}{\mu_\alpha^3}\right)^{\frac{1}{2}} v + t v \mu_\alpha^{-1}, \tag{22}$$

applying (19), (21) and the fact that

$$\frac{N_t}{t} \to \mu_\alpha^{-1} \tag{23}$$

in distribution, we conclude that

$$\lim_{t\to\infty} P\{S_{N_t} \leqslant x_t\} = \Phi(\gamma). \quad \square \tag{24}$$

When the Malthusian parameter does not exist, but G is in the "sub-exponential" class $\mathscr{S}$ defined in section IV.4, then we have the following result:

Theorem 2 A. *If* $m<1$ *and* $G \in \mathscr{S}$ *then*

$$\lim_{t\to\infty} \frac{\mu(x,t)}{\mu(t)} = (1-m) \sum_n m^n\, \Gamma^{*n}(x).$$

Proof. From lemmas IV.4.4 and IV.4.7, and (16) it follows that

$$\lim_{t\to\infty} P\{N_t = k\} = (1-m)\, m^k,$$

and the theorem then follows from (20). □

Remarks. (i) The fact that a limit exists for fixed x contrasts with theorem 2, where we had to allow x_t to grow with time. For further results along these lines, see Athreya-Ney (1971).

(ii) Recall that when $m<1$ and $G \in \mathscr{S}$, then theorem IV.10.2 tells us that

$$P\{Z(t)=1 \mid Z(t)>0\} \to 1 .$$

Since

$$P\{Z(t)>0\} \sim \frac{1-G(t)}{1-m} \sim \mu(t),$$

one can "roughly" interpret the limit probability $(1-m)m^k$ as the probability that the lone surviving particle has k ancestors.

(iii) It is possible to give a purely analytical proof of theorem 2, but so far this has required extra regularity assumptions (see e.g. Ney (1965b)). These methods also apply to the case when offspring particles of a common parent do not move independently.

(iv) If the motions of the offspring particles are not independent random variables, but have common marginal distribution $\Gamma(x)$, then formula (14) continues to hold, and hence theorems 2 and 2A are still valid.

1.3 Second Moments and Mean Square Convergence

We can do somewhat more than describe the mean. Let

$$P_t(x) = \frac{Z(x,t)}{Z(t)},$$

where $Z(t)$ = total number of particles at t. $P_t(x)$ is the *proportion* of particles to the left of x, and is of course a random variable. With probability one we have $P_t(-\infty)=0$, $P_t(\infty)=1$; and $P_t(\cdot)$ is non-decreasing and can be defined to be right continuous. Thus $P_t(x)$ is a random distribution function. In this section we will show that it converges in an appropriate sense to the normal distribution. Of course we will here limit ourselves to $m>1$, or else $Z(t) \to 0$ almost surely.

To this end we need analogs to theorem 2 for the second moments of $Z(x_t, t)$. Let $m_2 = f''(1)$,

$$\zeta(x,t) = \sum_n n^2 p_n(x,t) = E Z^2(x,t), \tag{25}$$

and

$$\rho(x,t) = \sum \sum n m P\{Z(x,t)=n, Z(t)=m\} = E(Z(x,t))(Z(t)).$$

We will prove

Theorem 3. *If* $m>1$, $m_2<\infty$, *and the hypotheses of theorem 2 are satisfied, then*

$$\zeta(x_t, t) \sim e^{2\alpha t} K^2 \Phi^2(\gamma) \tag{26.a}$$

and

$$\rho(x_t, t) \sim e^{2\alpha t} K^2 \Phi(\gamma), \tag{26.b}$$

where

$$K^2 = \left(\frac{m_2 \int_0^\infty e^{-2\alpha\tau} dG(\tau)}{1 - m \int_0^\infty e^{-2\alpha\tau} dG(\tau)} \right) \left(\frac{m-1}{\alpha m \mu_\alpha} \right)^2 .$$

Note that when $m > 1$, G_α automatically has all moments.

Proof. Using by now familiar methods we can differentiate through the generating function equation (4) and get

$$\begin{aligned} \zeta(x,t) = [1 - G(t)] D(x) + m_2 \int_0^t dG(\tau) \left[\int_{-\infty}^{\infty} \mu(x-y, t-\tau) d\Gamma(y) \right]^2 \\ + m \int_0^t dG(\tau) \int_{-\infty}^{\infty} \zeta(x-y, t-\tau) d\Gamma(y) . \end{aligned} \tag{27}$$

If we treat μ as known for the time being, and let

$$J(x,t) = [1 - G(t)] D(x) + m_2 \int_0^t dG(\tau) \left[\int_{-\infty}^{\infty} \mu(x-y, t-\tau) d\Gamma(y) \right]^2 , \tag{28}$$

then we can recognize (27) as a two-dimensional renewal equation in the unknown function $\zeta(x, t)$, namely

$$\zeta(x,t) = J(x,t) + m \int_0^t dG(\tau) \int_{-\infty}^{\infty} \zeta(x-y, t-\tau) d\Gamma(y) . \tag{29}$$

We multiply (29) through by $e^{-2\alpha t}$. Let

$$e^{-2\alpha t} \zeta(x,t) = \zeta_\alpha(x,t), \qquad \hat{m} = m \int_0^\infty e^{-2\alpha\tau} dG(\tau) ,$$

and

$$\hat{G}_\alpha(t) = \frac{m}{\hat{m}} \int_0^t e^{-2\alpha\tau} dG(\tau) .$$

Note that since $m \int_0^\infty e^{-\alpha\tau} dG(\tau) = 1$ and $\alpha > 0$. we must have $\hat{m} < 1$; also that $\hat{G}_\alpha(\cdot)$ is a proper distribution function. Thus

$$\zeta_\alpha(x,t) = e^{-2\alpha t} J(x,t) + \hat{m} \int_0^t d\hat{G}_\alpha(\tau) \int_{-\infty}^{\infty} \zeta_\alpha(x-y, t-\tau) d\Gamma(y) , \tag{30}$$

or

$$\zeta_\alpha(x,t) = [e^{-2\alpha t} J(x,t)] * \sum_{n=0}^{\infty} \hat{m}^n \hat{G}_\alpha^{*n}(t) \Gamma^{*n}(x) . \tag{31}$$

Now we let x_t be defined as in (10), and substitute this expression for x in (31). We observe that due to theorem 2

$$\lim_{t\to\infty} e^{-2\alpha t} J(x_t, t) = m_2 \frac{\hat{m}}{m} \left(\frac{m-1}{\alpha m \mu_\alpha}\right)^2 \Phi^2(\gamma). \tag{32}$$

Since $\hat{m}<1$, the series in (31) converges uniformly in x and t, and hence

$$\lim_{t\to\infty} \zeta_\alpha(x_t, t) = \frac{\hat{m}}{1-\hat{m}} \frac{m_2}{m} \left(\frac{m-1}{\alpha m \mu_\alpha}\right)^2 \Phi^2(\gamma), \tag{33}$$

proving (26.a).

The second part is proved similarly from an equation for $\rho(x, t)$. □

An immediate consequence of theorem 3 is

Theorem 4. *If $m>1$, $m_2<\infty$, and the second moment of Γ exists, then $Z(x_t, t)/[\mu(t)\Phi(\gamma)]$ converges in mean square to the same random variable W which is the limit of $Z(t)/\mu(t)$ in theorems IV.11.1 and 2.*

Proof. Write

$$E\left[\frac{Z(x_t, t)}{\mu(t)\Phi(\gamma)} - W\right]^2 \leqslant 2E\left[\frac{Z(x_t, t)}{\mu(t)\Phi(\gamma)} - \frac{Z(t)}{\mu(t)}\right]^2 + 2E\left[\frac{Z(t)}{\mu(t)} - W\right]^2,$$

square out the first term on right side and apply theorems 3 and IV.11.1. □

Finally, we add the obvious

Corollary. *If $p_0=0$, then under the conditions of theorem 4*

$$\frac{Z(x_t, t)}{Z(t)} \to \Phi(\gamma) \quad \text{in mean square as } t\to\infty.$$

(The condition $p_0=0$ is merely to assure $P\{Z(t)=0\}=0$.)

Remark (i). Observe that if the convergence in theorem 4 can be shown to be exponentially fast, then the convergence in that theorem and the corollary can be asserted to hold with probability 1. Further ideas on how to carry through such an argument can be obtained from the paper of Bellman and Harris (1952), where the analogous calculations are performed for the ordinary age-dependent process.

Remark (ii). The conclusion of theorem 4 and its corollary can be shown to hold without the independence assumption on offspring motion. In particular it suffices to assume that if a particle has n offspring which move by amounts $X_1, \ldots, X_n$, then the marginal distributions of the joint distribution of $X_1, \ldots, X_n$ are all given by $\Gamma(\cdot)$.

2. Cascades; Distributions of Generations

Several applications which at first sight are quite different, turn out to be merely reformulations of the of the last section.

2.1 Binary Nuclear Cascades

The objective is to describe the energy distribution of nucleons undergoing binary fission. For a discussion of the physical assumptions underlying this process, as well as references to the extensive physical literature we refer the reader to chapter 5 of Barucha-Reid (1960). For related discussion see also chapters III and VII of T. Harris (1963).

The process again starts with an initial "parent" particle at time $t=0$, which undergoes a collision after a random time, thus giving rise to two nucleons. Each of these then lives a random time, collides, and gives rise to two more particles. It is customary to assume that life-lengths of particles till collision are exponentially and independently distributed, though neither the exponentiallity nor the binary nature of particle production is essential to the kind of results we will give.

The essential information still to be specified is the probability law governing the energies of the particles. We let $H(E_1, E_2 | E_0)$ denote the joint cummulative distribution of the energies of the particles resulting from the collision of a parent particle of energy E_0. We assume that H satisfies "conservation of energy", i.e. that $P\{0 \leqslant E_1, 0 \leqslant E_2, E_1+E_2 \leqslant E_0\}=1$ and that it is symmetric in E_1, E_2, and homogeneous, i.e.

$$H(kE_1, kE_2 | kE_2) = H(E_1, E_2 | E_0) = H(E_2, E_1 | E_0)$$

for any positive constant k. (In the physics literature it is usually assumed that H has a density function h which is homogeneous of order two, i.e. such that $h(kE_1, kE_2 | kE_0) = (1/k^2) h(E_1, E_2 | E_0)$, which of course implies homogeneity of H.)

From this data one can derive an integral equation for the distribution and generating function of

$N(E, t | E_0) =$ the number of particles of energy at least E at time t, given that the cascade was initiated by a particle of energy E_0 at time 0.

Let $p_n(E, t | E_0) = P\{N(E, t | E_0) = n\}$. Arguing as in section 1, we can show that

$$p_n(E, t | E_0) = p_n\left(\frac{E}{E_0}, t \,\middle|\, 1\right)$$

which we write $\equiv p_n((E/E_0), t)$; and that

$$p_n(x,t)=\delta_{n1}\,e^{-\lambda t}+\int_0^t \lambda e^{-\lambda y}\,dy \int_x^1\int_x^1 H(du_1,du_2)\cdot\sum_{i=0}^{n} p_i\left(\frac{x}{u_1},t-y\right)p_{n-i}\left(\frac{x}{u_2},t-y\right), \tag{1}$$

where $H(x_1,x_2)\equiv H(x_1,x_2|1)$.

Introducing the generating function

$$Q(x,t,z)=\sum_{n=0}^{\infty} p_n(x,t)\,z^n,$$

and differentiating with respect to t (it is not hard to justify this), one obtains

$$\left(\frac{1}{\lambda}\frac{\partial}{\partial t}+1\right)Q(x,t,z)=\int_x^1\int_x^1 H(du_1,du_2)\,Q\left(\frac{x}{u_1},t,z\right)\cdot Q\left(\frac{x}{u_2},t,z\right). \tag{2}$$

This equation is well-known in cascade theory, and is called the Janossy G-equation. Existence and uniqueness theorems are given in Ney (1962a, b) very much along the lines of chapter IV.

As one might expect from the conservation of energy, the energies of the particles get smaller as they increase in number, and in fact for fixed x

$$p_0(x,t)\to 1 \quad \text{as } t\to\infty \tag{3}$$

(see Ney (1964)), and the mean $\mu(x,t)\to 0$. To describe the rate of decay of $N(x,t)\equiv N(x,t|1)$ it is natural to try to determine a function $y_t\to 0$ such that the *proportion* of particles existing at t and having energy at least y_t converges to a non-degenerate limit. To this end let X be a random variable with distribution

$$P\{X\leqslant u\}=H(u,\infty)=H(\infty,u),$$

and

$$\begin{aligned}\mu&=E(-\log X),\\ \sigma^2&=\operatorname{var}(-\log X).\end{aligned} \tag{4}$$

Define

$$y_t=\exp\{-2\lambda\mu t-\gamma[2\lambda t(\mu^2+\sigma^2)]^{\frac{1}{2}}\}. \tag{5}$$

Then we have

Theorem 1. *If the moments in (4) exist then*

$$\frac{N(y_t,t)}{N(0,t)}\to\Phi(\gamma) \quad \textit{in mean square} \tag{6}$$

as $t\to\infty$, *where* $\Phi(\cdot)$ *is the Gaussian distribution.*

Remark. Probability 1 convergence also holds, but this result has not been published.

Proof. A change of variables which replaces energy by |log (energy)| reduces the problem to a special case of the branching random walk in section 1. The multiplicative homogeneity assumption is transformed into the usual additive (spatial) homogeneity of the random walk. Making the appropriate notational identification

$$\frac{Z(x_t, t)}{Z(t)} = \frac{N(e^{-x_t}, t)}{N(0, t)},$$

$$\alpha = \lambda, \quad \mu_\alpha = \frac{1}{2\lambda}, \quad \sigma_\alpha^2 = \frac{1}{4\lambda^2}, \quad \Gamma(x) = P\{-\log X \leqslant x\}, \quad \nu = \mu, \quad \theta^2 = \sigma^2,$$

we see that $y_t = e^{-x_t}$, and theorem 1 follows from the corollary of theorem 1.4 and remark (ii) at the end of part 3 of the last section.

Analogs of theorems 1.1—1.4 can be proved similarly.

2.2 *Kolmogorov's "Rock-Crushing" Problem*

Several decades before any of the above work, Kolmogorov (1941) considered a discrete time version of the above model, which indicated the direction that results could be expected to take. Start with a "rock" of given mass (say unit mass), and then "strike" it so that it breaks into two parts in some random fashion, then break each of these again, etc. The sizes of the rocks will get smaller and smaller. It is shown, under suitable assumptions on the distribution of mass among the "offspring" pieces, that the ratio of the mean number of pieces having masses in an appropriate range, divided by the total mean number of pieces, converges to a Gaussian function. By a suitable change of variables, this model can again be identitied as a special case of the branching random walk.

2.3 *Distribution of Generations*

Consider an age-dependent branching process $Z(t)$ as defined in chapter IV. The generation of any particle alive at a given time is the number of ancestors he has. In general, of course, different particles alive at the same time need not be of the same generation, and it is of interest to determine the distribution of particles among generations. To this end let $Z_n(t)$ = the number of particles alive at t which are of the nth generation, and $Z(n, t) = \sum_{k=0}^{n} Z_k(t)$.

Theorem 2. *If* $m>1$ *and* α *(necessarily* >0*) is the Malthusian parameter, and*

$$n_t = \frac{t}{\mu_\alpha} + \gamma \frac{\sigma_\alpha^2}{\mu_\alpha^{\frac{3}{2}}} t^{\frac{1}{2}} + o(t^{\frac{1}{2}}),$$

then as $t \to \infty$

(i) $EZ(n_t, t)/\mu(t) \to \Phi(\gamma)$,

and

(ii) $Z(n_t, t)/Z(t) \to \Phi(v)$ *in mean square.*

Part (i) is just formula (1.19), and was proved by Samuels (1969). She also has a convergence in probability form of (ii), and a variety of related results on generation numbers. Special cases were first treated by Kharlamov (1969) and Bühler (1970). The latter (1971) has also studied other aspects of the family structure of branching processes.

Observe that theorem 2 can be regarded as a special case of theorems 1.2 and 1.4, with

$$\Gamma(x) = \begin{cases} 1 & x \geq 1, \\ 0 & x < 1, \end{cases}$$

in which case x_t just becomes n_t.

3. Branching Diffusions

There is another interesting model similar to that of the last sections in which particles reproduce according to a Galton-Watson branching process and move according to a standard Brownian motion in Euclidean space R_N. The particle lifetimes are taken to be exponentially distributed, and the entire process is then Markovian. It is thus possible to undertake a more refined analysis than in the case of the age-dependent process.

The elegant approach which follows is due to S. Watanabe (1965). For a similar analysis the reader is also referred to H. Conner (1967). The details of the calculations are similar in some respects to those of section 1; and we shall here only outline the main ideas.

To describe the state of the system adequately (so as to make it Markovian) we need to know the total number of particles Z_t in the system and their locations. Thus if we set

$$\mathbf{X}_t = \{X_t^1, X_t^2, \ldots, X_t^{Z_t}\}, \tag{1}$$

where X_t^j, $(j=1,2,\ldots,Z_t)$ denotes the location of the jth particle, then in view of the exponential life time and the Markovian nature of Brownian motion it is clear that $\{\mathbf{X}_t; t \geq 0\}$ is Markovian.

Let $f(s)=\sum_{n=0}^{\infty} p_n s^n$ be the offspring generating function and let the lifetime parameter be c. Then $\{Z_t; t\geqslant 0\}$ is a continuous time Markov branching process of the type discussed in chapter III and has as its infinitesimal generating function

$$u(s)=c[f(s)-s]. \tag{2}$$

If $m=f'(1)\leqslant 1$ then the process dies out (eventually) with probability one and so to make the problem interesting we shall assume $m>1$. In fact, let us assume that $p_0=0$ so that the process necessarily grows to ∞ as $t\to\infty$.

We wish to study the asymptotic behavior of the distribution of the locations of the particles in the system. For any set D let

$$Z_t^{\mathsf{D}} = \text{number of particles in } \mathsf{D} \text{ at } t = \sum_{j=1}^{Z_t} \chi_{\mathsf{D}}(X_t^j), \tag{3}$$

where $\chi_{\mathsf{D}}(\cdot)$ is the indicator function of the set D. If D is a bounded set then it can be shown that

$$\frac{EZ_t^{\mathsf{D}}}{EZ_t} \sim \text{constant}\cdot t^{-\frac{N}{2}} \tag{4}$$

(see complements). Thus one has to normalize the process to get a nondegenerate limit law, as is shown by the following

Theorem 1. *Let $f''(1-)<\infty$, $f'(1)>1$, $f(0)=0$. Then, for any bounded domain $\mathsf{D}\subset R_N$*

$$P_x\left\{\lim_{t\to\infty} (2\pi t)^{\frac{N}{2}} \frac{Z_t^{\mathsf{D}}}{Z_t} = |\mathsf{D}|\right\} = 1, \tag{5}$$

where $|\mathsf{D}|=$ the Lebesgue measure of D and P_x denotes the probability measure corresponding to starting with one particle at x.

Here are the main steps of the proof.

Step 1. Introduce the *characteristic functional* Y_t^{ξ} of the random distribution X_t

$$Y_t^{\xi} = \sum_{j=1}^{Z_t} e^{i(\xi\cdot X_t^j)}. \tag{6}$$

Show that

$$E_x Y_t^{\xi} = e^{at} e^{\frac{-|\xi|^2 t}{2}} e^{i(\xi\cdot x)}, \tag{7}$$

where $a=c(m-1)$, and $|\xi|$ is the Euclidean norm; and that

$$E_x Y_t^{\xi} Y_{t+s}^{\eta} = e^{i(\xi+\eta)\cdot x}\left[c f''(1)\exp\left\{\left(a-\frac{|\xi+\eta|^2}{2}\right)t+\left(a-\frac{|\eta|^2}{2}\right)s\right\}\right.$$

$$\times\int_0^t \exp\{(a+(\xi\cdot\eta)u)\}\,du$$

$$\left.+\exp\left\{\left(a-\frac{|\eta|^2}{2}\right)s+\left(a-\frac{|\xi+\eta|^2}{2}\right)t\right\}\right], \tag{8}$$

and

$$E_x(|Y_t^{\xi}|^2) = c f''(1)\, e^{at}\int_0^t e^{(a-|\xi|^2)u}\,du + e^{at}$$

$$\sim \frac{c f''(1)}{a-|\xi|^2}\, e^{(2a-|\xi|^2)t} \quad \text{if } a>|\xi|^2$$

$$\sim c f''(1)\, t\, e^{at} \quad \text{if } a=|\xi|^2 \tag{9}$$

$$\sim \left(\frac{c f''(1)}{|\xi|^2-a}+1\right) e^{at} \quad \text{if } a<|\xi|^2$$

as $t\to\infty$.

Step 2. Note from step 1 that $e^{i(\xi\cdot x)}$ is a right eigen functional of the mean operator of the process, and hence that

$$W_t^{\xi} = Y_t^{\xi}\, e^{-\left(a-\frac{|\xi|^2}{2}\right)t} \tag{10}$$

is a martingale. (See that next section for a general discussion of this point.) Furthermore if $|\xi|^2<a$ then $\sup_t E_x|W_t^{\xi}|^2<\infty$, and hence

$$\lim_{t\to\infty} W_t^{\xi} = W^{\xi} \quad \text{exists in mean square} \tag{11}$$

and almost surely. Also

$$E_x|W^{\xi}-W^{\eta}|^2 = O(|\xi-\eta|^2)\,.$$

Step 3. For any $L_1(R^N)$ function F, define the Fourier transform

$$\hat{F}(\xi) = \frac{1}{(2\pi)^N}\int_{R^N} e^{-i(\xi\cdot x)} F(x)\,dx\,.$$

Let $\mathscr{F}=\{F: F\in L_1(R^N),\ \hat{F}(\xi)=O(e^{-\alpha|\xi|^2})$ for some $\alpha>0\}$. Let D be a bounded open set in R_N such that $|\partial\,\mathsf{D}|=0$, and χ_{D} be the indicator function of the set D. Show that for every $\varepsilon>0$ there exist F_1, F_2 in $\mathscr{F}$ such that

$$F_1\leqslant \chi_{\mathsf{D}}\leqslant F_2 \quad \text{and} \quad \int_{R^N}(F_2-F_1)\,d\boldsymbol{x}<\varepsilon\,.$$

Step 4. Show that, under the hypothesis of theorem 1, for any F in $\mathscr{F}$

$$P_x\left\{\lim_{t\to\infty}\frac{\sum_{j=1}^{Z_t}F(X_t^j)}{\left(e^{at}t^{-\frac{N}{2}}\right)}=(2\pi)^{\frac{N}{2}}\hat{F}(0)\,W^0\right\}=1\,.$$

This is done as follows:

$$\begin{aligned}Z_t^F &\equiv \sum_{j=1}^{Z_t}F(X_t^j)=\sum_{j=1}^{Z_t}\int_{R^N}e^{i(\xi\cdot X_t^j)}\hat{F}(\xi)\,d\xi=\int_{R^N}Y_t^\xi\hat{F}(\xi)\,d\xi\\ &=\int_{|\xi|^2\le a-\varepsilon}W^\xi e^{\left(a-\frac{|\xi|^2}{2}\right)t}\hat{F}(\xi)\,d\xi+\int_{|\xi|^2\le a-\varepsilon}(W_t^\xi-W^\xi)\,e^{\left(a-\frac{|\xi|^2}{2}\right)t}\hat{F}(\xi)\,d\xi\\ &\quad+\int_{|\xi|^2>a-\varepsilon}Y_t^\xi\hat{F}(\xi)\,d\xi=I_1(t)+I_2(t)+I_3(t),\qquad\text{say},\end{aligned}$$

where $0<\varepsilon<a/2$.

Use the submartingale inequality for $|W_t^\xi|^2$, and Chebychev's inequality to show that

$$P_x\left\{\lim_{t\to\infty}\frac{I_3(t)}{\exp\left[\left(\frac{a}{2}+\varepsilon\right)t\right]}=0\right\}=1\,.$$

Next use step 2 to show that

$$P_x\left\{\lim_{t\to\infty}\frac{I_2(t)}{e^{at}t^{-\frac{N}{2}}}=0\right\}=1\,.$$

Finally make a change of variable from ξ to $\xi/\sqrt{t}$ in I_1, and use the continuity of W^ξ in ξ to show that

$$P_x\left\{\lim_{t\to\infty}\frac{I_1(t)}{e^{at}t^{\frac{-N}{2}}}=(2\pi)^{\frac{N}{2}}\hat{F}(0)\,W^0\right\}=1\,.$$

The reader may have noted that the methods are somewhat similar to those of section V.8.

Watanabe has also given another normalization in which one expands the set D at the rate of $\sqrt{t}$. The precise result is:

Under the hypothesis of theorem 1

$$P_x\left\{\lim_{t\to\infty}\frac{Z^{\sqrt{t}\,\mathsf{D}}}{Z_t}=\int_{\mathsf{D}}(2\pi)^{\frac{-N}{2}}e^{-\frac{|x|^2}{2}}\,dx\right\}=1\,.$$

Note the similarity of this result to theorem 1.4 and its corollary. Theorem 1 has also been extended to the case when the particles move according to a symmetric stable process in R^N rather than the standard Brownian motion, in which case

$$P_x\left\{\lim_{t\to\infty} t^{\frac{N}{\alpha}} \frac{Z_t^{\mathsf{D}}}{Z_t} = k|\mathsf{D}|\right\} = 1$$

and

$$P_x\left\{\lim_{t\to\infty} \frac{Z_t^{t^{\frac{1}{\alpha}}\mathsf{D}}}{Z_t} = \int_{\mathsf{D}} p_\alpha(1, x)\, dx\right\} = 1,$$

where α is the parameter of the stable process, $p_\alpha(t, x)$ is the transition density, and k is a constant independent of D. Using the same method, Watanabe has also settled the conjecture of Harris on the discrete time version of theorem 1.4 (chapter IV, p. 75 of his book).

The generalization of Watanabe's results to the case of arbitrary lifetimes, and Markov processes more general than Brownian motion or symmetric stable processes, is an open problem.

4. Martingale Methods

Examples of martingales have appeared many times in this book; in connection with single and multitype Galton-Watson processes, continuous time Markov branching processes, and branching diffusions. For a variety of other branching models, almost sure convergence is most easily established by finding a suitable martingale; and it is worth while to outline the general principle envolved.

Consider the discrete time, general state space process $\{Z_n; n=0, 1, \ldots\}$ constructed by T. Harris in chapter III of his book (1963). (The continuous time parameter case is quite similar.) This process

(i) is Markov with stationary transition probabilities;

(ii) has as its state space the collection of point distributions (of finite numbers of points) in an abstract set Ω of "types";

(iii) has a transition mechanism such that the distribution of Z_{n+1} given Z_n is the convolution of the measures $P(x_i, r, \mathsf{A})$, where

$$P\{x, r, \mathsf{A}\} = P\{\text{a particle located at } x \text{ gives rise to } r \text{ particles in the set } \mathsf{A} \text{ in the next generation}\},$$

and where x_i ranges over the point distribution Z_n.

Let $Z_n(\mathsf{A})$ = the number of particles in the set A at time n. Then the state of the process at time n is given by specifying $Z_n(\mathsf{A})$ for all subsets A of Ω.

This process is precisely the generalization to an arbitrary "type" space, of the multitype process of chapter V. The details of the construction are in chapter III of Harris (1963).

Let $M_n(x, \mathsf{A}) = E_x\{Z_n(\mathsf{A})\}$ = the expected number of particles in A at the nth generation, given that the process started with a single particle at x. (We assume that these moments exist.) Clearly $M_n(x, \mathsf{A})$ is a measure in A, and satisfies

$$M_{n+1}(x, \mathsf{A}) = \int_{\Omega} M_n(y, \mathsf{A}) M_1(x, dy). \tag{1}$$

A complex valued function f on Ω is a right eigenfunction with eigenvalue λ for the kernel $M_1(x, \mathsf{A})$ if

$$\int f(y) M_1(x, dy) = \lambda f(x). \tag{2}$$

The following theorem now tells us how to construct a martingale from the original process.

Theorem 1. *Let f be a right eigenfunction for M_1 with eigenvalue λ. Then the sequence*

$$Y_n = \lambda^{-n} \int_{\Omega} f(y) Z_n(dy) \tag{3}$$

is a martingale.

Proof.

$$E\{Y_1 | Z_0\} = \lambda^{-1} \int_{\Omega} Z_0(dx) \int_{\Omega} f(y) M_1(x, dy) = \lambda^{-1} \lambda \int_{\Omega} f(x) Z_0(dx) = Y_0.$$

By the Markov property and the stationarity of the transition probabilities

$$E\{Y_{n+1} | \mathbb{F}_n\} = Y_n,$$

where $\mathbb{F}_n$ is the σ-field generated by $Z_0, Z_1, \ldots, Z_n$. Clearly

$$\begin{aligned} E\{Y_{n+1} | Y_0, Y_1, \ldots, Y_n\} &= E\{E\{Y_{n+1} | \mathbb{F}_n\} | Y_0, Y_1, \ldots, Y_n\} \\ &= E\{Y_n | Y_0, Y_1, \ldots, Y_n\} = Y_n. \quad \square \end{aligned}$$

There is a variety of sufficient conditions for the almost sure convergence of these martingales. For example if $f(\cdot)$ is nonnegative, then necessarily λ is nonnegative, and hence Y_n is nonnegative. This implies that Y_n converges almost surely. A sufficient condition for the existence of a nonnegative eigenvalue is an analog of the Perron-Frobenius condition on the kernel $M_1(x, \mathsf{A})$. Another condition is $\sup_n E|Y_n|^\alpha < \infty$ for some $\alpha > 1$, usually $\alpha = 2$.

We have seen examples of the above for a finite type space in chapter V, and in a continuous time setting for branching Brownian motion in section 3 of this chapter. We will here illustrate the method with one more example, namely the branching random walk.

Let Ω be the real line, and let

$$P(x, r, \mathsf{A}) = \sum_{n=r}^{\infty} p_n \binom{n}{r} [\Gamma(\mathsf{A} - x)]^r [1 - \Gamma(\mathsf{A} - x)]^{n-r},$$

where $p_n \geqslant 0$, $\sum p_n = 1$, and $\Gamma(\cdot)$ is a measure on the Borel sets of the line. This process corresponds to the branching random walk of section 1 for the special case when particle lifetime $\equiv 1$; i.e. $G(t)=1$ for $t \geqslant 1$, $=0$ for $t<1$. (Here we use Γ to denote the measure associated with the distribution Γ of section 1.)

It is easy to see that now

$$M_1(x, \mathsf{A}) = m\Gamma(\mathsf{A}-x), \tag{4}$$

and hence eigenfunctions f for M_1 satisfy the equation

$$\int_R f(y)\Gamma(dy-x) = \lambda m^{-1} f(x);$$

or equivalently

$$\int_R f(x+y)\Gamma(dy) = \lambda m^{-1} f(x). \tag{5}$$

Now if we take $f_t(x)=e^{itx}$, $-\infty<t<\infty$, then we see that

$$\int_R f_t(x+y)\Gamma(dy) = [m\varphi(t)]\, m^{-1} f_t(x),$$

where $\varphi(t)$ is the characteristic function of $\Gamma(\cdot)$; and hence by (5), e^{itx} is an eigenfunction for M_1 with eigenvalue $\lambda = m\varphi(t)$. Hence by theorem 1 we have

Corollary 1[15]**.** *For any real t*

$$Y_n = [m\varphi(t)]^{-n} \int_R e^{itx} Z_n(dx) = [m\varphi(t)]^{-n} \sum_{j=1}^{Z_n} e^{itX_n^j},$$

where $X_n^1, \ldots, X_n^{Z_n}$, *are the locations of the* Z_n *particles at time n, is a martingale for the branching random walk with unit lifetime.*

There are of course other eigenfunctions. For example, if $\int x\Gamma(dx)=0$, then $f(x)=x$ is an eigenfunction with eigenvalue m. If Γ has its support on the negative line then $f(x)=e^{-\theta x}$, $0 \leqslant \theta < \infty$, is an eigenfunction with eigenvalue $m\psi(\theta)$, where $\psi(\theta) = \int_0^\infty e^{-\theta y}\Gamma(dy)$. In this case

$$Y_n = (m\psi(\theta))^{-n} \int_0^\infty e^{-\theta x} Z_n(dx)$$

is a nonnegative martingale sequence, and hence converges with probability one.

The analogs of these results hold for continuous time models. In particular, the ordinary one dimensional age-dependent branching process can be identified with a general branching process in continuous

[15] This result has been independently observed by A. Joffe (private communication).

time, where the 'type' is the age of an object. Harris has indicated that the so called reproductive age value $V(x)=e^{\alpha x}(1-G(x))^{-1}\int_x^\infty e^{-\alpha u}\,dG(u)$ is a right eigenfunction for the mean kernel $M_t(x, \mathsf{A})$ with eigenvalue $e^{\alpha t}$. (See Harris (1963).) The model discussed in section 1 of this chapter could also be put in the present general framework.

5. Branching Processes with Random Environments

5.1 Introduction

A basic feature of the branching models discussed previously is the time homogeneity of the offspring distribution. But there are many situations where the offspring distribution depends on an environment which changes with time. Smith and Wilkinson (1969) proposed a model with such a structure and discussed the extinction probability. Athreya and Karlin (1971) reformulated and generalized the Smith-Wilkinson model, treated the extinction problem, and gave extensions of the basic limit theorems of the Galton-Watson process. In this section we shall present a summary of these results.

5.2 Smith-Wilkinson Model

Assume that we start with Z_0 particles. Let us denote the population size at time n by Z_n. The transition from Z_n to Z_{n+1} takes place as follows. All the Z_n members of the nth generation reproduce according to the same offspring distribution with p.g.f. $\varphi_{\zeta_n}(s)$, where $\varphi_{\zeta_n}(s)$ is chosen at random from a collection Φ of p.g.f.'s according to some specified distribution. It is postulated that $\varphi_{\zeta_n}(s)$ for $n=0, 1, 2, \ldots$ are all stochastically independent and have a common distribution. The process $\{Z_n: n=0, 1, 2, \ldots\}$ is referred to by Smith and Wilkinson as a *Branching Process With Random Evironments.*

A moment's calculation yields the result that $\Pi_n(s)\equiv E(s^{Z_n}|Z_0=1)$, the p.g.f. of Z_n with $Z_0=1$, is given by

$$\Pi_n(s)=E\{\varphi_{\zeta_0}(\varphi_{\zeta_1}(\ldots\varphi_{\zeta_{n-1}}(s))\ldots)\} \tag{1}$$

where E refers to the expectation with respect to the random choices of the $\varphi_{\zeta_i}(s)$, $i=1, 2, \ldots, n$.

Smith and Wilkinson noted that the usual functional iteration property does not quite go over for the Π_n's, and more seriously that the particles no longer have independent lines of descent since one can get some information about one from the history of another via an inference about the unknown p.g.f.'s $\varphi_{\zeta_i}(s)$. This created complications. They overcame some of these by a clever analysis and established a number of results about extinction probability using fluctuation theory methods.

It was pointed out, however, by Athreya and Karlin, that once we condition on the entire environment sequence $\{\varphi_{\zeta_i}(s);\ i=0,1,2,\ldots\}$, the process *does* evolve like an ordinary Galton-Watson process, and the basic feature of independence of lines of descent is preserved under this conditioning. This enabled them to simplify the setting, drop the assumption of independence of $\{\varphi_{\zeta_i}(s)\}$, and to replace it by stationarity and ergodicity. The independence assumption was rather crucial to the Smith-Wilkinson arguments. It made the process $\{Z_n\}$ Markovian.

We shall now proceed to discuss the Athreya-Karlin model in some detail. While we shall give proofs of some of the results, for the most part we shall refer the reader to their original papers for details.

5.3 The Athreya-Karlin Model

To describe our model properly we need some technical preliminaries. Let $(\Omega, \mathbb{F}, P)$ be a probability space and $\mathscr{M}$ designate the collection of probability distributions on the nonnegative integers

$$\left\{\bar{p}=\{p_i\}_{i=0}^{\infty},\ \sum i p_i<\infty,\ 0\leqslant p_0+p_1<1\right\}. \tag{2}$$

Let $\zeta_i(\omega)$ for $i=0,1,2,\ldots$ be a sequence of mappings from $(\Omega, \mathbb{F}. P)$ into $(\mathscr{M}, \mathscr{B})$ where $\mathscr{B}$ is the Borel σ-algebra in $\mathscr{M}$ generated by the usual topology. For any such mapping $\zeta(\omega)$ from Ω to $\mathscr{M}$, define the associated p.g.f.

$$\varphi_\zeta(s)=\sum_{i=0}^{\infty} p_i(\zeta)s^i, \qquad |s|\leqslant 1, \tag{3}$$

where $\{p_i(\zeta)\}$, $i=0,1,2,\ldots$ is the probability distribution corresponding to ζ.

Let $\{Z_n(\omega);\, n=0,1,2,\ldots\}$ be a sequence of nonnegative integer valued random variables defined on $(\Omega, \mathbb{F}, P)$. As in earlier chapters, let $\sigma(\mathsf{D})$ be the sub σ-algebra of $\mathbb{F}$ generated by a given collection D of random variables on $(\Omega, \mathbb{F}, P)$. Set

$$\begin{aligned} \bar{\zeta}&=(\zeta_0,\zeta_1,\zeta_2,\ldots),\\ F_n(\bar{\zeta})&=\sigma(\zeta_0,\zeta_1,\ldots,\zeta_n),\\ F(\bar{\zeta})&=\sigma(\bar{\zeta}),\\ F_{n,z}(\bar{\zeta})&=\sigma(\zeta_0,\ldots,\zeta_n,Z_0,Z_1,\ldots,Z_n). \end{aligned} \tag{4}$$

We postulate that $\{Z_n\}$ satisfies the recurrence relation

$$E(s^{Z_{n+1}}|F_{n,z})=[\varphi_{\zeta_n}(s)]^{Z_n} \qquad \text{a.s.}, \tag{5}$$

and that for any set of integers $1 \leqslant n_1 < n_2 < \cdots < n_k$,

$$E(s_1^{Z_{n_1}} s_2^{Z_{n_2}} \ldots s_k^{Z_{n_k}} \mid F(\overline{\zeta}), Z_0 = m) = [E(s_1^{Z_{n_1}} s_2^{Z_{n_2}} \ldots s_k^{Z_{n_k}} \mid F(\overline{\zeta}), Z_0 = 1)]^m \quad \text{a.s.,} \tag{6}$$

where $|s_i| \leqslant 1, i = 1, 2, \ldots, k$.

The proof of the existence of a process satisfying these axioms is routine and parallels the constructions given in Harris (1963). The process $\{Z_n(\omega); n = 0, 1, 2, \ldots\}$ is again called a *branching process with random environments* and the process $\overline{\zeta} = (\zeta_0(\omega), \zeta_1(\omega), \zeta_2(\omega), \ldots)$ is called the *environmental process*. It must be clear that the transition from Z_n to Z_{n+1} is the same as discussed before.

An easy consequence of (5) and (6) is

Lemma 1.

$$E(s^{Z_{n+1}} \mid F(\overline{\zeta}), Z_0 = k) = [\varphi_{\zeta_0}(\varphi_{\zeta_1}(\ldots \varphi_{\zeta_n}(s) \ldots))]^k . \tag{7}$$

Although the process $\{Z_n\}$ is no longer Markov it is evident from (5), (6) and (7) that, conditioned on the environmental process $\overline{\zeta} = \{\zeta_i\}$, the process $\{Z_n\}$ is Markov and has independent lines of descent. This observation is the key to the simplicity of the entire situation.

Let

$$\begin{aligned} \mathsf{B} &= \{\omega : Z_n(\omega) = 0 \ \text{ for some } n\} , \\ q_k(\overline{\zeta}) &= P(\mathsf{B} \mid F(\overline{\zeta}), Z_0 = k) , \\ q_k &= P(\mathsf{B} \mid Z_0 = k) . \end{aligned} \tag{8}$$

We refer to B as the set of extinction and to $q_k, q_k(\overline{\zeta})$ as the unconditional and conditional probabilities of extinction. It is trivial from (7) that

$$q_k(\overline{\zeta}) = [q_1(\overline{\zeta})]^k$$

and

$$q_k = E(q_1(\overline{\zeta}))^k ,$$

thus making q_k a moment sequence. From now on we shall write $q(\overline{\zeta})$ for $q_1(\overline{\zeta})$.

Since the sequence of events $\mathsf{B}_n = \{\omega : Z_n(\omega) = 0\}$ increases to B we have the relation

$$q(\overline{\zeta}) = \lim_{n \to \infty} \varphi_{\zeta_0}(\ldots \varphi_{\zeta_n}(0) \ldots),$$

which in turn can be rewritten to yield the important functional equation

$$q(\overline{\zeta}) = \varphi_{\zeta_0}(q(T\overline{\zeta})) , \tag{9}$$

where T denotes the shift transformation $T\overline{\zeta} = (\zeta_1, \zeta_2, \ldots)$. The reader will notice that (9) is simply the stochastic analog of the equation $q = f(q)$ of I.5. It can easily be shown (see Athreya-Karlin (1971)) that $q(\overline{\zeta})$ is the minimal solution of (9), and that when $P(q(\overline{\zeta}) < 1) = 1$ it is the unique one with that property. That is, if $\tilde{q}(\overline{\zeta}(\omega))$ is any random variable mapping Ω to $[0,1]$ and satisfying

$$\tilde{q}(\overline{\zeta})=\varphi_{\zeta_0}(\tilde{q}(\overline{\zeta}))$$

then $P(q(\overline{\zeta})\leqslant\tilde{q}(\overline{\zeta}))=1$, and if $P(\tilde{q}(\overline{\zeta})<1)=1$, then $P(\tilde{q}(\overline{\zeta})=q(\overline{\zeta}))=1$.

It is immediate from (9) and our basic assumption $P(\zeta_0\in\mathcal{M})=1$, that the event $\{q(\overline{\zeta})=1\}$ is *shift invariant* i.e. the two events $\{\omega:q(\overline{\zeta})=1\}$ $=\{\omega:q(T\overline{\zeta})=1\}$ coincide with probability one.

In order to exploit this shift invariance we shall make the important assumption that the shift transformation T, is *stationary and ergodic.* (This is so in the case when the ζ_i are independently identically distributed (Smith-Wilkinson model).) This hypothesis plus shift invariance of the set $\{q(\overline{\zeta})=1\}$ at once yields the fact that

$$P(q(\overline{\zeta})=1)=0 \quad \text{or} \quad 1\,.$$

5.4 *Extinction Probabilities*

We now exhibit a necessary condition for $P(q(\overline{\zeta})<1)=1$. Write $a^+=\max(a,0)$; $a^-=\max(-a,0)$.

Theorem 1. *Suppose that*

$$E(\log\varphi'_{\zeta_0}(1-))^+<\infty \quad \textit{and} \quad P(q(\overline{\zeta})<1)=1\,.$$

Then

(i) $E(\log\varphi'_{\zeta_0}(1))^-<\infty$ *and* $E\log\varphi'_{\zeta_0}(1)>0$.

Furthermore

(ii) $E|\log((1-q(\overline{\zeta}))/(1-q(T\overline{\zeta})))|<\infty$, $E\log((1-q(\overline{\zeta}))/(1-q(T\overline{\zeta})))=0$.

Proof. We shall make repeated use of the ergodic theorem.

From the basic functional equation (9) satisfied by $q(\overline{\zeta})$ and the hypothesis $P(q(\overline{\zeta})<1)=1$ we get

$$h(\overline{\zeta})=f(\overline{\zeta})+h(T\overline{\zeta})\,, \tag{10}$$

where

$$0\leqslant h(\overline{\zeta})=-\log(1-q(\overline{\zeta}))<\infty \qquad \text{a.s.}\,,$$

and

$$f(\overline{\zeta})=-\log\left(\frac{(1-\varphi_{\zeta_0}(T\overline{\zeta}))}{1-q(T\overline{\zeta})}\right).$$

Iteration of the above yields

$$h(\overline{\zeta})=f(\overline{\zeta})+f(T\overline{\zeta})+\cdots+f(T^n\overline{\zeta})+h(T^{n+1}\overline{\zeta})\,, \tag{11}$$

and since $h(\cdot)$ is nonnegative it follows that

$$\sum_{i=0}^{n} f(T^i\overline{\zeta})\leqslant h(\overline{\zeta})\,.$$

Breaking this up into positive and negative parts we rewrite it as

$$\frac{1}{n}\sum_{i=0}^{n} f^{+}(T^{i}\bar{\zeta}) - \frac{1}{n}\sum_{i=0}^{n} f^{-}(T^{i}\bar{\zeta}) \leqslant \frac{1}{n} h(\bar{\zeta}).$$

But

$$\begin{aligned}
0 &\leqslant E(f^{-}(\bar{\zeta})) = E(-f(\bar{\zeta}); f(\bar{\zeta}) \leqslant 0) \\
&= E\left(\log\left(\frac{1-\varphi_{\zeta_0}(q(T\bar{\zeta}))}{1-q(T\bar{\zeta})}\right); \frac{1-\varphi_{\zeta_0}(q(T\bar{\zeta}))}{1-q(T\bar{\zeta})} \geqslant 1\right) \\
&\leqslant E(\log \varphi'_{\zeta_0}(1); \varphi_{\zeta_0}(1) \geqslant 1) \\
&= E(\log \varphi'_{\zeta_0}(1))^{+} < \infty .
\end{aligned}$$

Therefore, applying the ergodic theorem to $f^{-}(\bar{\zeta})$ yields

$$0 \leqslant \limsup \frac{1}{n}\sum_{i=0}^{n} f^{+}(T^{i}\bar{\zeta}) \leqslant E(\log \varphi'_{\zeta_0}(1))^{+} < \infty, \quad \text{a.s.}$$

Because of the nonnegativity of f^{+}, the converse of the ergodic theorem implies that

$$E f^{+}(\bar{\zeta}) < \infty , \tag{12}$$

and hence that $E|f(\bar{\zeta})| < \infty$. To finish (ii) we have to show that $E(f(\bar{\zeta})) = 0$. This is also a consequence of the ergodic theorem. In fact, note first from (11) that,

$$\lim_{n\to\infty} \frac{1}{n}\sum_{i=0}^{n} f(T^{i}\bar{\zeta}) = \lim_{n\to\infty}\left[\frac{1}{n} h(\bar{\zeta}) - \frac{1}{n} h(T^{n+1}\bar{\zeta})\right],$$

where the left hand side exists and equals $E f(\bar{\zeta})$ by the ergodic theorem. Clearly, $\lim (1/n) h(\bar{\zeta}) = 0$. It follows that $\lim (1/n) h(T^{n+1}\bar{\zeta})$ exists and by stationarity has the same distribution as $\lim (1/n) h(\bar{\zeta})$ which is $\equiv 0$. This proves that $E f(\bar{\zeta}) = 0$.

We next turn to (i). By hypothesis, $E(\log \varphi'_{\zeta_0}(1))^{+} < \infty$. We shall show that $E(\log \varphi'_{\zeta_0}(1))^{-} < E(\log \varphi'_{\zeta_0}(1))^{+}$. Observe that

$$\begin{aligned}
E(\log \varphi'_{\zeta_0}(1))^{-} &= E(-\log \varphi'_{\zeta_0}(1); \varphi'_{\zeta_0}(1) \leqslant 1) \\
&\leqslant E\left[-\log\left(\frac{1-\varphi_{\zeta_0}(q(T\bar{\zeta}))}{1-q(T\bar{\zeta})}\right); \varphi'_{\zeta_0}(1) \leqslant 1\right] \\
&\leqslant E(f(\bar{\zeta}); f(\bar{\zeta}) \geqslant 0) \\
&= E f^{+}(\bar{\zeta}) \leqslant E(\log \varphi'_{\zeta_0}(1))^{+} .
\end{aligned}$$

If $E(\log\varphi'_{\zeta_0}(1))^- = E(\log\varphi'_{\zeta_0}(1))^+$ then $E\log\varphi'_{\zeta_0}(1)=0$. Since also $E f(\overline{\zeta})=0$, it follows that

$$E\left[\log\varphi'_{\zeta_0}(1)-\log\frac{1-\varphi_{\zeta_0}(q(T\overline{\zeta}))}{1-q(\overline{\zeta})}\right]=0\,.$$

But the quantity in brackets is always nonnegative and can be zero only if $q(\overline{\zeta})=1$. Thus, we get the conclusion $P(q(\overline{\zeta})=1)=1$ which contradicts the hypothesis $P(q(\overline{\zeta})<1)=1$. □

Stating the conclusion in contrapositive form yields

Corollary 1. *If* $E(\log\varphi'_{\zeta_0}(1))^+<\infty$ *and* $E(\log\varphi'_{\zeta_0}(1))^+ \leqslant E(\log\varphi'_{\zeta_0}(1))^- \leqslant\infty$ *then* $P(q(\overline{\zeta})=1)=1$.

Before proceeding to the converse of theorem 1 we shall extract an integrability condition from theorem 1 (ii).

Theorem 2. *Assume that the hypothesis of theorem 1 holds and that there exist* $\varepsilon>0$ *and* $c>0$ *such that*

$$\inf_{\zeta_0} P\{q(T\overline{\zeta})\leqslant 1-\varepsilon\,|\,\zeta_0\}>c \qquad \text{a.s.} \tag{13}$$

Then, $E(-\log(1-\varphi_{\zeta_0}(0)))<\infty$.

Proof. Note that when $\{\zeta_i\}$ are independent, or when ζ_i form an irreducible finite Markov chain, the additional hypothesis in theorem 2 is clearly satisfied. Smith (1968) obtained this result for the Smith-Wilkinson models. The proof of theorem 2 just involves noting that

$$\begin{aligned}\infty > E\left|\log\frac{1-q(\overline{\zeta})}{1-q(T\overline{\zeta})}\right| &\geqslant E\left[\left|\log\left(\frac{1-\varphi_{\zeta_0}(q(T\overline{\zeta}))}{1-q(T\overline{\zeta})}\right)\right|;1-q(T\overline{\zeta})\geqslant\varepsilon\right]\\ &\geqslant E[|\log(1-\varphi_{\zeta_0}(0))|;1-q(T\overline{\zeta})\geqslant\varepsilon]-|\log\varepsilon|\\ &\geqslant c\,E|\log(1-\varphi_{\zeta_0}(0))|-|\log\varepsilon|\,. \qquad \square\end{aligned}$$

Now turning to the converse of theorem 1 we have the following sufficient condition for $P(q(\overline{\zeta})<1)=1$.

Theorem 3. *Suppose* $E(-\log(1-\varphi_{\zeta_0}(0)))<\infty$ *and* $E(\log\varphi'_{\zeta_0}(1))^- < E(\log\varphi'_{\zeta_0}(1))^+\leqslant\infty$. *Then* $P(q(\overline{\zeta})<1)=1$.

Proof. Let $Y_n(\overline{\zeta})=\varphi_{\zeta_0}(\varphi_{\zeta_1}(\dots\varphi_{\zeta_n}(0)))$. It is clear from the definition of the shift transformation that

$$Y_n(\overline{\zeta})=\varphi_{\zeta_0}(Y_{n-1}(T\overline{\zeta}))\,. \tag{14}$$

Since the φ_{ζ_i} are nontrivial with probability one, it follows that $P\{Y_n(\overline{\zeta})<1 \text{ for all } n\}=1$. Thus,

$$0\leqslant -\log(1-Y_n(\overline{\zeta}))<\infty \quad \text{for all } n \qquad \text{a.s.}$$

If we let $\mu_n = E(-\log(1-Y_n(\bar{\zeta})))$, then $\mu_0 < \infty$ by hypothesis. We will use induction and the identity (derived from (14))

$$-\log(1-Y_n(\bar{\zeta})) = -\log\left(\frac{1-\varphi_{\zeta_0}(Y_{n-1}(T\bar{\zeta}))}{1-Y_{n-1}(T\bar{\zeta})}\right) - \log(1-Y_{n-1}(T\bar{\zeta})). \tag{15}$$

Note that $(1-\varphi_{\zeta 0}(x))/(1-x)$ is increasing in x in $[0,1]$, and so

$$-\log\varphi'_{\zeta_0}(1-) \leqslant -\log\left(\frac{1-\varphi_{\zeta_0}(Y_{n-1}(T\bar{\zeta}))}{1-Y_{n-1}(T\bar{\zeta})}\right) \leqslant -\log(1-\varphi_{\zeta_0}(0)). \tag{16}$$

Thus, $\mu_n \leqslant (n+1)\mu_0 < \infty$. If

$$\theta_n = E\left(-\log\left(\frac{1-\varphi_{\zeta_0}(Y_{n-1}(T\bar{\zeta}))}{1-Y_{n-1}(T\bar{\zeta})}\right)\right),$$

then $-\infty \leqslant \theta_n < \infty$ by hypothesis and (16). Now using (15) we get

$$\mu_n = \sum_{j=1}^{n} \theta_j + \mu_0. \tag{17}$$

Assume that $P(q(\bar{\zeta})<1) \neq 1$. Then by the zero-one law $P(q(\bar{\zeta})<1)=0$ and $Y_{n-1}(T\bar{\zeta})\uparrow 1$ a.s. This fact implies (by monotone convergence) that $\mu_n \uparrow \infty$ and $\theta_n \downarrow \theta = E(-\log\varphi'_{\zeta_0}(1)) < 0$. This clearly violates (17). □

We summarize the above results:

1. If $E(\log\varphi'_{\zeta_0}(1))^+ < \infty$,

then

a) $E(\log\varphi'_{\zeta_0}(1))^+ \leqslant E(\log\varphi'_{\zeta_0}(1))^- \leqslant \infty$ implies $P(q(\bar{\zeta})=1)=1$;

while

b) $E(\log\varphi'_{\zeta_0}(1))^+ > E(\log\varphi'_{\zeta_0}(1))^-$, $E\{-\log(1-\varphi_{\zeta_0}(0))\} < \infty$ imply $P(q(\bar{\zeta})=1)=0$.

2. If $E(\log\varphi'_{\zeta_0}(1))^+ = \infty$, then $E\{-\log(1-\varphi_{\zeta_0}(0))\} < \infty$ implies $P(q(\bar{\zeta})=1)=0$.

The unresolved cases are:

3. $E(\log\varphi'_{\zeta_0}(1))^+ = \infty$ and $E(\log\varphi'_{\zeta_0}(1))^- = \infty$;
4. $\infty \geqslant E(\log\varphi'_{\zeta_0}(1))^+ > E(\log\varphi'_{\zeta_0}(1))^-$ and $E\{-\log(1-\varphi_{\zeta_0}(0))\} = \infty$.

A known feature of the classical Galton-Watson process is that it either explodes or dies out. That is, $P(\mathsf{A} \cup \mathsf{B}) = 1$ where $\mathsf{A} = \{\omega : Z_n \to \infty\}$. The proof depended on the fact that the Galton-Watson process is Markovian. Even though B.P.R.E. is not Markov in general, this property does hold for B.P.R.E. The proof is nontrivial and may be found in Athreya-Karlin (1971), where the reader will also find results about rates of convergence of $\varphi_{\zeta_0}(\ldots\varphi_{\zeta_n}(0)\ldots)$ to $q(\bar{\zeta})$ similar to those in section 11 of chapter I.

5.5. Limit Theorems

The B.P.R.E. $\{Z_n\}$ will be labeled supercritical, critical or subcritical, as $E(\log\varphi'_{\zeta_0}(1))>0$, $=0$ or <0, respectively. In the supercritical case, under the extra moment condition $E\{-\log(1-\varphi_{\zeta_0}(0))\}<\infty$, we know that $P(q(\bar{\zeta})<1)=1$ and the process explodes with positive probability. The martingale theorem (I.10.1) here has a natural extension.

Theorem 4. *Let* $W_n=Z_nP_n^{-1}$ *where (for* $n\geqslant 1$*)* $P_n=\prod_{j=0}^{n-1}\varphi'_{\zeta_j}(1)$ *and* $P_0=Z_0=1$. *Then, the family* $\{W_n, F_{n,z}(\bar{\zeta}); n=0,1,2,\dots\}$ *is a nonnegative martingale, and hence* $\lim_{n\to\infty} W_n=W$ *exists* a.s. *Suppose, in addition, that*

$$E\left\{(\varphi'_{\zeta_0}(1))^{-1}\sum_{j=2}^{\infty} p_{\zeta_0}(j)\,j\log j\right\}<\infty.$$

Then,

(i) $\lim_{n\to\infty} E(e^{-uW_n}|F(\bar{\zeta}))\equiv\psi(u,\bar{\zeta})$ *is the unique solution of the functional equation*

$$\psi(u,\bar{\zeta})=\varphi_{\zeta_0}\left(\psi\left(\frac{u}{\varphi'_{\zeta_0}(1)},T\bar{\zeta}\right)\right)\quad \text{a.s.}$$

among those solutions which are of the form $\int_0^\infty e^{-u}dH(x,\bar{\zeta})$, *where* $H(x,\bar{\zeta})$ *is a probability distribution on* $[0,\infty)$ *with* $H(0+,\bar{\zeta})<1$ *and* $\int_0^\infty x\,dH(x,\bar{\zeta})=1$.

(ii) $E(W|F(\bar{\zeta}))=1$.

(iii) $P(W=0|F(\bar{\zeta}))=q(\bar{\zeta})$ a.s.

The Yaglom theorem for the subcritical case in section I.8 also has an extension here. We need an extra hypothesis about the ζ's. We say that $\bar{\zeta}\equiv\{\zeta_n\}$ is an *exchangeable process* if the vectors $(\zeta_i,\zeta_{i+1},\dots,\zeta_{i+n})$ and $(\zeta_{i+n},\zeta_{i+n-1},\dots,\zeta_i)$ are identically distributed for each $i\geqslant 0$ and $n\geqslant 0$.

Theorem 5. *Assume that* $\bar{\zeta}$ *is exchangeable and that* $E(\log\varphi'_{\zeta_0}(1))<0$. *Then the sequence* $\{P\{Z_n=j|Z_n\neq 0, F(\bar{\zeta})\}; j=1,2,\dots\}$ *(of random distributions) converges weakly to a (random) proper distribution* $v_j(\bar{\zeta})$.

At the moment there is no appropriate generalization of Kolmogorov's exponential limit law for the critical case.

The proofs of the above theorems are too involved to be given here. We refer the reader to Athreya-Karlin (1971) for details. The multidimensional generalization of the B.P.R.E. has also been briefly treated in that paper. See also Kaplan (1972b, c) and Savits (1972).

6. Continuous State Branching Processes

6.1 Introduction

The state spaces of the branching models we have studied have all been discrete; namely the non-negative integers, or lattice points in p-dimensions. There are analogous processes on continuous state spaces which preserve the basic features of the branching property. These were first introduced by Jirina (1958); and related models have been studied by Lamperti (1967 a, c), by Watanabe (1968 b) and by Seneta and Vere-Jones (1969). Some of these treat continuous and some discrete time versions. We shall here present a brief account of Lamperti's work.

The process will be defined in terms of its transition function. A one parameter family $\{P_t(x,E); t \geqslant 0\}$ of functions will be called a *continuous branching function* (C.B. function) if it satisfies the following conditions:

(i) $P_t(x, \mathsf{E})$ is defined for $t \geqslant 0$, $x \geqslant 0$, and E a Borel subset of the half line $[0, \infty)$.

(ii) For fixed t and x, $P_t(x, \mathsf{E})$, as a function of E, is a probability measure; and for fixed E, $P_t(x, \mathsf{E})$ is jointly measurable in x and t.

(iii) The Chapman-Kolmogorov equation holds:

$$\int_0^\infty P_t(u, \mathsf{E}) P_s(x, du) = P_{t+s}(x, \mathsf{E}).$$

(iv) For any $x, y, t \geqslant 0$, $\{P_t\}$ satisfies

$$P_t(x+y, \mathsf{E}) = \int P_t(x, \mathsf{E}-u) P_t(y, du) \tag{1}$$

for each E.

(v) There exist $t > 0$ and $x > 0$ such that $P_t(x, \{0\}) < 1$.

Let us examine these conditions. Assumptions (i) to (iii) are simply the definition of a Markov transition function. Postulate (iv) is precisely the basic branching feature, namely the *additive property:* if we start at $x+y$ then the resulting process is equivalent to the sum of two independent processes, one starting at x and the other at y. Finally, (v) rules out the trivial case that, starting at any x, the process instantly dies and stays there.

We shall check that a C.B. function does give rise to a strong Markov process on the nonnegative reals. This process will be called a *continuous state space branching processes* or simply a *C.B. process.*

These processes apart from being of interest in their own right also arise as limit processes of sequences of Galton-Watson processes in which the initial population size goes to ∞ and the time scale shrinks in a specified manner.

We will develop the basic properties of C.B. functions, and then discuss some limit theorems and open problems.

6.2 Existence, Construction and Some Basic Properties of C.B. Processes

Let $\{P_t(x, \mathsf{E})\}$ be a C.B. function. From (1) it is clear that for each (t, x) the measure $P_t(x, \mathsf{E})$ is infinitely divisible. If we denote the Laplace transform of $P_t(x, \mathsf{E})$, by

$$\psi_t(x,\lambda) = \int_0^\infty e^{-\lambda y} P_t(x,dy), \qquad \lambda \geqslant 0,$$

then we see from (1) that $\psi_t(x,\lambda)$ satisfies the functional equation

$$\psi_t(x+y,\lambda) = \psi_t(x,\lambda)\,\psi_t(y,\lambda). \tag{2}$$

We know from (ii) that for fixed t and λ, $\psi_t(x,\lambda)$ is a measurable function of x. It is also clear that $\psi_t(x,\lambda)$ is monotone decreasing in x, and so we may conclude from (2) that for each t and λ there exists a number $\psi_t(\lambda)$ such that

$$\psi_t(x,\lambda) = e^{-x\psi_t(\lambda)}. \tag{3}$$

The Chapman-Kolmogorov equation (iii) now transforms into

$$\psi_{t+s}(\lambda) = \psi_t(\psi_s(\lambda)). \tag{4}$$

This is the analog of the functional iteration property of the generating function of a Galton-Watson process.

To go from a C.B. function $\{P_t\}$ to a Markov process corresponding to it, one uses the tools of semigroup theory (see Dynkin (1965)). For any bounded measurable function f on $[0, \infty)$, define

$$(T_t f)(x) = \int_0^\infty f(y) P_t(x,dy). \tag{5}$$

This defines a family $\{T_t; t \geqslant 0\}$ of bounded linear operators of unit norm on the Banach space $\boldsymbol{B}$ of bounded measurable functions on $[0, \infty)$ (with norm $\|f\| = \sup_{x \geq 0} |f(x)|$).

The Chapman-Kolmogorov equation yields the relation

$$T_{t+s} = T_t T_s, \tag{6}$$

and this makes the family $\{T_t\}$ a semigroup.

Theorem 1. (i) T_t *maps the subspace* C_0 *of all continuous functions on* $[0, \infty)$ *that vanish at* ∞ *into itself.*

(ii) $\{T_t\}$ *is a strongly continuous semigroup; i.e. for each* f *in* C_0, $T_t f \to f$ *uniformly in* x *as* $t \downarrow 0$.

A general result from the theory of Markov process now assures us that there exists a unique strong Markov process $\{X(t,\omega); t \geqslant 0\}$ with $\{P_t\}$ as its transition function (see Dynkin (1965)).

Proof of (i). Let $f(x)$ be a continuous function on $[0, \infty)$ vanishing at ∞.

By the Stone-Weirstrass approximation theorem, finite linear combinations of the form $\sum_{i=1}^{N} a_i e^{-\lambda_i x}$ (where the a_i's are real and the λ_i's are $\geqslant 0$) are dense in C_0. Moreover, T_t is a bounded linear operator with norm 1. Therefore to show that

$$\lim_{y \to x} (T_t f)(y) = f(x) \quad \text{for each finite } x, \tag{7}$$

and

$$\lim_{x \to \infty} (T_t f)(x) = 0 \tag{8}$$

for each f in C_0, it suffices to do the same for functions of the form $f(x) = e^{-\lambda x}$ for some $\lambda > 0$. For this special f, $T_t f(x) = \exp\{-x \psi_t(\lambda)\}$, implying (7). As for (8), we use for the first time the postulate (v) in the definition of a C.B. function to assert the existence a pair (t_0, x_0), $x_0 > 0$, $t_0 > 0$, such that $P_{t_0}(x_0, \{0\}) < 1$. This can be shown (we leave it as an exercise) to imply that $P_t(x, \{0\}) < 1$ for all t, $x > 0$, which in turn assures that $\psi_t(\lambda) > 0$ for each $\lambda > 0$. Now (8) is immediate. This proves (i).

The proof of (ii) needs a little more functional analysis than we want to develop here and we refer the reader to Lamperti's paper for the proof.

Let us derive some elementary properties of C.B. processes $\{X(t); t \geqslant 0\}$. We can use (3) and (4) to compute the moments when they exist. Noting that $\psi_t(0) = 0$ we get from (3)

$$E(X(t) | X(0) = x) = -(e^{-x\psi_t(\lambda)})'|_{\lambda=0} = x \psi_t'(0+), \tag{9}$$

where the prime denotes differention with respect to λ. Now (4) yields the functional equation

$$\psi_{t+s}'(0+) = \psi_t'(0+) \psi_s'(0+), \tag{10}$$

so that if $\psi_t'(0)$ is finite for any t, then it must be so for all t; and so the mean must be of the form $e^{\alpha t}$, for some constant α.

This suggests that $\{X(t) e^{-\alpha t}\}$ is a nonnegative martingale. An open problem is to generalize the convergence theorems I.10.1 and I.10.2.

The second moment can be evaluated similarly, when it exists. The expressions for the first and second moments resemble the analogous ones for Galton-Watson process. They are of the form:

$$E(X(t) | X(0) = x) = x e^{\alpha t}, \tag{11}$$

$$E(X^2(t) | X(0) = x) = \begin{cases} x^2 + \beta x t & \text{if } \alpha = 0, \\ x^2 e^{2\alpha t} + \left(\dfrac{\beta}{\alpha}\right) x (e^{2\alpha t} - e^{\alpha t}) & \text{if } \alpha \neq 0. \end{cases} \tag{12}$$

These formulas yield

$$\lim_{t\to 0} t^{-1}[E(X(t)-x) \mid X(0)=x)] = \alpha x,$$

$$\lim_{t\to 0} t^{-1}[E((X(t)-x)^2 \mid X(0)=x)] = \beta x.$$

This suggests that if $X(t)$ is a diffusion process then αx and βx should be the corresponding diffusion coefficients, i.e., that the backward (Kolmogorov) equation satisfied by $X(t)$ is

$$\frac{\partial u}{\partial t} = \beta x \frac{\partial^2 u}{\partial x^2} + \alpha x \frac{\partial u}{\partial x}. \tag{13}$$

This is, in fact, true and a proof may be found in Lamperti's paper. The corresponding $\psi_t(\lambda)$ for this diffusion is given by

$$\psi_t(\lambda) = \begin{cases} \dfrac{\lambda}{1 + \dfrac{\beta t \lambda}{2}} & \text{if } \alpha = 0, \\[2ex] \dfrac{\lambda e^{\alpha t}}{1 - \dfrac{\beta\lambda}{2\alpha}(1-e^{\alpha t})} & \text{if } \alpha \neq 0. \end{cases} \tag{14}$$

Noting that for any C.B. process (3) is the same as saying

$$E(e^{-\lambda X(t)} \mid X(0)=x) = e^{-x\psi_t(\lambda)},$$

we see that

$$P_t(x,\{0\}) = P\{X(t)=0 \mid X(0)=x\} = \exp\{-x \lim_{\lambda\to\infty} \psi_t(\lambda)\}.$$

Thus, if $X(t)$ is a diffusion with (13) as its backward equation then

$$P_t(x,\{0\}) = \begin{cases} \exp\left\{-\dfrac{2x}{\beta t}\right\} & \text{if } \alpha = 0, \\[2ex] \exp\left\{\dfrac{2\alpha e^{\alpha t} x}{\beta(e^{\alpha t}-1)}\right\} & \text{if } \alpha \neq 0. \end{cases} \tag{15}$$

If we set $q(x) = P\{X(t)=0$ for some $t \mid X(0)=x\}$ = the probability of extinction of the process, then (15) yields

$$q(x) = \lim_{t\to\infty} P_t(x,\{0\}) = \begin{cases} 1 & \text{if } \alpha \leqslant 0, \\[1ex] \exp\left\{-\dfrac{2\alpha}{\beta} x\right\} & \text{if } \alpha > 0. \end{cases} \tag{16}$$

An example of a C.B. process which is not a diffusion, is the purely discontinuous process which evolves as follows: if $X(0)=x$ then the process sits there an exponential length of time τ with mean inversely proportion to x, and then jumps to a new position $x+Y$ where Y is a nonnegative random variable whose distribution does not depend on x or τ; then the process evolves the same way from $X(\tau+0)=x+Y$. This is the continuous state space analog of the continuous time Markov branching process discussed in Chapter III. The process can explode here too, and the non-explosion hypothesis resembles the one in Chapter III Namely: $P_x\{X(t)<\infty\}=1$ for all $x,t\geqslant 0$ if and only if

$$\int_0^{\delta} \frac{1}{[\varphi(\lambda)-e^{-\lambda}]}\,d\lambda = -\infty \quad \text{for each } \delta>0\,,$$

where $\varphi(\lambda)=E(e^{-\lambda Y})$.

Notice that these purely discontinuous processes cannot move to the left since the jump Y is always nonnegative. Thus any process for which absorption into zero in finite time is possible, i.e. $P_t(x,\{0\})>0$ for some $t,x>0$ cannot be of the purely discontinuous type.

Lamperti describes the construction of the most general C.B. process and shows that every such process can be obtained by a *random time change* from a process with stationary independent increments which cannot jump to the left. This is a generalization of a similar result in III.11. We shall merely summarize his result.

Let $\{X(t,\omega)\}$ be any C.B. process. Define

$$J(\omega)=\sup\{t\colon X(t,\omega)>0\}$$

and

$$Y(t,\omega)=\int_0^t X(u,\omega)\,du\,.$$

Then $Y(\cdot,\omega)$ is strictly increasing for $t<J(\omega)$, and therefore has a unique inverse $Y^{-1}(t,\omega)$ for $t<K(\omega)=\int_0^\infty X(u,\omega)\,du$. Now define

$$Z(t,\omega)=\begin{cases} X(Y^{-1}(t,\omega)) & \text{for } t<K(\omega)\,,\\ 0 & \text{for } t\geqslant K(\omega)\,. \end{cases} \tag{17}$$

Then $\{Z(t,\omega);t\geqslant 0\}$ is a process with stationary independent increments which cannot jump to the left.

By essentially reversing the above argument one can get X from Z.

As a corollary to his construction Lamperti shows that the only diffusions among C.B. processes are those whose backward equations are given by (13).

6.3 A Characterization of C.B. Processes

Lamperti has also obtained the following characterization of C.B. processes, as limits of sequences of discrete time Galton-Watson processes.

Let $\{Z_n^{(r)}\}$, $n=0,1,2,\ldots$, $r=1,2,\ldots$ be a sequence of Galton-Watson processes with offspring distribution $\{p_k^{(r)}\}$. Define

$$X_t^{(r)} = \frac{Z_{[rt]}^{(r)}}{b_r} \quad \text{for } 0 \leqslant t < \infty, \tag{18}$$

where the b_r's are nonnegative. Then, the following two results yield the desired characterization.

Theorem 2. *Suppose the process $\{X_t^{(r)}\}$ converges to $\{X_t\}$ in the sense of finite dimensional distributions. Suppose $Z_0^{(r)} \to \infty$ and that $P(X_t=0\} < 1$ for some $t>0$. Then* (i) $b_r \to \infty$, (ii) $\lim_{r\to\infty} (Z_0^{(r)}/b_r) = c$ *exists and is* >0, *and* (iii) *$\{X_t\}$ is a C.B. process with $X_0 = c$.*

Theorem 3. *Let $\{X_t\}$ be any C.B. process with $X_0=c$. Then there exists a sequence of branching process $\{Z_n^{(r)}\}$ with $Z_0^{(r)} \to \infty$ such that $\{X_t^{(r)}\}$ as defined in (18), with $b_r = Z_0^{(r)}/c$, converge to $\{X_t\}$ in the same sense as in theorem 2.*

A characterization of all limit processes has also been obtained in the case when the definition in (18) is changed to

$$X_t^{(r)} = \frac{Z_{[rt]}^{(r)} - a_r}{b_r}, \tag{19}$$

where $\{a_r\}$, $\{b_r\}$ and $\{Z_0^{(r)}\}$ behave appropriately as $r \to \infty$. That is, one centers the process at some a_r which need not be zero.

7. Immigration

Thus far we have considered populations which have existed in "isolation"; that is, which have grown or declined purely according to the multiplicative laws of a branching process. A useful and realistic modification of this scheme is the addition of the possibility of an immigration into the population from an outside source, or of an emmigration out of the population.

From the point of view of applications to biological or ecological processes, such emmigration—immigration factors are clearly of great importance. From a mathematical viewpoint we will see that the immigration process sheds light on one of the basic limit theorems of the Galton-Watson process, and also leads to an interesting new limit law.

7.1 Stationary Immigration Rate

Consider an ordinary Galton-Watson process with generating function f. Assume that at the time of birth of the nth generation, that is, at time n, there is an immigration of Y_n particles into the population. The Galton-Watson process with immigration (G.W.I. process) is defined by a sequence of random variables $\{Z_n\}$ which are determined by the relation

$$Z_{n+1} = X_1^{(n)} + \cdots + X_{Z_n}^{(n)} + Y_{n+1}, \qquad n \geqslant 1,\ Z_0 = Y_0, \tag{1}$$

where $X_1^{(n)}, X_2^{(n)}, \ldots$ are independent and identically distributed with generating function $f(s)$, the $Y_0, Y_1, \ldots$ are i.i.d. with generating function $h(s)$, and the X's and Y's are independent.

A similar process can be constructed in continuous time by immigrating a Poisson or compound Poisson process into a continuous time Markov branching process of the type constructed in chapter III. Results for such processes which parallel to some degree those of the present section, were obtained by Sevastyanov (1957).

Let $g_n(s) = E s^{Z_n}$, $0 \leqslant s \leqslant 1$. Then $g_{n+1}(s) = h(s) g_n[f(s)]$, $n \geqslant 0$, and hence $g_n(s) = \prod_{k=0}^{n} h[f_k(s)]$. Furthermore

1. $\lim g_n(s) = g(s)$ exists, and satisfies
2. $g(s) = h(s) g(f(s))$.

When does the process $\{Z_n\}$ have a proper limit distribution? The following result (see e.g. Heathcote (1965), Foster (1969)) gives the answer in the most elementary cases.

Theorem 1. (i) *If* $m = f'(1) > 1$, *or* $m = 1$ *and* $f''(1) < \infty$, *then for* $0 < k < \infty$

$$\lim_{n \to \infty} P\{Z_n = k\} = 0.$$

(ii) *If* $0 < h'(1) < \infty$, *and* $m < 1$, *then*

$$\lim_{n \to \infty} P\{Z_n = k\} = \rho_k \quad \textit{exist,}$$

and $\{\rho_k, k \geqslant 0\}$ *is a probability function.*

Proof. If $h'(1) < \infty$ then (as $k \to \infty$)

$$1 - h[f_k(s)] = h'(1)\,[1 - f_k(s)] + o[1 - f_k(s)]. \tag{2}$$

Thus

$$g(s) \equiv \prod_{k=0}^{\infty} h[f_k(s)] > 0 \tag{3}$$

if and only if

$$\sum [1 - f_k(s)] < \infty. \tag{4}$$

By the results of sections I.5, I.9 and I.11, the sum (4) diverges under (i) and converges under (ii). If $h'(1) = \infty$ and $m > 1$ then $f_k(s) \to q$, and hence $\sum (1 - h[f_k(s)]) = \infty$. If $h'(1) = \infty$ and $m = 1$ then $1 - h[f_k(s)] \geqslant M[1 - f_k(s)] \sim (2/\sigma^2) M k$, where M is a constant, and hence the sum again diverges. Taking logarithms and applying the dominated converges theorem to the product below, one observes that under (ii)

$$\lim_{s \uparrow 1} g(s) = \lim_{n \to \infty} g[f_n(0)] = \lim_{n \to \infty} \prod_{k=0}^{\infty} h[f_{n+k}(0)] = 1;$$

and the continuity theorem for generating functions implies the last assertion of the theorem. □

The complete answer is given in theorem 2, proved by Foster and Williamson (1971). For earlier work in this direction, see also Heathcote (1966) and Seneta (1968 c).

Theorem 2. *For a G.W.I. process with stationary immigration and* $m \leqslant 1$, *a necessary and sufficient condition for* $\{Z_n\}$ *to converge in distribution to a proper random variable is*

$$\int_0^1 \frac{1 - h(s)}{f(s) - s} \, ds < \infty. \tag{5}$$

In the supercritical case it is not surprising that immigration has little effect on the limiting behavior of the process. We state the most general known result.

Theorem 3 (Seneta (1970 b)). *If* $m > 1$, *then there exists a sequence of constants* $\{C_n\}$ *such that* $\{Z_n / C_n\}$ *converges with probability one to a random variable* V. *If* $E \log Y_1 < \infty$ *then* $P\{V < \infty\} = 1$, *and* V *has an absolutely continuous distribution on* $(0, \infty)$. *If* $E \log Y_1 = \infty$, *then* $P\{V < \infty\} = 0$.

The constants C_n are the same as in theorem I.10.3. For the proof one observes that e^{-Z_n/C_n} is a submartingale. The details parallel quite closely those of the proof of theorem I.10.3.

In the critical case, the same normalization as for the ordinary G.W. process leads to a proper limit law. The comparison of the result with that for the G.W. process is here particularly interesting.

Theorem 4 (Foster (1969), Seneta (1970d)). *If* $m=1$, $f''(1)=\sigma^2<\infty$, *and* $0<h'(1)<\infty$, *then* $2Z_n/n\sigma^2$ *converges in distribution to a random variable with density*

$$w(x)=\frac{1}{\Gamma\left(\frac{2h'(1)}{\sigma^2}\right)}x^{\frac{2h'(1)}{\sigma^2}-1}e^{-x}. \tag{6}$$

When $h'(1)=\sigma^2/2$, then $w(x)=e^{-x}$. Thus recalling the limit law of section I.9 we observe that

Corollary. *In the critical case, the effect (on the limit law of* Z_n/n*) of conditioning on non-extinction is the same as immigrating at an average rate of* $\sigma^2/2$.

The mathematically useful device of conditioning on non-extinction is physically unappealing since it involves conditioning on an event whose probability goes to zero. It is thus reassuring to find that in the above sense it is equivalent to an operation which is physically quite meaningful; namely immigration at a specified rate.

The proof of theorem 4 involves estimating $\prod_{k=1}^{n} h[f_k(s)]$ in terms of the estimates of $f_k(s)$ obtained in section I.9; and is very similar to the proof of the exponential limit law in the conditioned case. For details see Foster (1969), where similar results for multi-type processes are also developed.

7.2 Time-Variable Immigration Rates

When the immigration into the nth generation is a random variable with generating function $h_n(s)$, then

$$Es^{Z_n}=\prod_{k=0}^{n} h_{n-k}[f_k(s)].$$

Foster and Williamson (1971) have considered the question of when limit laws exist for such a process. In the critical case they show that when $\sigma^2<\infty$, then $P\{Z_n=k\}\to 0$ no matter how the h_k's are chosen. They also give a sufficient condition for the convergence of the normalized sequence Z_n/n.

7.3 A Branching Process with a Reflecting Barrier at Zero

From the point of view of applications it would be desirable to be able to treat processes where the immigration rate depends on the population size. In general this is a difficult and open problem. A partial result is in Foster (1969, 1971), in a very special case.

He considers a model of a critical process in which there is no immigration unless the population size hits zero, at which time one particle is immigrated into the population. (The precise number immigrated is not crucial, and could be a random variable with any fixed distribution.) In this case there results an interesting new limit law.

Theorem 5. *For the "reflecting" immigration process described above, if* $f'''(1) < \infty$, *then* $(\log Z_n / \log n)$ *converges in distribution to a random variable with uniform distribution on* $[0,1]$.

As indicated before it may also be desireable to introduce an emmigration factor into the process. This usually adds no new mathematical difficulties, however, since the process with emmigration can frequently be constructed from the original one by suitable modification of the particle production distribution.

For example, if each particle emmigrates (independently) with probability $1-p$, then the new process is an ordinary Galton-Watson process with generating function $f(1-p+ps)$.

8. Instability

We have derived many different limit laws through a variety of devices such as normalizing factors, conditioning, immigration. These constructions were necessary because the unadulturated Galton-Watson process is unstable (chapter I), i.e.

$$\lim_{n\to\infty} P\{Z_n = k\} = 0 \quad \text{for all } k \geqslant 1 . \tag{1}$$

One can inquire whether it is possible to produce a limit distribution for Z_n which does not have all its mass at 0 or ∞, when the particle production distribution is allowed to vary suitably from generation to generation. The answer is essentially no. The obvious exception is if exactly one particle is produced by each parent. The following theorem, due to Church (1967) states that (1) will hold unless the generating function governing particle production in the nth generation, say $h_n(s)$, converges sufficiently rapidly to the degenerate generating function s.

Theorem. *Let* $\Pi_n(s) \equiv E(s^{Z_{n+1}} | Z_0 = 1)$. *Then,* $\Pi_n(s) = h_0(h_1(h_2 \ldots (h_n(s) \ldots)$, *and* $\lim_{n\to\infty} \Pi_n(s) = \Pi(s)$ *exists for* $0 \leqslant s < 1$. *Further,* $\Pi(s)$ *is strictly increasing in* $[0,1)$ *if and only if*

$$\sum_j (1 - h_j'(0)) < \infty .$$

Corollary. *For the* $\{Z_n\}$ *process* $\lim_{n\to\infty} P\{Z_n=k\}=a_k$ *exists for all* k. *But* $a_k=0$ *for all* $k\geqslant 1$, *if and only if*

$$\sum_j (1-h_j'(0))=\infty.$$

A proof of this theorem may be found in Church (1967) and also in Athreya and Karlin (1971) who used it to establish a similar instability for branching processes with random environments.

Complements and Problems VI

1. It is reasonable to guess that the condition $f''(1)<\infty$ in theorem 3.1 is not crucial (compare with section 1).

2. Proof of (3.4). Hint: Use (3.3) to observe that $EZ_t^D/EZ_t=P\{$a brownian particle is in D at $t\}$.

3. In the definition of a C.B. function (section 6) show that condition (v) implies that

$$P_t(x,\{0\})<1 \quad \text{for all } t>0, \quad x>0.$$

4. Use theorem 7.2 to show that if $m<1$, then a necessary and sufficient condition for the process with stationary immigration rate (section 7.2) to be positive recurrent is $E\log Y_n<\infty$. This was first proved directly by Heathcote (1966).

5. *Further recent work on immigration:*

Asymptotic results for the transition function of a G.W.I. process, along the lines of the Karlin, McGregor spectral theorems have been obtained by Pakes (1971).

Stationary measures for the G.W.I. process have been studied by Seneta (1969a), (1971).

M. Pinsky (1971), has considered continuous state space processes with immigration.

Bibliography[16]

Papers or books written before about 1962 are included in the bibliography only if they are specifically referred to in the book. We have attempted to make the bibliography fairly complete only from the period of publication of T. Harris's book (1963). The latter reference has a very good bibliography of work up to 1962.

The following abbreviations of frequently referred to journals will be used:

AMS: The Annals of Mathematical Statistics
BAMS: Bulletin of the American Mathematical Society
Berk. Symp.: Proceedings of the Berkeley Symposium on Mathematical Statistics and Probability
IJM: Illinois Journal of Mathematics
JAP: Journal of Applied Probability
JMAA: Journal of Mathematical Analysis & Applications
JMKU: Journal of Mathematics Kyoto University
JRSS: Journal of the Royal Statistical Society (Series B)
PAMS: Proceedings of the American Mathematical Society
PJA: Proceedings of the Japan Academy
TAMS: Transactions of the American Mathematical Society
TPA: Theory of Probability & its Applications
ZW: Zeitschrift für Wahrscheinlichkeitstheorie und verwandte Gebiete.

Adke, S. R.: (1964) The maximum population size in the first N generations of a branching process. Biometrics **20**, 649–651.

Ahfors, L. V.: (1953) Complex Analysis. New York: McGraw-Hill.

Athreya, K. B.: (1968) Some results on multitype continuous time Markov branching processes. AMS **39**, 347–357.

Athreya, K. B.: (1969a) On the equivalence of conditions on a branching process in continuous time and on its offspring distribution. JMKU **9**, 41–53.

Athreya, K. B.: (1969b) Limit theorems for multitype continuous time Markov branching processes. I. The case of an eigenvector linear functional. ZW **12**, 320–332.

Athreya, K. B.: (1969c) Limit theorems for multitype continuous time Markov branching processes. II. The case of an arbitrary linear functional. ZW **13**, 204–214.

Athreya, K. B.: (1969d) On the supercritical one dimensional age dependent branching processes. AMS **40**, 743–763.

Athreya, K. B.: (1969e) On a characteristic property of Polya's urn. Studia Sci. Math. Hungar. **4**, 31–35.

Athreya, K. B.: (1970) A simple proof of a result of Kesten and Stigum on supercritical multitype Galton-Watson branching processes. AMS **41**, 195–202.

Athreya, K. B.: (1971a) A note on a functional equation arising in Galton-Watson branching processes. JAP **8**, 589–598.

---

[16] Papers having more than one author are listed only once, under the name of the author appearing first on the paper.

Athreya, K. B.: (1971b) Some refinements in the theory of supercritical multitype branching processes. ZW **20**, 47–57.

Athreya, K. B.: (1971c) On the absolute continuity of the limit random variable in the supercritical Galton-Watson branching process. PAMS **30**, 563–565.

Athreya, K. B., Karlin, S.: (1967) Limit theorems for the split times of branching processes. J. Math. Mech. **17**, 257–277.

Ahreya, K. B., Karlin, S.: (1968) Embedding of urn schemes into continuous time Markov branching processes and related limit theorems. AMS **39**, 1801–1817.

Athreya, K. B., Karlin, S.: (1970) Branching processes with random environments. PAMS **76**, 865–870.

Athreya, K. B., Karlin, S.: (1972a) On branching processes in random environments: I. Extinction probability. AMS **42**, 1499–1520.

Athreya, K. B., Karlin, S.: (1972b) On branching processes in random environments: II. Limit theorems. AMS **42**, 1843–1858.

Athreya, K. B., Ney, P.: (1970) The local limit theorem and some related aspects of super-critical branching processes. TAMS **152**, 233–251.

Athreya, K. B., Ney, P.: (1971) Limit theorems for the means of branching random walks. To appear in the proceedings of the Sixth Prague Conference on Information Theory.

Badalbaev, I. S.: (1970) The properties of an estimate of a regulating parameter of a branching stochastic process. (Russ.) Stochastic Processes and Related Problems. Part I (Russ.), Tashkent: Izdat. "Fan" Uzbek. SSR, 31–42.

Bartoszyński, R.: (1967a) Some limit properties of generalized branching processes. (Russ. summ.) Bull. Acad. Polon. Sci. Sér. Sci. Math. Astronom. Phys. **15**, 157–160.

Bartoszyński, R.: (1967b) A limit property of certain branching processes. (Loose Russ. Summ.) Bull. Acad. Polon. Sci. Sér. Sci. Math. Astronom. Phys. **15**, 615–618.

Bartoszyński, R.: (1969) Branching processes and models of epidemics. Dissertations Math. Rozprawy Mat. 61, 51 pp.

Baum, L. F., Katz, M.: (1965) Convergence rates in the law of large numbers. TAMS **120**, 109–123.

Bellman, R., Harris, T. E.: (1952) On age-dependent binary branching processes. Ann. of Math. **55**, 280–295.

Bharucha-Reid, A. T.: (1960) Elements of the Theory of Markov Processes and Their Applications. New York: McGraw-Hill.

Bharucha-Reid, A. T.: (1965) Markov branching processes and semigroups of operators. JMAA **12**, 513–536.

Billingsley, P.: (1968) Convergence of Probability Measures. New York: Wiley.

Bircher, J. J.: (1969) An age-dependent branching process with arbitrary state space. Ph. D. Thesis, State Univ. of N. Y. at Buffalo.

Bircher, J. J., Mode, C. J.: (1970) On the foundations of age-dependent branching processes with arbitrary state space. JMAA **32**, 435–444.

Bircher, J. J., Mode, C. J.: (1971a) An age-dependent branching process with arbitrary state space I. JMAA **36**, 41–59.

Bircher, J. J., Mode, C. J.: (1971b) An age-dependent branching process with arbitrary state space II, JMAA **36**, 227–240.

Bishir, J.: (1962) Maximum population size in a branching process. Biometrics **18**, 394–403.

Blackwell, D., Kendall, D.: (1964) The Martin boundary for Polya's urn scheme and an application to stochastic population growth. JAP **1**, 284–296.

Boyd, A. V.: (1971) Formal power series and total progeny in a branching process. JMAA **34**, 565–566.

Brown, B. J., Heyde, C. C.: (1971) An invariance principle and some convergence rate results for branching processes. ZW **20**, 271–278.

Bühler, W. J.: (1967) Slowly branching processes. AMS **38**, 919–921.

Bühler, W. J.: (1968) Some results on the behavior of branching processes. TPA **13**, 52–64.

Bühler, W. J.: (1969) Ein zentraler Grenzwertsatz für Verzweigungsprozesse. (Engl. summ.) ZW **11**, 139–141.

Bühler, W. J.: (1970) Generations and degree of relationship in supercritical Markov branching processes. ZW **18**, 141–152.

Bühler, W. J.: (1971) The distribution of generations and other aspects of the family structure of branching processes. To appear in Sixth Berkeley Symposium.

Čerkasov, I. D.: (1963) On the theory of one-dimensional branching random processes with continuous time. (Russ.-Eng. summary.) Bul. Inst. Politehn. Iasi (N. S.) **9** (13), fasc. 1–2, 51–58.

Čerkasov, I. D.: (1966) Factorial moments and probabilities of degeneration of a nonhomogeneous branching process. Rev. Roumaine Math. Pures Appl. **11**, 979–987.

Chistyakov, V. P.: (1957) Local limit theorems in the theory of branching random processes. TPA **2**, 345–363.

Chistyakov, V. P.: (1959) Generalization of a theorem for branching stochastic processes. TPA **4**, 103–106.

Chistyakov, V. P.: (1960) Transitional phenomena in branching processes with *n* types of particles. Soviet Math. Dokl. **1**, 300–302.

Chistyakov, V. P.: (1961) Transient phenomena in branching stochastic processes. TPA **6**, 27–41.

Chistyakov, V. P.: (1964) A theorem on sums of independent positive random variables and its applications to branching random processes. TPA **9**, 640–648.

Chistyakov, V. P.: (1970a) Some limit theorems for branching processes with immigration. TPA **15**, 241–257.

Chistyakov, V. P.: (1970b) Some limit theorems for branching processes with particles of final type. TPA **15**, 515–521.

Chistyakov, V. P.: (1970c) On convergence of branching processes to diffusion processes. TPA **15**, 707–710.

Chistyakov, V. P., Markova, N. P.: (1962) Some theorems for nonhomogeneous branching processes. (Russ.) Dokl. Akad. Nauk SSSR **147**, 317–320.

Chistyakov, V. P., Savin, A. A.: (1962) Some theorems for branching processes with several types of particles. TPA **7**, 93–100.

Chover, J., Ney, P.: (1965) A non-linear integral equation and its applications to critical branching processes. J. Math. Mech. **14**, 723–735.

Chover, J., Ney, P.: (1968) The non-linear renewal equation. J. D'Analyse Math. **21**, 381–413.

Chover, J., Ney, P., Wainger, S.: (1969) Functions of probability measures. Technical report, Univ. of Wisconsin. Modified version to appear in J. D'Analyse Math. (1972a).

Chover, J., Ney, P., Wainger, S.: (1972b) Degeneracy properties of subcritical branching processes. To appear in AMS.

Chung, K. L.: (1967) Markov Chains with Stationary Transition Probabilities, 2nd Edition, New York: Springer.

Chung, K. L.: (1968) A Course In Probability Theory. New York: Harcourt, Brace & World.

Church, J. D.: (1967) Composition limit theorems for probability generating functions. Math. Research Center, report 732.
Coddington, E. A., Levinson, N.: (1955) Theory of Ordinary Differential Equations. New York: McGraw-Hill.
Conner, H. E.: (1961) A limit theorem for a position-dependent branching process. JMAA **3**, 560–591.
Conner, H. E.: (1964) Extinction probabilities for age—and position—dependent branching processes. SIAM J. Appl. Math. **12**, 899–909.
Conner, H. E.: (1966) Limiting behavior for age and position-dependent branching processes. JMAA **13**, 265–295.
Conner, H. E.: (1967a) A note on limit theorems for Markov branching processes. PAMS **18**, 76–86.
Conner, H. E.: (1967b) Asymptotic behavior of averaging-processes for a branching process of restricted brownian particles. JMAA **20**, 464–479.
Crump, K. S.: (1968) Some generalized age-dependent branching processes. Thesis, Montana State Univ.
Crump, K. S.: (1970) Migratory populations in branching processes. JAP **7**, 565–572.
Crump, K. S.: (1970) On systems of renewal equations. JMAA **30**, 425–434.
Crump, K. S., Mode, C. J.: (1968) A general age-dependent branching process. I. JMAA **24**, 494–508.
Crump, K. S., Mode, C. J.: (1969a) A general age-dependent branching process. II. JMAA **25**, 8–17.
Crump, K. S., Mode, C. J.: (1969b) An age-dependent branching process with correlations among sister cells. JAP **6**, 205–210.
Daley, D. J.: (1968) Extinction conditions for certain bisexual Galton-Watson branching processes. ZW **9**, 315–322.
Darling, D. A.: (1970) The Galton-Watson process with infinite mean. JAP **7**, 455–456.
Davis, A. W.: (1965) On the theory of birth, death and diffusion processes. JAP **2**, 293–322.
Davis, A. W.: (1967a) Branching diffusion processes with no absorbing boundaries. I. JMAA **18**, 276–296.
Davis, A. W.: (1967b) Branching diffusion processes with no absorbing boundaries. II. JMAA **19**, 1–25.
Dayanithy, K.: (1971) Spectralities of branching processes. ZW **20**, 279–307.
Doney, R. A.: (1971) The total progeny of a branching process. JAP **8**, 407–412.
Doney, R. A.: (1972) Age-dependent birth and death processes. ZW **22**, 69–90.
Doob, J. L.: (1953) Stochastic Processes. New York: Wiley.
Dubuc, S.: (1969) Positive harmonic functions of branching processes. PAMS **21**, 324–326.
Dubuc, S.: (1970) La Fonction de Green d'un processus de Galton-Watson. Studia Math. **34**, 69–87.
Dubuc, S.: (1971) Problemes relatifs a l'itération de fonctions suggérés par les processus en cascade. Ann. Inst. Fourier Grenoble **21**, 171–251.
Durham, S. D.: (1971a) Limit theorems for a general critical branching process. JAP **8**, 1–16.
Durham, S. D.: (1971b) A problem concerning generalized age-dependent branching processes with immigration. AMS **42**, 1121–1123.
Dwass, M.: (1969) The total progeny in a branching process and a related random walk. JAP **6**, 682–686.
Dynkin, E. B.: (1965) Markov Processes. New York: Academic Press.

Erickson R. V.: (1971) On the existence of absolute moments for the extinction time of a Galton-Watson process. AMS **42**, 1124–1128.

Fahady, K.S., Quine, M.P., Vere-Jones, D.: (1971) Heavy traffic approximations for the Galton-Watson process. Adv. App. Prob. **3**, 282–300.

Feller, W.: (1966) An Introduction to Probability Theory and Its Applications, Vol. II. New York: Wiley.

Feller, W.: (1968) An Introduction to Probability Theory and Its Applications, Vol. I, 3rd Ed. New York: Wiley.

Fisher, R.A.: (1930) The genetical Theory of Natural Selections. Oxford Univ. Press; New York: Dover 1958.

Foster, J.: (1969) Branching processes involving immigration. Ph. D. Thesis, Univ. of Wisconsin.

Foster, J. A.: (1971) A limit theorem for a branching process with state-dependent immigration. AMS **42**, 1773–1776.

Foster, J., Williamson, J. A.: (1971) Limit theorems for the Galton-Watson process with time-dependent immigration. ZW **20**, 227–235.

Freedman, D.A.: (1965) Bernard Friedman's urn. AMS **36**, 956–970.

Galton, F.: (1873) Problem 4001. Educational Times 1 April, 17.

Galton, F.: (1889) Natural Inheritance, 2nd American ed. (An earlier ed. appeared in 1869) App. F, 241–248. London: Macmillan and Co.

Galton, F.: (1891) Hereditary Genius, 2nd American ed. New York: D. Appleton and Co.

Gnedenko, B.V., Kolmogorov, A.N.: (1954) Limit Distributions for Sums of Independent Random Variables. Cambridge, Mass: Addison-Wesley.

Goldstein, M.I.: (1971 a) Critical age-dependent branching processes: Single and multitype. ZW **17**, 74–88.

Goldstein, M.I.: (1971 b) A uniform limit theorem and a conjecture in multitype age-dependent branching processes. (Centre de Recherches Mathématiques, Université de Montreal, tech. report 107).

Goodman, L.A.: (1967) The probabilities of extinction for birth-and-death processes that are age-dependent or phase-dependent. Biometrika **54**, 579–596.

Goodman, L. A.: (1968 a) Stochastic models for population growth of the sexes. Biometrika **55**, 469–487.

Goodman, L.A.: (1968 b) How to minimize or maximize the probabilities of extinction in a Galton-Watson process and in some related multiplicative population processes. AMS **39**, 1700–1710.

Haezendonck, J.: (1968) Construction d'un processus de branchement continu discernable. C. R. Acad. Sci. Paris Ser. A–B 267, A940–A942.

Harris, T.E.: (1948) Branching Processes. AMS **19**, 474–494.

Harris, T.E.: (1951) Some Mathematical Models for Branching Processes. 2nd Berk. Symp. 305–328.

Harris, T.E.: (1963) The Theory of Branching Processes. Berlin: Springer.

Heathcote, C.R.: (1965) A branching process allowing immigration. JRSS **27**, 138–143.

Heathcote, C.R.: (1966) Corrections and comments on the paper "A branching process allowing immigration". JRSS **28**, 213–217.

Heathcote, C.R., Seneta, E.: (1966) Inequalities for branching processes. JAP **3**, 261–267.

Heathcote, C.R., Seneta, E.: (1967) Correction: "Inequalities for branching processes". JAP **4**, 215.

Heathcote, C.R., Seneta, E., Vere-Jones, D.: (1967) A refinement of two theorems in the theory of branching processes. TPA **12**, 297–301.

Heyde, C.C.: (1970a) A rate of convergence result for the supercritical Galton-Watson process. JAP **7**, 451–454.

Heyde, C.C.: (1970b) Extension of a result of Seneta for the supercritical Galton-Watson process. AMS **41**, 739–742.

Heyde, C.C.: (1971a) Some central-limit analogues for supercritical Galton-Watson processes. JAP **8**, 52–59.

Heyde, C.C.: (1971b) Some almost sure convergence theorems for branching processes. ZW **20**, 189–192.

Hu Ti-ho: (1964) The invarience principle and its applications to branching processes. (Chinese, Engl. Summ.) Acta. Sci. Natur. Univ. Pekinensis **10**, 1–27.

Ikeda, N., Nagasawa, M., Watanabe, S.: (1965) On branching Markov processes. PJA **41**, 816–821.

Ikeda, N., Nagasawa, M., Watanabe, S.: (1966a) Fundamental equations of branching Markov processes. PJA **42**, 252–257.

Ikeda, N., Nagasawa, M., Watanabe, S.: (1966b) A construction of Markov processes by piecing out. PJA **42**, 370–375.

Ikeda, N., Nagasawa, M., Watanabe, S.: (1966c) A construction of branching Markov processes. PJA **42**, 380–384.

Ikeda, N., Nagasawa, M., Watanabe, S.: (1966d) Transformation of branching Markov processes. PJA **42**, 719–724.

Ikeda, N., Nagasawa, M., Watanabe, S.: (1966e) On branching semigroups, I. PJA **42**, 1016–1021.

Ikeda, N., Nagasawa, M., Watanabe, S.: (1966f) On branching semigroups, II. PJA **42**, 1022–1026.

Ikeda, N., Nagasawa, M., Watanabe, S.: (1968a) Branching Markov processes, I. JMKU **8**, 233–278.

Ikeda, N., Nagasawa, M., Watanabe, S.: (1968b) Branching Markov Processes, II. JMKU **8**, 365–410.

Ikeda, N., Nagasawa, M., Watanabe, S.: (1969) Branching Markov processes, III. JMKU **9**, 95–160.

Ikeda, N., Nagasawa, M., Watanabe, S.: (1971) Correction to "Branching Markov Processes, II." JMKU **11**, 195–196.

Ikeda, N., Watanabe, S.: On the uniqueness and non-uniqueness of a class of non-linear equations and explosion problem for branching processes. Technical report.

Imai, H.: (1967) On the branching transport processes. (Jap. Eng. summ.) Proc. Inst. Stat. Math. **15**, 11–21.

Imai, H.: (1968) Notes on a local limit theorem for discrete time Galton-Watson branching processes. Ann. Inst. Statist. Math. **20**, 391–410.

Imai, H.: (1970) On an asymptotic form of the individual distribution of the Galton-Watson process with mean one. (Japanese Engl. summ.) Proc. Inst. Statist. Math. **17**, 99–107.

Jagers, P.: (1967) Integrals of branching processes. Biometrika **54**, 263–271.

Jagers, P.: (1968) Age-dependent branching processes allowing immigration. TPA **13**, 225–236.

Jagers, P.: (1969a) The proportions of individuals of different kinds in two-type populations. A branching process problem arising in biology. JAP **6**, 249–260.

Jagers, P.: (1969b) A general stochastic model for population development. Skand. Aktuarietidskr. **1–2**, 84–103.

Jagers, P.: (1969c) Renewal theory and the almost sure convergence of branching processes. Ark. Mat. **7**, 495–504.

Jagers, P.: (1970) The composition of branching populations: A mathematical result and its application to determine the incidence of death in cell proliferation. Math. Biosci. **8**, 227–238.

Jagers, P.: (1971) Diffusion approximations of branching processes. AMS **42**, 2074–2078.

Jirina, M.: (1957) The asymptotic behavior of branching stochastic processes. Czechoslovak. Math. J. **7**, 130–153. Russian. Engl. translation in: Selected Translations in Mathematical Statistics and Probability, Vol. 2, 87–107.

Jirina, M.: (1958) Stochastic branching processes with continuous statespace. (Russ. Summ.) Czechoslovak. Math. J. **8**, 292–313.

Jirina, M.: (1960) The Asymptotic Behavior of Branching Stochastic Processes. New York: Gordon and Breach.

Jirina, M.: (1962) Branching processes with measure-valued states. Trans. Third Prague Conf. Info. Theory., Stat. Dec. Funct., & Random Processes. 333–357.

Jirina, M.: (1966) Asymptotic behavior of measure-valued branching processes. (Czech. summ.) Rozpravy Československé Akad. Věd. **76**, No. 3, 55 pp.

Jirina, M.: (1967) General branching processes with continuous time parameter. Fifth Berk. Symp. (Berkeley, Calif. 1965/66) Vol II: Contributions to probability theory, Part 1, 389–399.

Jirina, M.: (1969) On Feller's branching diffusion processes. (Czech. summ.) Časopis Pěst. Mat. **94**, 84–90; 107.

Jirina, M.: (1970) A simplified proof of the Sevastyanov theorem on branching processes. (French summ.) Ann. Inst. H. Poincaré Sect. B. (N.S.) **6**, 1–7.

Jirina, M.: (1971) Diffusion branching processes with several types of particles. ZW **18**, 34–46.

Joffe, A.: (1967) On the Galton-Watson branching processes with mean less than one. AMS **38**, 264–266.

Joffe, A., Spitzer, F.: (1967) On multitype branching processes with $\rho \leqq 1$. JMAA **19**, 409–430.

Kaplan, N.: (1972a) A theorem on compositions of random probability generating functions and applications to branching processes with random environments. JAP **9**, 1–12.

Kaplan, N.: (1972b) A continous time Markov branching model with random environments. To appear in JAP.

Kaplan, N.: (1972c) Some results about multidimensional branching processes with random environments. To appear in AMS.

Karlin, S.: (1966) A First Course In Stochastic Processes. New York-London: Academic Press.

Karlin, S.: (1967) Local limit laws for the supercritical continuous time simple branching processes. (Unpublished manuscript.)

Karlin, S.: (1968a) Equilibrium behavior of population genetic models with non-random mating, II. Pedigrees, homozygosity and stochastic models. JAP **5**, 487–566.

Karlin, S.: (1968b) Branching Processes. Mathematics of the Decision Sciences, Part 2, 195–234. Providence, R.I.: American Mathematical Society.

Karlin, S., McGregor, J.: (1964) Direct product branching processes and related Markov chains. Proc. Nat. Acad. Sci. U.S.A. **51**, 598–602.

Karlin, S., McGregor, J.: (1965) Direct product branching processes and related induced Markov chains. I. Calculations of rates of approach to homozygosity. Proc. Internat. Res. Sem. Stat. Lab., Univ. of Calif., Berkeley, Calif., 1963. 111–145. New York: Springer.

Karlin, S., McGregor, J.: (1966a) Spectral theory of branching processes. I. The case of discrete spectrum. ZW **5,** 6–33.
Karlin, S., McGregor, J.: (1966b) Spectral theory of branching processes. II. The case of continuous spectrum. ZW **5,** 34–54.
Karlin, S., McGregor, J.: (1967a) Uniqueness of stationary measures for branching processes and applications. Fifth Berk. Sym. (Berkeley, Calif. 1965/66) Vol II. Contributions to prob. theory. Part 2, 243–254.
Karlin, S., McGregor, J.: (1967b) Properties of the stationary measure of the critical case simple branching process. AMS **38**, 977–991.
Karlin, S., McGregor, J.: (1968a) Embeddability of discrete time simple branching processes into continuous time branching processes. TAMS **132,** 115–136.
Karlin, S., McGregor, J.: (1968b) Embedding iterates of analytic functions with two fixed points into continuous groups. TAMS **132,** 137–145.
Karlin, S., McGregor, J.: (1968c) On the spectral representation of branching processes with mean one. JMAA **21,** 485–495.
Kawazu, K., Watanabe, S.: (1971) Branching processes with immigration and related limit theorems. TPA **16,** 34–51.
Kemeny, J.G., Snell, J.L., Knapp, A.W.: (1966) Denumerable Markov Chains. Princeton, N.J.: Van Nostrand.
Kendall, D.C.: (1951) Some problems in the theory of queues. JRSS **13,** 151–185.
Kendall, D.G.: (1966a) Branching processes since 1873. J. London Math. Soc. **41,** 385–406.
Kendall, D.G.: (1966b) On super-critical branching processes with a positive chance of extinction. Research papers in statistics (Festschrift J. Neyman) 157–165. London: Wiley.
Kesten, H.: (1970) Quadratic transformations: A model for population growth. Adv. App. Prob. **2,** 1–82; 179–228.
Kesten, H.: (1971) Some non-linear stochastic growth models. BAMS **77,** 492–511.
Kesten, H.: (1972) Limit theorems for stochastic growth models. (To appear).
Kesten, H., Ney, P., Spitzer, F.: (1966) The Galton-Watson process with mean one and finite varience. TPA **11,** 513–540.
Kesten, H., Stigum, B.P.: (1966a) A limit theorem for multidimensional Galton-Watson processes. AMS **37**, 1211–1223.
Kesten, H., Stigum, B.P.: (1966b) Additional limit theorems for indecomposable multidimensional Galton-Watson processes. AMS **37**, 1463–1481.
Kesten, H., Stigum, B.P.: (1967) Limit theorems for decomposible multi-dimensional Galton-Watson processes. JMAA **17**, 309–338.
Kharlamov, B.P.: (1968) On properties of branching processes with an arbitrary set of particle types. TPA **13,** 84–98.
Kharlamov, B.P.: (1969a) On the generation numbers of particles in a branching process with overlapping generations. TPA **14,** 44–50.
Kharlamov, B.P.: (1969b) The numbers of generations in a branching process with an arbitrary set of particle types. TPA **14,** 432–449.
Kingman, J.F.C.: (1963) Continuous time Markov processes. Proc. London. Math. Soc. **13,** 593–604.
Kingman, J.F.C.: (1965) Stationary measures for branching processes. PAMS **16,** 245–247.
Kolda, S.: (1969) Branching processes with a denumerable set of types of particles. (Czech. Eng. summ.) Časopis Pěst. Mat. **94,** 168–193.
Kolmogorov, A.: (1938) Zur Lösung einer biologischen Aufgabe. Izvestiya nauchno—issledovatelskogo instituta matematiki i mechaniki pri Tomskom Gosudarstvennom Universitete **2,** 1–6.

Kolmogorov, A.: (1941) Über das logarithmisch normale Verteilungsgesetz der Dimensionen der Teilchen bei Zerstückelung. Dokl. Acad. Nauk. SSSR **31,** 99–101.

Kolmogorov, A., Dmitriev, N.: (1947) Branching stochastic processes. Dokl. Acad. Nauk. SSSR **56,** 5–8.

Kuczma, M.: (1968) Functional Equations In A Single Variable. Warszawa, PWN-Polish Scientific Publishers.

Kurtz, T.: (1970) Approximation of age-dependent, multitype, branching processes. AMS **41,** 363–368.

Lamperti, J.: (1967a) Continuous state branching processes, BAMS **73,** 382–386.

Lamperti, J.: (1967b) Limiting distributions for branching processes. Fifth Berk. Sym. (Berkeley, Calif. 1965/66) Vol. II: Contributions to prob. theory, Part 2, 225–241.

Lamperti, J.: (1967c) The limit of a sequence of branching processes. ZW **7,** 271–288.

Lamperti, J.: (1970) Maximal branching processes and 'long range percolation'. JAP **7,** 89–98.

Lamperti, J., Ney, P.: (1968) Conditioned branching processes and their limiting diffusions. TPA **13,** 128–139.

Levinson, N.: (1959) Limiting theorems for Galton-Watson branching processes. IJM **3,** 554–565.

Levinson, N.: (1960) Limiting theorems for age-dependent branching processes. IJM **4,** 100–118.

Levy, P.: (1937) Theorie de l'addition des variables aléatoires. Paris: Gauthier-Villars.

Lipow, C.: (1971) Two branching models with generating functions dependent on population size. University of Wisconsin, Ph. D. Thesis.

Loeve, M.M.: (1963) Probability Theory, 3rd Ed. Princeton, N.J.: Van Nostrand.

Lootgieter, Jean-Claude: (1969) Processus de Galton-Watson de moyenne 1. C.R. Acad. Sci. Paris Ser. A–B, 268, A817–A818.

Mandl, P.: (1965) Age structure of Markov branching processes. (Czech, Russian and German summ.), Časopis Pěst. Mat. **90,** 353–360.

Martin-Löf, A.: (1966) A limit theorem for the size of the nth generation of an age-dependent branching process JMAA **15,** 273–279.

Mode, C.J.: (1966a) Some multidimensional branching processes as motivated by a class of problems in mathematical genetics I. Bull. Math. Biophys. **28,** 25–50.

Mode, C.J.: (1966b) Some multidimensional branching processes as motivated by a class of problems in mathematical genetics II. Bull. Math. Biophys. **28,** 181–190.

Mode, C.J.: (1968a) A multidimensional age-dependent branching process with applications to natural selection, I. Math. Biosci. **3,** 1–18.

Mode, C.J.: (1968b) A multidimensional age-dependent branching process with applications to natural selection, II. Math. Biosci. **3,** 231–247.

Mode, C. J.: (1969a) Lag time in cell division from the point of view of the Bellman-Harris process. Math. Biosci. **5**, 341–345.

Mode, C. J.: (1969b) Applications of generalized multi-type age-dependent branching processes in population genetics. Bull. Math. Biophys. **31**, 575–589.

Mode, C.J.: (1971) Multitype Branching Processes. N.Y. American Elsevier.

Mode, C. J.: (1972) Multi-type age-dependent branching processes and cell cycle analysis. To appear in Math. Biosci.

Mode, C. J., Nair, K. A.: (1969) On the distribution of the W-random variable in a general age-dependent branching process. JMAA **28**, 636–646.

Mode, C. J., Nair, K. A.: (1971a) A multi-dimensional age-dependent branching processes—subcritical case. JMAA **34**, 567–577.

Mode, C. J., Nair, K. A.: (1971b) The reducible multi-dimensional age-dependent branching process. JMAA **33**, 131–139.

Moy, Shu-teh C.: (1967a) Extensions of a limit theorem of Everett, Ulam, and Harris on multitype branching processes to a branching process with countably many types. AMS **38,** 992–999.

Moy, Shu-teh C.: (1967b) Ergodic properties of expectation matricies of a branching process with countably many types. J. Math. Mech. **16,** 1207–1225.

Moyal, J. E.: (1962a) Multiplicative population chains. Proc. Roy. Soc. London Ser. A 266, 518–526.

Moyal, J. E.: (1962b) The general theory of stochastic population processes. Acta. Math. **108,** 1–31.

Moyal, J. E.: (1964) Multicative population processes. JAP. **1**, 267–283.

Mullikin, T. W.: (1963) Limiting distributions for critical multitype branching processes with discrete time. TAMS **106,** 469–494.

Mullikin, T. W.: (1968) Branching processes in neutron transport theory. Probabilistic Methods in Applied Mathematics, Vol 1, 199–281. New York: Academic Press.

Nagaev, A. V.: (1961) More exact statements of certain theorems in the theory of branching processes. (Russian), Trudy Taškent. Gos. Univ. **189,** 55–63.

Nagaev, A. V.: (1967) On estimating the expected number of direct descendants of a particle in a branching process. TPA **12,** 314–320.

Nagaev, A. V., Badalbaev, I.: (1967) A refinement of certain theorems on branching random processes (Russ.-Engl. summ.). Litovsk. Mat. Sb. **7,** 129–136.

Nagaev, S. V., Muhamedhanova, R.: (1966a) Transition phenomena in branching random processes with discrete time. (Russ.) Limit Thms. Statist. Inference (Russian) Tashkent: Izdat. "Fan". 83–89.

Nagaev, S. V., Muhamedhanova, R.: (1966b) Certain limit theorems in the theory of branching random processes. (Russ.) Limit Thms. Statist. Interference, Tashkent: Izdat. "Fan" 90–113.

Nagaev, S. V., Muhamedhanova, R.: (1968) Certain remarks apropos of earlier published limit theorems in the theory of branching processes. (Russ.) Probabilistic Models and Duality Control (Russ.). Tashkent: Izdat. "Fan" Uzbek. SSR. 46–49.

Nagasawa, Masao: (1968) Construction of branching Markov processes with age and sign. Kōdai Math. Sem. Rep. **20**, 469–508.

Neuts, Marcel F.: (1969) The queue with Poisson input and general service times, treated as a branching process. Duke Math. J. **36**, 215–231.

Neveu, J.: (1964) Chaines de Markov et theorie du potential. Ann. Fac. Sci. Univ. Clermont-Ferrand, No. **24**, 37–89.

Ney, P. E.: (1964a) Generalized branching processes I: Existence and uniqueness theorems. IJM **8**, 316–331.

Ney, P. E.: (1964b) Generalized branching processes II: Asymptotic theory. IJM **8**, 332–350.

Ney, P. E.: (1964c) Ratio limit theorems for cascade processes. ZW **3**, 32–49.

Ney, P.: (1965a) The limit distribution of a binary cascade process. JMAA **10**, 30–36.

Ney, P.: (1965b) The convergence of a random distribution function associated with a branching process. JMAA **12**, 316–327.

Ney, P.: (1967) Critical multi-type degenerate branching processings. Technical report, Univ. of Wisc.

Oguru, Y.: (1969/70) Spectral representation for branching processes on the real half time. Publ. Res. Inst. Math. Sci. **5**, 423–441.

Pakes, A. G.: (1970a) An asymptotic result for a subcritical branching process with immigration. Bull. Austral. Math. Soc. **2**, 223–228.

Pakes, A. G.: (1970b) On a theorem of Quine and Seneta for the Galton-Watson process with immigration. To appear in Austral. J. Stat.

Pakes, A. G.: (1971a) On the Critical Galton-Watson process with immigration. J. Austral. Math. Soc. **12**, 476–482.

Pakes, A. G.: (1971b) Further results on the Critical Galton-Watson process with immigration. To appear in J. Austral. Math. Soc.

Pakes, A. G.: (1971c) A branching process with state dependent immigration component. Adv. Appl. Prob. **3**, 301–314.

Pakes, A. G.: (1971d) Branching processes with immigration. JAP **8**, 32–42.

Papangelou, F.: (1968) A lemma on the Galton-Watson process and some of its consequences. PAMS **19**, 1169–1479.

Pinsky, M. A.: (1971) Limit theorems for continuous state branching processes with immigration. Northwestern University Report.

Pollak, E.: (1969) Bounds for certain branching processes. JAP **6**, 201–204.

Pollak, E.: (1971) On survival probabilities and extinction times for some branching processes. JAP **8**, 633–654.

Puri, P. S.: (1967) Some limit theorems on branching processes related to development of biological populations. Math. Biosci. **1**, 77–94.

Quine, M. P.: (1970) The multitype Galton-Watson process with immigration. JAP **7**, 411–422.

Quine, M. P., Seneta, E.: (1969) A limit theorem for the Galton-Watson process with immigration. Austral. J. Stat. **11**, 166–173.

Radcliffe, J.: (1972) An immigration super-critical branching diffusion process. JAP **9**, 13–23.

Radcliffe, J., Staff, P. J.: (1970) Immigration-migration-death processes with multiple latent roots. Math. Biosci. **8**, 279–290.

Reynolds, J. F.: (1970) A theorem on discrete time branching processes allowing immigration. JAP **7**, 446–450.

Rhyzhov, Y. M., Skorokhod, A. W.: (1970) Homogeneous branching processes with a finite number of types and continuously varying mass. TPA **15**, 704–707.

Rubin, H., Vere-Jones, D.: (1968) Domains of attraction for the subcritical Galton-Watson branching process. JAP **5**, 216–219.

Ryan, T. A., Jr.: (1968) On age-dependent branching processes. Ph. D. dissertation, Cornell University.

Samuels, M. L.: (1971) Distribution of the branching process population among generations. JAP **8**, 655–667.

Sankoff, D.: (1971) Branching process with terminal types: applications to context free grammers. JAP **8**, 233–240.

Savage, I. R., Shimi, I. N. A.: (1969) A branching process without rebranching. AMS **40**, 1850–1851.

Savits, T.: (1969) The explosion problem for branching Markov processes. Osaka J. Math. **6**, 375–395.

Savits, T. H.: (1972) Branching Markov processes in a random environment. Princeton Univ. Tech. report.

Sawyer, S. A.: (1970) A formula for semigroups, with an application to branching diffusion processes. TAMS **152**, 1–38.

Selivanov, B. I.: (1969) Certain explicit formulas in the theory of random branching processes with discrete time and one type of particles. TPA **14**, 336–342.

Seneta, E.: (1967a) The Galton-Watson process with mean one. JAP **4**, 489–495.

Seneta, E.: (1967b) On the transient behavior of a Poisson branching processes. J. Austral. Math. Soc. **7**, 465–480.

Seneta, E.: (1968a) On asymptotic properties of subcritical branching processes. J. Austral. Math. Soc. **8**, 671–682.

Seneta, E.: (1968b) On recent theorems concerning the supercritical Galton-Watson process. AMS **39**, 2098–2102.

Seneta, E.: (1968c) The stationary distribution of a branching process allowing immigration: A remark on the critical case. JRSS **30**, 176–179.

Seneta, E.: (1969a) Functional equations and the Galton-Watson process. Adv. Appl. Prob. **1**, 1–42.

Seneta, E.: (1969b) Some second-order properties of the Galton-Watson extinction time distribution. Sankhyā Ser. A **31**, 75–78.

Seneta, E.: (1970a) A note on the supercritical Galton-Watson process with immigration. Math. Biosci. **6**, 305–312.

Seneta, E.: (1970b) On the supercritical Galton-Watson process with immigation. Math. Biosci. **7**, 9–14.

Seneta, E.: (1970c) Population growth and the multi-type Galton-Watson process. Nature **225**, 776.

Seneta, E.: (1970d) An explicit-limit theorem for the critical Galton-Watson process with immigration. JRSS **32**, 149–152.

Seneta, E.: (1971) On invariant measures for simple branching processes. JAP **8**, 43–51.

Seneta, E., Vere-Jones, D.: (1966) On quasi-stationary distributions in discrete-time Markov chains with a denumerable infinity of states. JAP **3**, 403–434.

Seneta, E., Vere-Jones, D.: (1968) On the asymptotic behavior of subcritical branching processes with continuous state space. ZW **10**, 212–225.

Seneta, E., Vere-Jones, D.: (1969) On a problem of M. Jiřina concerning continuous state branching processes. Czechoslovak Math. J. **19** (94) 277–283.

Sevastyanov, B. A.: (1951) The theory of branching random processes. Uspehi Mat. Nauk. **6**, 47–99 (Russ.).

Sevastyanov, B. A.: (1964) Age-dependent branching processes. TPA **9**, 521–537.

Sevastyanov, B. A.: (1964) On the general definition of branching stochastic processes. (Russ.) Winter School in Thy. of Prob. and Math. Statistics held in Uzgarod, 217–220, Kiev: Izdat. Akad. Nauk Ukrain. SSR.

Sevastyanov, B. A.: (1967a) Asymptotic behavior of the probability of non-extinction of a critical branching process. TPA **12**, 152–154.

Sevastyanov, B. A.: (1967b) Regularity of branching processes (Russ.). Mat. Zametki **1**, 53–62.

Sevastyanov, B. A.: (1968a) Theory of branching processes (Russ.). Theory of Probability, Mathematical Statist., Theoret. Cybernet. 1967 (Russ.), Akad. Nauk SSSR Vsesojuz. Inst. Naucn. i. Tehn. Informacii, Moscow, 5–46.

Sevastyanov, B. A.: (1968b) Branching processes (Russ.). Mat. Zametki **4**, 239–251.

Sevastyanov, B. A.: (1968c) Limit theorems for age-dependent branching processes. TPA **13**, 237–259.

Sevastyanov, B. A.: (1968d) Renewal type equations and moments of branching processes (Russ.). Mat. Zametki **3**, 3–14.

Sevastyanov, B. A.: (1971) Branching Processes. Moscow: Nauka. (Russ.)

Siddiqui, M. M.: (1967) Strong law of large numbers for branching processes. SIAM J. Appl. Math. **15**, 893–897.

Silverstein, M. L.: (1968a) Markov processes with creation of particles. ZW **9**, 235–257.

Silverstein, M. L.: (1968b) A new approach to local times. Jour. Math. and Mech. **17**, 1023–1054.

Silverstein, M. L.: (1969) Continuous state branching semigroups. ZW **14**, 96–112.

Skorohod, A. V.: (1967) Branching diffusion processes. TPA **9**, 445–449.

Slack, R. S.: (1968) A branching process with mean one and possibly infinite varience. ZW **9**, 139–145.

Smith, W. L.: (1968) Necessary conditions for almost sure extinction of a branching process with random environment. AMS **39**, 2136–2140.

Smith, W. L., Wilkinson, W. E.: (1969) On branching processes in random environments. AMS **40**, 814–827.

Snow, R. N.: (1959a) *N*-dimensional age-dependent branching processes. Preliminary Rep. I: Formulation. AMS Notices **6**, 616.

Snow, R. N.: (1959b) *N*-dimensional age-dependent branching processes. Preliminary Rep. II: Asymptotic behavior, irreducible case. AMS Notices **6**, 616–617.

Spitzer, F.: (1964) Principles of Random Walk. Princeton, New Jersey: Van Nostrand.

Spitzer, F.: (1967) Two explicit Martin boundary constructions. Symposium on Prob. Methods in Analysis (1966), Lecture Notes in Mathematics, Berlin-Heidelberg-New York: Springer.

Srinivasan, S. K.: (1969) Stochastic Theory and Cascade Processes. New York: American Elsevier.

Stam, A. J.: (1966) On a conjecture by Harris. ZW **5**, 202–206.

Stigum, B. P.: (1966) A theorem on the Galton-Watson process. AMS **37**, 695–698.

Stratton, H. H., Jr., Tucker, H. G.: (1964) Limit distributions of a branching process. AMS **35**, 557–565.

Szasz, D.: (1967) On the general branching process with continuous time parameter. Studia Sci. Math. Hungar. **2**, 227–247.

Urbanik, K.: (1962/63) The limiting behavior of indecomposable branching processes. Studia Math. **22**, 109–125.

Vinogradov, O. P.: (1964) On an age-dependent branching process. TPA **9**, 131–136.

Watanabe, S.: (1965) On the branching process for Brownian particles with an absorbing boundary. JMKU **4**, 385–398.

Watanabe, S.: (1967a) Branching Brownian motion. Unpublished mimeograph, Stanford University.

Watanabe, S.: (1967b) Limit theorem for a class of branching processes. (Markov Processes and Potential Theory.) (Proc. Sympos. Math. Res. Center, Madison, Wisconsin) N. Y.: Wiley 205–232.

Watanabe, S.: (1968) A limit theorem of branching processes and continuous state branching processes. JMKU **8**, 141–167.

Watanabe, S.: (1969) On two dimensional Markov processes with branching property. TAMS **136**, 447–466.

Watson, H. W.: (1873) Solution to problem 4001. Educational Times, 1. Aug., 115–116

Watson, H. W., Galton, F.: (1874) On the probability of the extinction of families. J. Anthropol. Inst. Great B. and Ireland **4**, 138–144.

Waugh, W. A. O'N.: (1968) Age-dependent branching processes under a condition of ultimate extinction. Biometrika **55**, 291–296.

Weiner, H. J.: (1965) Asymptotic properties of an age dependent branching process. AMS **36**, 1165–1568.

Weiner, H. J.: (1965) An integral equation in age dependent branching processes. AMS **36**, 1569–1573.

Weiner, H. J.: (1966a) Monotone convergence of moments in age dependent branching processes. AMS **37**, 1806–1808.

Weiner, H. J.: (1966b) Applications of the age distribution in age dependent branching processes. JAP **3**, 179–201.

Weiner, H. J.: (1966c) On age dependent branching processes. JAP **3**, 383–402.

Weiner, H. J.: (1969) Sums of lifetimes in age dependent branching processes. JAP **6**, 195–200.

Weiner, H. J.: (1970) On a multi-type critical age-dependent branching processes. JAP **7**, 523–543.

Weissner, E. W.: (1971) Multitype branching processes in random environments. JAP **8**, 17–31.

Whittle, P.: (1964) A branching process in which individuals have variable lifetimes. Biometrika **51**, 262–264.

Wilkinson, W. E.: (1967) Branching processes in stochastic environments. Univ. of North. Car., mimeo ser. no. 544.

Wilkinson, W. E.: (1969) On calculating extinction probabilities for branching processes in random environments. JAP **6**, 478–492.

Yaglom, A. M.: (1947) Certain limit theorems of the theory of branching processes. Dokl. Acad. Nauk SSSR **56**, 795–798.

Yang, Y. S.: (1972) On branching processes allowing immigration. JPA **9**, 24–31.

Zubkov, A. N.: (1970) A degeneracy condition for a bounded branching process. (Russ.) Mat. Zametki **8**, 9–18. English translation in Math. Notes **8**, 472–477

Bibliography added in proof

Dayanithy, K.: (1972) Representation of branching processes. The critical case. ZW **22**, 268–292.

Erickson, K. B.: (1972) Self annihilating branching processes. To appear.

Hering, H.: (1972) Critical Markov branching processes with general set of types. TAMS **160**, 185–202.

Kaplan, N., Karlin, S.: (1972) Criteria for extinction of certain population growth processes with interesting types. To appear in JAP.

Pakes, A. G.: (1971e) Some limit theorems for the total progeny of a branching process. Adv. Appl. Prob. **3**, 176–192.

Waugh, W. A. O'N.: (1972) The apparent "lag phase" in a stochastic population model in which there is no variation in the conditions of growth. Technical report, Univ. of Toronto.

List of Symbols

(in order of appearance)

Referencing Scheme

"Theorem II.8.3" refers to theorem 3 of section 8 in chapter II; "(II.8.3)" refers to expression (3) of section 8 in chapter II. If the reference is in the same chapter, then the chapter number is dropped; if in the same section then the chapter and section number are dropped.

Author Index

Subject Index

Die Grundlehren der mathematischen Wissenschaften in Einzeldarstellungen mit besonderer Berücksichtigung der Anwendungsgebiete

Eine Auswahl

74. Boerner: Darstellungen von Gruppen
76. Tricomi: Vorlesungen über Orthogonalreihen
77. Behnke/Sommer: Theorie der analytischen Funktionen einer komplexen Veränderlichen
78. Lorenzen: Einführung in die operative Logik und Mathematik
86. Richter: Wahrscheinlichkeitstheorie
87. van der Waerden: Mathematische Statistik
94. Funk: Variationsrechnung und ihre Anwendung in Physik und Technik
97. Greub: Linear Algebra
99. Cassels: An Introduction to the Geometry of Numbers
104. Chung: Markov Chains with Stationary Transition Probabilities
107. Köthe: Topologische lineare Räume
114. MacLane: Homology
116. Hörmander: Linear Partial Differential Operators
117. O'Meara: Introduction to Quadratic Forms
120. Collatz: Funktionalanalysis und numerische Mathematik
121./122. Dynkin: Markov Processes
123. Yosida: Functional Analysis
124. Morgenstern: Einführung in die Wahrscheinlichkeitsrechnung und mathematische Statistik
125. Itô/McKean jr.: Diffusion Processes and their sample Paths
126. Lehto/Virtanen: Quasikonforme Abbildungen
127. Hermes: Enumerability, Decidability, Computability
128. Braun/Koecher: Jordan-Algebren
129. Nikodým: The Mathematical Apparatus for Quantum Theories
130. Morrey jr.: Multiple Integrals in the Calculus of Variations
131. Hirzebruch: Topological Methods in Algebraic Geometry
132. Kato: Perturbation Theory for Linear Operators
133. Haupt/Künneth: Geometrische Ordnungen
134. Huppert: Endliche Gruppen I
135. Handbook for Automatic Computation. Vol. 1/Part a: Rutishauser: Description of ALGOL 60
136. Greub: Multilinear Algebra
137. Handbook for Automatic Computation. Vol. 1/Part b: Grau/Hill/Langmaack: Translation of ALGOL 60
138. Hahn: Stability of Motion
139. Mathematische Hilfsmittel des Ingenieurs. 1. Teil
140. Mathematische Hilfsmittel des Ingenieurs. 2. Teil
141. Mathematische Hilfsmittel des Ingenieurs. 3. Teil
142. Mathematische Hilfsmittel des Ingenieurs. 4. Teil
143. Schur/Grunsky: Vorlesungen über Invariantentheorie
144. Weil: Basic Number Theory
145. Butzer/Berens: Semi-Groups of Operators and Approximation
146. Treves: Locally Convex Spaces and Linear Partial Differential Equations
147. Lamotke: Semisimpliziale algebraische Topologie

148. Chandrasekharan: Introduction to Analytic Number Theory
149. Sario/Oikawa: Capacity Functions
150. Iosifescu/Theodorescu: Random Processes and Learning
151. Mandl: Analytical Treatment of One-dimensional Markov Processes
152. Hewitt/Ross: Abstract Harmonic Analysis. Vol. 2: Structure and Analysis for Compact Groups. Analysis on Locally Compact Abelian Groups
153. Federer: Geometric Measure Theory
154. Singer: Bases in Banach Spaces I
155. Müller: Foundations of the Mathematical Theory of Electromagnetic Waves
156. van der Waerden: Mathematical Statistics
157. Prohorov/Rozanov: Probability Theory
159. Köthe: Topological Vector Spaces I
160. Agrest/Maksimov: Theory of Incomplete Cylindrical Functions and Their Applications
161. Bhatia/Szegö: Stability Theory of Dynamical Systems
162. Nevanlinna: Analytic Functions
163. Stoer/Witzgall: Convexity and Optimization in Finite Dimensions I
164. Sario/Nakai: Classification Theory of Riemann Surfaces
165. Mitrinović/Vasić: Analytic Inequalities
166. Grothendieck/Dieudonné: Eléments de Géométrie Algébrique I
167. Chandrasekharan: Arithmetical Functions
168. Palamodov: Linear Differential Operators with Constant Coefficients
170. Lions: Optimal Control Systems Governed by Partial Differential Equations
171. Singer: Best Approximation in Normed Linear Spaces by Elements of Linear Subspaces
172. Bühlmann: Mathematical Methods in Risk Theory
173. F. Maeda/S. Maeda: Theory of Symmetric Lattices
174. Stiefel/Scheifele: Linear and Regular Celestial Mechanics. Perturbed Two-body Motion—Numerical Methods—Canonical Theory
175. Larsen: An Introduction of the Theory of Multipliers
176. Grauert/Remmert: Analytische Stellenalgebren
177. Flügge: Practical Quantum Mechanics I
178. Flügge: Practical Quantum Mechanics II
179. Giraud: Cohomologie non abélienne
180. Landkoff: Foundations of Modern Potential Theory
181. Lions/Magenes: Non-Homogeneous Boundary Value Problems and Applications I
182. Lions/Magenes: Non-Homogeneous Boundary Value Problems and Applications II
183. Lions/Magenes: Non-Homogeneous Boundary Value Problems and Applications III. In preparation
184. Rosenblatt: Markov Processes. Structure and Asymptotic Behavior
185. Rubinowicz: Sommerfeldsche Polynommethode
186. Wilkinson/Reinsch: Handbook for Automatic Computation II, Linear Algebra
187. Siegel/Moser: Lectures on Celestial Mechanics
188. Warner: Harmonic Analysis on Semi-Simple Lie Groups I
189. Warner: Harmonic Analysis on Semi-Simple Lie Groups II
190. Faith: Algebra: Rings, Modules, and Categories I. In preparation
191. Faith: Algebra: Rings, Modules, and Categories II. In preparation
192. Maltsev: Algebraic Systems. In preparation
193. Pólya/Szegö: Problems and Theorems in Analysis. Vol. 1
194. Igusa: Theta Functions
195. Berberian: Baer *-Rings
196. Athreya/Ney: Branching Processes